Biology and freedom

Biology and freedom

An essay on the implications of human ethology

S. A. BARNETT

The Australian National University

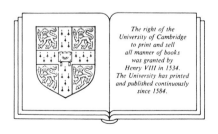

*The right of the
University of Cambridge
to print and sell
all manner of books
was granted by
Henry VIII in 1534.
The University has printed
and published continuously
since 1584.*

CAMBRIDGE UNIVERSITY PRESS

Cambridge

New York   New Rochelle   Melbourne   Sydney

Published by the Press Syndicate of the University of Cambridge
The Pitt Building, Trumpington Street, Cambridge CB2 1RP
32 East 57th Street, New York, NY 10022, USA
10 Stamford Road, Oakleigh, Melbourne 3166, Australia

First published 1988

Printed in Great Britain at the University Press, Cambridge

British Library cataloguing in publication data
Barnett, S.A.
Biology and freedom : an essay on the
implications of human ethology
1. Sociobiology
I. Title
304.5   GN365.9

Library of Congress cataloguing in publication data
Barnett, S.A. (Samuel Anthony), 1915-
Biology and freedom : an essay on the implications of human
ethology / S.A. Barnett.
p.   cm.
Bibliography: p.
Includes indexes.
ISBN 0 521 35316 5
1. Human behavior. 2. Social evolution. 3. Social psychology.
4. Human biology–Social aspects. I. Title.
[DNLM: 1. Aggression. 2. Communication. 3. Ethology.
4. Evolution. 5. Social Behavior. HM 106 B261b]
HM106.B268 1988
304.2′7-dc 19   87-35182   CP

ISBN 0 521 35316 5

Is human nature incurably depraved?
If it is, reading this book will be
waste of time.

G.B. Shaw

Contents

Contents

x *Contents*

Illustrations

Preface

No book can ever be finished. While working on it we learn just enough to find it immature the moment we turn away from it.

K.R. Popper[1]

On her deathbed, the poet, Gertrude Stein, turned to a friend and asked, 'What is the answer?' When the friend, unable to speak, made no reply, she said, 'In that case, what is the question?'[2] In the late twentieth century, the first question that we face is not in doubt. Modern technology and modern politics combine to enable us to commit mass suicide, and we have to decide whether to do so.[3] This book is written in the belief that we and our descendants will choose survival, not self-destruction, and that, in human society, there will be not only change but progress.

Like our ability to destroy ourselves, future progress depends in part on our use of scientific knowledge. The traditional image of science has two faces, both benign: science helps us to know the world; and it gives us power. But science is also a source of myths.[4] When modern physics began, it became easy to think of the universe as a system of which all parts obey Newton's laws of motion. Yet this conclusion does not follow from Newtonian mechanics: it is an assumption. And indeed the whole notion was overthrown by Einstein's theory of relativity.

During the past century, biology, not physics, has become a leading source of our pictures both of the world and of ourselves. It has provided descriptions of the human species each founded on ideas, of narrow scope, derived from the work of specialists. The present book has grown from attempts to answer questions about these images. Are we doing ourselves justice when we equate human societies with those of baboons, or human intelligence with that of a dog salivating at the sound of a bell? To answer, the images, or models, must be put in a much larger picture than is usually provided.

Such a task presents severe difficulties, for the modern biological images of humanity are not merely biological.[5] Some are supposed to tell us how we can plan ideal communities, how we should bring up our children, and how we should treat mental illness. Others propose biological explanations for our inhumanity to our fellows, or for our greed and lust for power, with the seeming implication that these disagreeable characteristics are part of our nature and beyond remedy. Confident proposals on these lines are presented to the public, by all the available means, as the last word of science. Catch-phrases resembling advertisers' slogans, such as the territorial imperative, the naked ape and the selfish gene, enter the language, and each brings a message.

Whatever their authors' intentions, the images of humanity, represented by such phrases, provoke cynicism about the future. Accepting these portraits is likely to weaken our resolve to cope with present dangers: they make it seem pointless to try to change things for the better. The images also conceal what is valid and useful in the sciences on which they are supposed to be based.

They are, however, also beneficial. They oblige us to face fundamental questions, factual and moral, about ourselves. We then have to examine, side by side, diverse kinds of knowledge: for example, findings on dominance among animals and on the great variety of human social systems. We have to ask whether the theory of evolution has any bearing on, say, Machiavellianism in politics or on the concept of original sin; and whether laboratory experiments on the effects of reward and punishment tell us anything useful about why we work, or about the enigma of free will.

When writing on such themes, it is easy to fall back on vague generalization or mere conjecture. One objective of this book is to show that there are plenty of relevant facts, to describe some of them, and to say where others can be found. The examples used are those most significant for 'western' readers. Although the fundamental questions are universal, an audience in China or India or the USSR would need a different book.

Many of these fundamentals appear repeatedly, in widely separated chapters. I hope therefore that readers will find it convenient to read the book straight through. Some will wish to go more deeply into special topics, or to confirm that the strange statements and events mentioned are correctly reported. For them there are notes and a bibliography. There is also an appendix on the meanings and uses of key words.

The fashionable myths of today will fade, but crucial questions that they raise will remain. This book is above all about such questions. It suggests some answers. But the important thing is for readers to be ready to debate the questions, and to act when they have reached conclusions for themselves.

Karl Popper's resigned comment, quoted in the epigraph, applies with special force to a book such as this. But others will do better.

S.A. Barnett

Acknowledgements

While I was writing this book I had helpful correspondence or conversations with many kind people, none of whom knew in full what I was up to. Among them were Marjorie Grene, R.B. Joynson, F.P. Lisovsky, I.S. McLean, Miranda Robertson, R.V. Short, Beryl Smalley, J.J.C. Smart, R.E. Ulrich.

Part 1

An introduction

1 Four portraits

Is there not a warfare to man upon earth?
And are not his days like the days of a hireling?

Job: 7:1.

In the west of Sicily, at Segesta, there is an ancient Greek temple of extraordinary perfection. This elaborately designed structure, and the nearby theatre, were put up, at great cost in work and stone, by the people of a small and poor community. Both were results of millennia of learning to make and to use tools, and of centuries of devising methods of measurement and calculation. These skills had been transmitted, from generation to generation, by teaching and imitation and by the use of languages that conveyed both facts and abstract ideas.

The temple at Segesta can be matched by the works of human communities in all lands and in every period. Even before agriculture began, our ancestors made carvings in rock and paintings on stone or bark; and the gatherer–hunters that remain still do so. As well, they compose or repeat legends and myths. Members of the species that produce these monuments and myths also describe themselves. As A.A. Kwapong has said, 'Every generation and every people have their own mythology on the origins and nature of man.'[1]

This book examines current myths that are sometimes held out as explaining the whole of human existence, including the temples we build, the pictures we paint and the stories we invent about ourselves. At first these myths seem to carry all the authority of modern science; but, when they are tested against actual scientific findings, they prove to be deficient in important ways. They do, however, represent attitudes which have appeared, again and again, for millennia.

The story begins near the time of the building of the temple at Segesta, when Plato (about 428–347 BC) proposed one of the most influential of all images of humanity. Plato wrote in a period of war and upheaval, when the people around him seemed to be driven mainly by greed and selfishness. These failings he held to be common to all human beings. In his most important political work, *The Republic*, he therefore proposed a state ruled by guardians who were to be bred for superiority and public spirit. He remarks on the need to breed horses and dogs selectively, in order to prevent degeneration, and states that the same principle applies to human beings. The guardians must combine fierceness in defending the state with kindness to the people they rule; and again he draws a canine analogy: well bred dogs are gentle toward the people they know, but hostile to strangers. And so, he tells us, 'the character we give our guardians is not contrary to nature'.[2]

Here we have themes that are still debated: that human beings are inherently violent and self-seeking; that only a few can be trusted to take any part in managing the state; that these few differ from the rest by their heredity; and that we can find evidence from animals that supports such ideas.

The form of the debate has varied; but, in the two millennia since the classical age of Greece, the most important new idea was that proposed by the English physician's son, Charles Darwin (1809–1882). Darwin was the leading exponent of a fundamental biological principle, that of evolution by the impersonal, amoral process of natural selection. This principle he and others daringly applied, not only to all plants and animals, but also

The Doric temple at Segesta in Sicily, of the fifth century BC, although unfinished, illustrates the perfection of classical design. Such temples were built by many communities, none of them wealthy.

to the human species. Humanity came to be seen, not as made in God's image, but as a product – one among many – of a ruthless struggle for survival.

Darwinism has become both a central feature of biological theory and also an inescapable part of modern accounts of our species. As a result, we are accustomed to thinking of ourselves as part of nature, as a product of evolution by natural selection, and as subject to natural law. Darwin had written that 'man bears in his bodily structure clear traces of his descent from some lower form' and that, despite the vast difference in 'mental power' between human beings and animals, 'there is no fundamental difference between man and the higher mammals in their mental faculties.'[3]

Today the theory of natural selection is much more elaborate than Darwin's version. One of the leading contributions was by an English mathematician, R.A. Fisher, whose *Genetical Theory of Natural Selection* was first published in 1930. This work is more than its title implies: the last five chapters are 'preliminary enquiries designed to examine whether man, in all his aspects, falls within the scope of a naturalistic theory of evolution.'[4] Fisher compares the 'societies' of the eusocial insects, such as ants and bees, with those of human beings. He emphasizes an analogy between the 'castes' of insects and specialization in human communities. His particular concern is with the differences in fertility observed in an extreme form among eusocial species of which the 'workers' are sterile. Among human beings, the rich are held to be genetically superior to the rest of us, and 'destructive consequences' are said to result from their relative infertility. Fisher attributes the decline and fall of past civilizations to the failure of the rich to breed. He does not, however, propose a complete biological image of humanity.

> A naturalistic view of Man provides no means of putting on an objective basis those mental valuations, moral and aesthetic, to which Man attaches such high importance; it cannot, therefore, be used to throw doubt upon these valuations.[5]

A grander design for biology has been proposed by an influential school of biological scientists, under the heading of sociobiology. A leader of this school, the American authority on ants, E.O. Wilson, presents a vision of humanity as a species fashioned by natural selection to behave in predetermined ways. He asks us to

> consider man in the free spirit of natural history, as though we were zoologists from another planet . . . In this macroscopic view the humanities and social sciences shrink to specialized branches of biology; history, biography, and fiction are the research protocols of human ethology; and anthropology and sociology together constitute the social biology of a single primate species.

Accordingly, Wilson predicts a future in which the sciences of behavior absorb the humanities and the social sciences – the disciplines that are usually regarded as the main sources of our systematic knowledge about ourselves. In this way, he tells us, the inevitable planned society of the twenty-first century will perhaps 'steer its members past . . . stresses and conflicts'. Wilson, however, seems doubtful whether the outcome will be welcome: 'When . . . the social sciences come to full flower, the result might be hard to accept'.

Wilson also writes that the 'dream of a culture predesigned for happiness will surely have to wait'.[6] The dreamer to whom he refers is the leading member of a prominent school of American psychologists, B.F. Skinner. Their image of humanity is based not on the theory of evolution but on laboratory experiments. Skinner himself expresses few doubts about future possibilities. What is needed, he says, is to concentrate on human *behavior*. All that we do is a result, not of our wishes or intentions or of choosing between possible actions, but of our learning the consequences – the pleasures and pains – that have resulted from previous actions. Traditional views on the springs of human action must be given up. 'There are wonderful possibilities', he writes, but he agrees that the resulting world is hard to imagine, a world

> in which people live together without quarrelling, maintain them-
> selves by producing [what] they need, enjoy themselves and contri-
> bute to the enjoyment of others in art, music, literature and
> games . . . and come to know themselves accurately and, therefore,
> manage themselves effectively. Yet all this is possible . . .[7]

The quotations in the preceding paragraphs represent powerful elements in modern thought. From these themes, the evolutionary and the behavioristic, have emerged four portraits of humanity. The first, which I call *Homo pugnax*, originated in its modern form in the nineteenth century. It was based on the idea that animals are in a continual state of strife, and on the presumption that human beings, since they too are a product of evolution, are incessantly violent. Since 1950, a similar belief has been energetically revived. Aggression, the associated buzz-word, has been indiscriminately applied to a great range of human activities, from homicide to hunting.

The second image, *Homo egoisticus*, is a more intellectual cousin of *H. pugnax*, and is a product of sociobiology. There is less preoccupation with violence. The central presumption is that our social lives and our attitudes and beliefs can all be explained by the action of natural selection: that is, by their role in ensuring our survival or the survival of our descendants. Hence we are irremediably selfish and competitive.

These two images present humanity as puppets jerked by evolutionary strings. What then becomes of human intelligence? The creators of *H. pugnax* and *H. egoisticus* acknowledge that people can adapt their behavior to circumstances, but they take little notice of this ability. The disregard is rather surprising. Most of them are academics dedicated to using their intelligence in elaborate ways, and to encouraging students to develop theirs. They also acknowledge that they are themselves human. Yet these writers seem to represent themselves as, to a large extent, instinct-bound automata.

The two alternatives concentrate not on intelligence but on our ability to develop new habits. One we owe largely to the Russian physiologist, I.P. Pavlov (1849–1936). He and his colleagues revived and strengthened an image of humanity that originally became important in the seventeenth century: the human being as a machine. The experiments on conditional reflexes, begun by Pavlov at the beginning of the present century, were designed to give a physiological or mechanistic account of animal and human intelligence. *Homo pavlovi* presents human action as a result of 'conditioning': our conduct is described as analogous to that of a dog, restrained in a harness, and salivating at the sound of a bell.

During the first decades of the twentieth century an alternative but related movement began to have a far-reaching influence, especially in the USA. Its goal was the prediction and control of behavior. In the hope of making this possible, consciousness, feelings and beliefs were rejected as proper studies. Psychologists of this school do not tie their animals down, but the behavior they record is strictly confined to a single operation, such as pressing a lever. The experimenter controls what the animal does by regulating what the animal experiences when the operation is performed: this may be a reward, such as food; punishment, such as electric shock; or relief from pain. The procedure is sometimes called behavioral engineering. Findings from this method have been used to propose applications in teaching, psychotherapy and even the design of a new kind of society. Their image I call *Homo operans*.

All four portraits are usually held out as products of modern science. Yet all seem to make, or to imply, moral judgments on human beings: each presents our usual conduct as narrowly limited and unpleasant – violent, selfish, stupid or greedy. Such a verdict on humanity is not part of science. A similar outlook is found in the works of influential thinkers of all periods. Chapter 2 is therefore on the historical context of cynical and misanthropic attitudes in European thought. Biological interpretations of the human species can be judged only if this context is taken into account.

The misanthropy of the past was often supported by animal analogies; and the modern biological images depend on such comparisons. Hence they raise questions about the use of comparisons and the roles of analogy and metaphor in our thinking. This, our last introductory topic, comes in Chapter 3.

To assess the special features of *Homo pugnax*, we need authentic knowledge on animal societies, and on their similarities to and differences from our own. We also need exact information on all the variety of activities that people discuss under the heading of aggression, from bird song to war. These are the topics of Chapters 4 and 5. To judge *Homo egoisticus*, we need a true account of what we know of evolution, of the interaction of environment and heredity, of our sexual and other behavior, and of the political role of Darwinian ideas. These I discuss in Chapters 6 to 9. For *H. pavlovi* and *H. operans* we need to examine present knowledge of habit formation and learning, and also the scope of 'behavioral engineering' in education, in the factory and in therapy; we also face questions such as the roles in our lives of exploration, curiosity and play; and we ask the apparently naive question, why do we work? These come in Chapters 10 and 11.

Each biological image is an attempt to *reduce* human existence to a simple principle, or to explain human action by a single idea, such as natural selection or conditional reflexes. Chapter 12 therefore shows further the strange, sometimes ethically deplorable, consequences of such a doctrine. The images are usually determinist: they imply that all we do is fixed by our genes or by mechanical responses to reward or punishment; hence they seem to deprive us of moral responsibility. Chapter 12 also discusses the ancient question: are we free to make choices?

Once we have rejected reduction and determinism, we are free, in Chapters 13 and 14, to accept human language, music, and communication generally, in all their complexity. These are maintained by our unique ability to keep or to alter traditions and skills by teaching and imitation. The human species has the means to transmit information, from generation to generation, in a way that allows rapid progress. But is further progress possible? That is the question of the final chapter.

2 The pessimistic tradition

Is heav'n unkind to man, and man alone?
Alexander Pope

The misanthropic theme in European thought can be illustrated from the words of a few powerful writers, whose views reappear, variously transformed, in current writings on human nature. I begin with the most remarkable of them all.

Ancient pessimism
The whole of western philosophy has been described as a series of footnotes to Plato; but even this understates his achievement. Plato wrote profoundly not only on a great range of philosophical questions, but also on history, anthropology, sociology and politics. K.R. Popper describes him as one of the first social scientists and 'by far the most influential'.[1]

Plato was a member of one of the wealthy families that ruled Athens in the fifth century BC. In his early years he fought for Athens in her ruinous wars with neighboring city states. In about 387 he founded the Academy, the earliest European university on which we have detailed information. There he attracted a remarkable group of students and teachers; some of the former became prominent generals or statesmen. Women, too, it seems, were among his students. The Academy came to have a substantial political influence, partly through its pupils and partly by giving advice to governments on their legal codes and their constitutions.

Mathematics, especially geometry, was central in the teaching of the Academy; and this leads to a question we still have to face. Geometry begins with definitions or axioms (for example, that parallel lines never meet); these are treated as self-evident. From them proofs or theorems follow. It was assumed by the Greeks (and by many later philosophers) that theorems give us sure knowledge about real things. 'It thus appeared possible to discover things about the actual world by first noticing what is self-evident and then using deduction.'[2] But deductive proofs, by themselves, can be misleading. The Greeks discovered this too. The famous paradoxes with which Zeno the Eleatic (495–345 BC) teased the Athenian philosophers included the 'dichotomy' and the 'Achilles'. Consider a person, P, who wishes to move from A to B; before P can reach B, he or she must reach the half-way mark; P must then reach the next half-way mark, then the next and so on: whatever distance remains, there is another half-distance to be covered. Each half-distance takes a certain time; there is an infinite number of half-distances; movement from A to B must therefore take an infinite time. Hence all movement is impossible. Similarly, Achilles gives a tortoise a start, runs to overtake it, and reaches the point where the tortoise set off; but the tortoise is still ahead and Achilles must now reach the next point achieved by the tortoise . . . The tortoise is still ahead and always will be. Both proofs lead to conclusions that contradict experience. To learn about the real world we may use mathematical hypotheses, but our conclusions must depend on what we observe and sometimes measure.

Biology, like the other natural sciences, rests on observation of the real world. But in our time sociobiologists have revived Plato's method: they make assumptions, or propose axioms, on how evolution comes about; they compute the consequences of these assumptions; and then they tell us what human beings do, or must be expected to do.

In their pessimism too they follow the Greeks. Belief in powerful forces beyond human control greatly influenced Greek thought.[3] The Greeks acknowledged the success that sometimes rewards human effort, but even that has its dangers. In a famous myth, the scientist, Daedalus, achieves flight, but only at a cost: for his son, Icarus, hang-gliding over the Mediterranean, flies too near the sun: the wax of his wings melts; he falls, and drowns in the wine-dark sea.[4]

Hence we have the rather grim concept of moira, sometimes inadequately translated as destiny. 'Moira', writes Bury, 'meant a fixed order in the universe' which demanded 'a philosophy of resignation'.[5] Hope of a better future is illusory. Natural selection, in one aspect, is – despite its implication of change – a modern counterpart of moira, when it is used to postulate a fixed human nature imposed by our evolutionary past.

The idea of moira is part of Plato's background. He advocated change, but only to a point: his ideal society, once achieved, was to be static. It was still to be a small city-state, or polis. As Bertrand Russell has said, 'It has always been correct to praise Plato, but not to understand him.'[6] Until recently his *Republic* has been treated as expressing profound moral insight, but today condemnation of Plato's politics is more usual. In his ideal state the ruling guardians undergo an exacting training from 18 to 36 years of age. They are then qualified for a subordinate role in government. Such training is open only to those who have been bred for superiority, in accordance with what we now call a eugenic program. Moreover, the children belong to the community, not to the parents.

The guardians have absolute authority and maintain it by deception. The third book of *The Republic* describes an elaborate myth designed to ensure that the members of the three classes, the guardians, the soldiers, and the husbandmen and craftsmen, accept their lot.

> The rich man in his castle,
>   The poor man at his gate,
> God made them, high or lowly,
>   And ordered their estate.

For the nineteenth-century poet this confident statement reflected divine law. But, for Plato, similar assertions were to be concocted to enable the guardians to preserve the social order. People were to be told that the gods had ordained their status, 'for the fostering of such a belief will make them care more for the city and for one another'.[7]

In the twentieth century, this sounds like a modern totalitarian system, especially that of Nazi Germany, founded on an explicit ideology, maintained by a clique, bolstered by calculated lying, and making constant use of presumptions concerning genetically based superiority. But such a comparison is misleading, for several reasons. The guardians are not to live in luxury or to run the state for their own benefit: if they derive satisfaction from their role, it is from a feeling of superiority and of duties properly performed. Plato's ideal state is designed for the benefit of all, not of a privileged minority, even though most of the people are held to be incapable of managing its affairs. Moreover, there is an element of communism: the guardians, at least, live a communal existence, with little private property; nobody is rich.

Plato's other writings reveal him as much more than a reactionary, authoritarian elitist. One of the central themes is exactly opposed to arbitrary rule by a chosen few. Free men are urged to seek truth by open argument. This, elenchos, has come down to us as the Sokratic method. (But Sokrates wrote nothing; most of what we know of him we learn from Plato.) In the *Phaedo*, the dialogue in which the death of Sokrates is described, Sokrates warns his hearers against becoming misologists: no worse thing can happen to a man, he says, than to become a hater of reasoning or of rational debate. Another passage is on the impossibility of attaining certainty on some philosophical questions, such as the outcome of death. Yet, we are told, a man should be held a coward if he refuses to examine such questions on every side.

> For he should persevere until he has achieved one of two things: either he should discover, or be taught the truth about them; or, if this be impossible, I would have him take the best and most irrefragable of human theories, and let this be the raft upon which he sails through life – not without risk, as I admit.[8]

Plato's continued influence is partly due to his use of argument among contrasted characters as a means of trying out new and often antithetical ideas. His dialogues are dramas in which there is a clash of personalities and principles. It is therefore appropriate to quote a famous example of Platonic irony. In the *Apology*, an admirer of Sokrates asks the Delphic oracle whether anyone is wiser than Sokrates. The divinely inspired oracle replies: nobody. So Sokrates, who tries to question every presumption, looks for somebody wiser than himself, in order to prove the god wrong. But politicians, poets and artisans all turn out to be in fact quite ignorant, though wise in their own estimation. Sokrates concludes that he is wiser than the others only through his awareness of his own ignorance.[9] Plato and his colleagues were, no doubt, often authoritarian, elitist and misanthropic; but they were much more than that, and they continue to give us food for thought even today.

Misanthropy and Machiavellianism

Early in the postmedieval period, two men helped to found a modern political science that revived classical misanthropy and pessimism. Niccolò Machiavelli (1469–1527) and Thomas Hobbes (1588–1679) both resembled Plato also in being much influenced by the political instability of their times.

As a young man Machiavelli saw the disintegration of the Italian city states and their subjugation by the French and the Spanish. He played a part in public affairs for about 14 years from 1498, when Florence was a republic, and he held a senior appointment in the Florentine civil service. But, on the fall of the republic to the aristocratic rule of the Medici, he was dismissed, imprisoned and tortured. On his release he took to writing.[10]

The entry in the *Oxford English Dictionary (OED)* under 'Machiavellian' includes 'preferring expedience to morality; practicing duplicity, especially in statecraft'; Machiavelli himself has commonly been described as an odious character, and his famous essay, *The Prince*, was said to have been inspired by the devil. After his death he was condemned by the Inquisition – an honour, J.A. Mazzeo suggests, that was hardly deserved. Kenneth Bock states that Machiavelli 'stripped people of all their cherished humane qualities and reduced them to the status of animals governed most effectively by force and fraud.'[11]

As might be expected, these judgments are incomplete. Machiavelli was a scholar of great ability and strong views. He was, however, certainly pessimistic about the scope of human morality. Near the beginning of his most important work, the *Discourses*, he writes:

> All those who have written upon civil institutions demonstrate (and history is full of examples to support them) that whosoever desires to found a state and give it laws, must start by assuming that all

men are bad and ever ready to display their vicious nature, whenever they may find an occasion for it. If their evil disposition remains concealed for a time, . . . we must assume that it lacked occasion to show itself.[12]

Whether he is writing of republican government or of a model supreme ruler, this passage represents a central Machiavellian theme.

In *The Prince* he describes the method an autocrat needs to rule a state effectively; and he asks which a ruler should choose, to be loved or to be feared. He answers that, if he cannot be both, he should choose to be feared.[13] The rule of law and the maintenance of custom depend on rewards and punishments. And, in a passage on how princes should honour their word, he writes that there are two ways of fighting: by law or by force.

> The first way is natural to men, and the second to beasts. But as the first way often proves inadequate one must needs have recourse to the second. So a prince must understand how to make a nice use of the beast and the man. . . . A prince . . . should learn from the fox and the lion; because the lion is defenceless against traps and a fox is defenceless against wolves. Therefore one must be a fox in order to recognize traps, and a lion to frighten off wolves.[14]

A prudent ruler, he adds, should not keep his word when it would be inconvenient to do so: instead he should imitate the fox, for history is full of examples of the success of this method.

Nonetheless, Machiavelli also asserts the value to a ruler of having the good will of his subjects. There is no glorification of war. The aim is always the welfare of the citizens of a city state. Above all, in a period of unending warfare, Machiavelli represents a passionate longing for peace and stability with which we, in the twentieth century, can immediately sympathize. But he differed from us, and resembled the classical authors he admired, in having no notion of continuous progress. 'For Machiavelli, as for Plato, change meant corruption.'[15] The stable society he sought was based on the Roman Republic; it was to conform to an ideal design; and this design was to meet the demands made by an unchanging (and disagreeable) human nature.

The war of all against all

Thomas Hobbes, the most notable of English political scientists, is the philosopher most often cited in discussions of biological images of humanity. The adjective 'Hobbesian' usually refers to his misanthropy. For him, human existence is an incessant struggle for wealth and power. He holds that this view of humankind is evident to anyone who merely looks at his own motives. Unlike Machiavelli, he does not base his conclusions on a study of history or on any kind of empirical evidence.

His most important work is *Leviathan*.[16] This is the name he gives to a sovereign, or apparatus of state, with supreme power. (In the Book of Job, Leviathan is 'king over all the sons of pride'.) A community united by such a rule, whether monarchical or republican, is a commonwealth. He attributes the English civil war to the division of power among king, lords and commons; such calamities may be prevented, he says, if there is a single source of authority.

Hobbes arrives at the need for a single, sovereign power by elaborate argument. Like Plato, he was greatly impressed by geometry, and he believed that the axioms of a mathematical system enable us to reach novel conclusions about the real world by pure reasoning, just as some modern Darwinists and economists rely on mathematics to reach conclusions about human societies.[17] The political system proposed by Hobbes has a correspondingly formal, impersonal character. As in Euclid's theorems, there is much emphasis on rigorous definition. Hobbes agrees with Plato that an understanding of political theory, arrived at by logical argument, can ensure a peaceful, orderly society. He likens his Leviathan to the rulers of Plato's republic. But, unlike Plato or Machiavelli, he pays little attention to ways of persuading people to accept the sovereign's rule. It is one thing to have settled axioms and definitions, quite another to agree on what should be done in the ungeometrical, disorderly world of reality.

Hobbes supposes that natural men once waged an unremitting war of all against all. Man has the right to preserve his life and, indeed, self-preservation is a law of nature. (Hobbes does not distinguish clearly between what is natural and what is right – another similarity to some modern writers.) The famous phrase, 'nasty, brutish, and short', refers to the lives of human beings in this primitive condition. Hence the notion of the survival of the fittest, applied to our species, did not have to wait for Darwin. Hobbes also anticipated Malthus in regarding war as a result not only of the lust for power, but also of over-population. Perpetual conflict results in private quarreling, and also in war and a constant state of fear.

To improve their condition, Hobbes tells us, people must cooperate, just as bees and ants combine for the good of the colony. But, unlike most writers who use analogies from the animal kingdom, he emphasizes the differences of these insects from ourselves. The bees of a colony work together by nature; they do not compete, nor do they criticize the administration. (He might have added that they have no government to criticize.) Human societies, in contrast, are not natural but artificial. People enter into a covenant with the sovereign, and so give up some of their former autonomy. The covenant is entirely imaginary; but, at a time when attempts were made to annul the 'divine right' of kings, it was a convenient fiction.

In constructing his political system Hobbes presumes that nature has a lawful structure that can be found out by rational human beings; and he applies this principle, not only to the universe of physics, but also to human conduct. René Descartes (1546–1650) had already expounded the concept of the animal or human body as a machine. Hobbes, in effect, reduces moral behavior to mechanics. His analogy is with the laws of motion, which can be expressed in formal, geometric terms.

> For seeing life is but a motion of limbs, . . . why may we not say, that all *automata* . . . have an artificial life? For what is the *heart*, but a spring; and the *nerves*, but so many strings; and the *joints*, but so many wheels . . . ?[18]

But the people of such a world, though indeed prone to error and folly, can learn to behave rationally. To ensure that they do so, rational conduct should be instilled by servants of the sovereign who are allowed to teach only what the sovereign approves. Surrender to the authority of the sovereign is itself a rational act, and this makes possible escape from unremitting conflict. In thus emphasizing our rationality, Hobbes is unlike the modern biologists who present humanity as driven by fixed impulsions imposed by natural selection. This part of his argument also seems to imply that we are capable of choosing between alternative policies: we can take his advice on the organization of the state, or reject it.

In the stable society to which Hobbes looks forward, the interests of the sovereign and of the common people are supposed to be, in the end, the same. There is no allowance for a conflict of interest between economic classes. Private ownership and the inheritance of property are accepted as necessary features of a political system. Like Machiavelli, Hobbes emphasizes the importance of power in state affairs, rather than rights. In a list of human characteristics, he puts first 'a general inclination' for 'a perpetual and restless desire of power, . . . that ceaseth only in death.'[19] Human beings are, however, not inherently combative, but rationally so; they lust for power, not so much for its own sake but to achieve a comfortable life.

This, in an abbreviated form, is the message with which Hobbes leaves us. A reader may ask whether his conduct matched his philosophy. On this I merely quote R.S. Peters: 'there was never any suggestion that he was anything but a man of integrity, honest and kindly in all his dealings.'[20]

Modern misanthropy

Our last pinnacles of pessimism belong almost to our own time. F.W. Nietzsche (1844–1900) derives his importance from his popularity, not from novel contributions to philosophy or political science. In this he resembles some present-day writers on human biology. His writings had become famous by the end of the nineteenth century; and later they were required reading for Hitler's 'National Socialists'.[21]

Nietzsche was a man of great charm; he was also a notable classical scholar, and became a professor of philology in Basel when only 25; but his health was bad, and ten years later he retired. In 1888 he became insane.

He has much in common (while sane) with Machiavelli and Hobbes; but his views are more extreme and even more explicit than theirs, though much less well argued. His outlook is expressed in the titles of two of his works, *The Will to Power* and *Beyond Good and Evil*; but his most widely read book is *Thus Spake Zarathustra*. These works are marked by scorn for conventional morals and indifference to the welfare of the majority. Admiration is reserved for men of power, or *Übermensch* (especially Napoleon Buonaparte). His heroes are, like Plato's guardians, superior by their biological nature; but, unlike the guardians, they recognize no responsibility to lesser folk. Their objective, achieved if necessary by ruthless struggle, is domination. And they deserve to dominate, because most men, and all women, are contemptible.

Nietzsche was overwhelmed by what he saw as the imperfections of the people around him: he refers to them as though they had been badly constructed by a manufacturer. (Perhaps this included himself.) For Nietzsche, concern for others represents decadence: sympathy is a pathological condition. The assertion of moral principles is only a means by which a man of power can enhance his control over others. Ordinary, weak, imperfect people must be deceived and prevented from combining to impose their demands on the strong. As we see later, some sociobiological writings (no doubt unintentionally) seem to echo this doctrine.

Nietzsche, though he wrote against the current of his times, was not an entirely isolated figure; and in the present century he has been followed by other writers, among them, Oswald Spengler (1880–1936). Spengler, too, was for a time immensely popular, and he is still read. According to J.P. Stern, he was, and remains, the mainstay of his German publishers.[22] Early in this century Spengler had a considerable following in the literary world, despite (or because of) his pessimism. In *The Decline of the West*, first published in 1918 and many times reprinted, he joins those who compare a civilization to an organism: like a living animal, it is born, grows, becomes strong, then ages, and finally dies. All civilizations are fated to end in this way. All aspects of a culture – art, science, politics, social life and religion – decline together. (Historians are almost unanimous in rejecting this hypothesis.) Spengler's gloom about the future was strengthened by his views on race: the prospect of increasing marriage of whites with people of other colors horrified him.

Spengler is important, not for his original contributions to scholarship, but for his large following. The central feature of politics, for him as for Nietzsche, is the lust for power; and, as we see in Chapter 9, like Nietzsche he helped to inspire Adolf Hitler's 'National Socialism'. He is a leading twentieth-century representative of that disgust with, and contempt for, the human species that has been such a persistent element in European thought.

A central theme of this book is that such pessimism is neither rational nor moral. The facts of biology and of history justify a guarded optimism. The apparent support that pessimism receives from modern biology is spurious. In the ordinary affairs of most societies, misanthropy takes second place to its opposite. Most of the time, most of us cooperate with and help others. If, however, we become convinced that we are depraved by nature, we shall be less likely to succumb to the temptation of behaving morally.

But to justify even this moderately buoyant position, we must first dispose of the 'Beast in Man'. We begin to do so in the next chapter.

3   Animals and analogy

My mistress' eyes are nothing like the sun.
Shakespeare

In the preceding chapters I quote influential writers who liken human beings
– including themselves – to insects, rats, dogs, monkeys and other animals.
We meet such comparisons throughout this book. Whenever they appear,
we have to decide what, if anything, they tell us about ourselves.

In human communication generally, animals have many roles: they may
help to present a vision of the world (a cosmology or a religion); they may
stand for, or be contrasted with, what we expect or demand of ourselves;
they may be invoked to support moral sermonizing; and they may be used

Animals in religious myth and symbolism. Egyptian cat of the first century AD.

in literary imagery to make what we say more dramatic, forceful, or entertaining. They have a powerful hold on us. Indeed our languages are full of concealed comparisons with other species. Words such as sheepish, capricious or vacillating rarely call up an image of the animals from whose names they are derived. I have heard a zoologist, at a scientific meeting, describe how a small bird *doggedly* built a nest. These are examples of buried metaphor.

A metaphor may be defined as a figure of speech 'in which a name or descriptive term is transferred to some object to which it is not properly applicable'.[1] But a lexical definition does not at all convey the power of metaphor. It can give us quite new insight or interest in phenomena or ideas. Unfortunately, as Shakespeare implies in this chapter's epigraph, it can also be misleading. More than a century ago, F. Max-Müller (1823–1900) wrote:

> Whenever any word, . . . first used metaphorically, is used without a clear conception of the steps that led from its original to its metaphorical meaning, there is danger of mythology . . . or, if I may say so, we have diseased language . . .[2]

To avoid 'diseased language', and to assess biological images of humanity, we need to be alert to the kinds of analogical or other metaphorical statements in which the images are clothed.

Animalitarianism

But first, there is more to say on the influence of animals on our ideas, and on the uses of animals as symbols. The importance of animals, for people who lived before even writing was invented, is shown in the paintings and engravings made by paleolithic artists. More than 30 000 years ago people began to make marvellous representations of animals on the walls and ceilings of caves. Often the pictures are diagrammatic, but some allow us to identify the species. The meaning of these works has been much debated: they have been said to represent art for art's sake, magic to ensure success in hunting, fertility symbols, and home decoration. In different times and places they may have had all these roles, or none.[3] What we know for certain is the dominance of portrayals of animals. Our paleolithic forebears were, it seems, not landscape painters.

Modern preliterate societies have elaborate languages and concepts of nature, in which animals are similarly prominent.[4] The gatherer-hunter aboriginals of Australia are among those who identify their tribal group with particular animal (or plant) species: these species are totems, and are described by anthropologists as sacred; they stand for ideas separate from the uses or dangers presented by the animals. In particular, they often represent an individual's descent or family and are protected by taboos.

A member of the flying fox totem who eats flying fox 'is supposed to develop such acute constipation that the excrement inside him eventually bursts out in the most horrible ulcerations'.[5]

Sometimes, however, the religious attitude toward animals is clearly related to economics. The pastoral Nuer of the Sudan depend on cattle for food (milk and blood), for fuel (dried dung), for hides, and for much else: cattle are their wealth.[6] Cattle are classified, in immense detail of color and structure, by means of an elaborate vocabulary; and each person has a name taken from the words in which his or her oxen or cows are described. The color patterns of the cattle are named by analogy with wild animals that have similar appearances; but wild animals are, for the Nuer, of only secondary importance.

Another African group, the Lele of Zaire, though cultivators, are also hunters, and have an intricate classification of wild animals: this is part of the knowledge that enables them to hunt efficiently.[7] They too have an extensive system of food taboos. Among the prohibited animals are tortoises and baboons. Both are regarded as anomalous: the tortoise lays eggs like a bird, but walks on four legs; baboons are said to stand up like a man, and to wash.

When societies became literate, they retained powerful animal symbolism. From ancient Egypt, the hawk-faced god, Horus, is familiar; so are the sphinxes, with rams' heads, that guard the tombs of pharaohs. In Hinduism, are Nandi, the bull, Ganesa, the god with an elephant's head, and Hanuman, the deified monkey. About them stories with strange ramifications have been told, evidently, for millennia. In literate societies we also find real animals used as exemplars.[8] Democritus, a Greek of the fifth century BC, regarded human beings as pupils of animals – of the spider in weaving, the swallow in architecture, and the nightingale in singing. Diogenes, a century later, held animals to be happier, healthier and stronger than human beings because, above all, they possess no property. (This was long before the idea of animal territory was proposed.)

The most influential writer on this theme was Pliny the Elder (about AD 23–79) who, in his *Natural History*, contrasted the blissful existence of animals with that of a humanity subject to ambition, avarice and aggression: only human beings are (he says) without instinctive knowledge, and suffer accordingly. The duration of Pliny's influence is shown in the writings of Montaigne (1533–1592), whose essays, famous in his time, are still read. His account of the characters of animals is taken from Pliny; but he is especially concerned with the nobility and virtue of animals – a theme that goes back even before the days of Democritus.[9]

'Go to the ant, thou sluggard; consider her ways, and be wise.' This familiar advice, from *Proverbs* of an early Iron Age civilization, is matched by similar exhortations in the still earlier Hindu *Panchatantra*, from which the Greek, Aesop (sixth century BC), may have derived his fables.[10] We retain from him the wolf in sheep's clothing, the sour grapes and other everyday phrases whose primary meanings are often hardly remembered. In these moralities, animals are commonly held out, not as examples, but as symbols of human folly. In *Physiologus*, a much copied work of animal lore of the fourth century AD, the ape and the wild ass both represent the devil. And in the European middle ages a conspicuous theme of many Bestiaries is the ape as a sinner, especially as one with excessive and ill-gotten wealth.[11] It has been supposed that the name of the demon of wealth, Mammon (which is also the Aramaic word for riches), stems from the Arabic, *maimum* (monkey). 'Ye cannot serve God and Mammon', says Matthew; but I do not suggest that this admonition should be used to castigate the modern writers who say that we are just like baboons.

A modern ecologist, asked to comment on the relationship of human beings with animals, would probably reply in terms of biological economics: the quantity, say, of protein to be derived from hunting, the influence of

Animals in religious myth and symbolism. Nandi bull, south India.

agricultural practices on the populations of pests and on the resulting losses of food; guano as a fertilizer. . . It would be easy to give a long list on these lines. But, as the preceding pages show, it would be very incomplete. In our speech and our art animals may stand for our tribal relationships, our moral principles, or our beliefs about the world or the transcendental.

Scientific analogy

Such metaphors, however important in human communication, fall outside the natural sciences. Has metaphor any place at all in scientific writing? Charles Singer, the historian of science, relegates metaphor to the science of the middle ages. He describes the use of analogy by medieval 'natural philosophers' as an intellectual weapon.[12] (The metaphor is Singer's.) Writers of that period learned from ancient philosophers that there are four elements (earth, air, fire, water) and that the body has four humors (blood, phlegm, yellow bile, black bile). Hence they supposed there must be four principal organs of the body. Similarly, the year has four seasons; so there must be four ages of man. Singer describes an elaborate system of physical and physiological fours. He also remarks that a modern scientist may base experiments on analogy, but 'buries deep among the debris of his abandoned working hypotheses the memory of the analogical processes he has used.' Above all, he 'never adduces analogy as a proof of his conclusions' – a statement that suggests W.S. Gilbert's sceptical question, 'What, never?'

But analogy is a kind of metaphor, and it is here that we find an obvious use for metaphor in science. The *OED*, under analogy, gives 'equivalency or likeness of relations' and 'reasoning based on the assumption that if things have some similar attributes, they will have other similar attributes'. Similarly, the philosopher, David Hume (1711–1776), in a passage on the reason of animals, writes that analogy 'leads us to expect from any cause the same events, which we have observed to result from similar causes'. But he also warns us against 'weak' analogies.[13]

In the most extreme (or 'strongest') type of analogy, two or more systems may both be defined by the same equation. Isaac Newton (1642–1727) explained the motion of the moon by treating it as a projectile like a cannon-ball. Pierre-Simon Laplace (1749–1827) demonstrated the analogy between the 'flow' of electric current, of liquids in a conduit and of heat along a conductor. The analogies of the physical sciences extend to biological systems. We take it for granted that the lens of an eye operates by the same rules as one made of glass, and that the mechanics of Newton applies to animal skeletons as well as levers designed by engineers.[14]

Even biology, despite the bewildering diversity of several million species, has unifying theories. Modern genetics, with its genes and dominant and recessive characters, was founded by breeding peas, flies and mice, but its principles apply to all sexually reproducing organisms, including ourselves: we can infer, say, certain effects of cousin marriage among human beings from observations on quite remote species.

Hence the findings of genetics provide useful analogies. The other unifying theory in biology, that of evolution, is a quite different case, for it was *founded* on an analogy with human society: the incessant strife described by Hobbes became, in the hands of Malthus, the struggle to survive in human populations always tending to breed beyond the resources of the environment. Darwin transferred this struggle to the lives of all organisms, and so was led to his theory of natural selection. Later he added another analogy, when he linked the changes that take place during the domestication of plants and animals with the much greater changes of evolution: Darwin equated *selection* of seeds or bulls, by cultivators or stockbreeders, with the effects of variation in fertility and mortality in natural populations.

Genetics and evolution therefore exemplify two contrasted roles of analogy in science: in the first, wide-ranging conclusions are drawn from established theory; in the second, analogy precedes theory and helps it into existence.

Such differing functions for analogy are not confined to far-reaching theories: they appear also in current ideas on the ways in which we rear, and think of, young children. Some comparisons of our own infants with those of other species are helpful. As an instance, mammals such as tree shrews, which secrete very concentrated milk, feed their young only at long intervals – 24 hours or more. Monkeys and apes have dilute milk, and their young have small, frequent meals. Human milk changes in composition during a feed but is always dilute. Hence a simple analogy holds between human and monkey milk. In some communities, infants are conventionally fed, from a bottle delivering rather strong milk, at four-hour intervals. Many mothers have nonetheless become impressed with the advantages of breast-feeding, but have been upset by their infants' protests when breast milk is offered only every four hours. The remedy is to do as apes do, and to feed the infant at shorter intervals, or even on demand.[15]

Other analogies concerning infants are of a different kind. 'Social smiling' occurs from about one to six months. Described objectively, as if performed by some remote species, it is a symmetrical contraction of the risorius muscles (at the corners of the mouth), usually made in response to a human face or face-like pattern. An ethologist may draw an analogy between it and the grimaces (say) of an infant monkey, and so classify it as a social signal typical of the species, *Homo sapiens*. An infant's crying, too, may be classified in this way. Of the three kinds of crying, one, the hunger cry, resembles, in its function, certain sounds made by other mammals; so does the pain cry.[16]

But a child also cries when it wants to be picked up and cuddled. In the last sentence I give up objective language, and discard the analogy with other species: instead, I write of the infant as a human being. Yet the sentence is not meaningless: on the contrary, it is common sense.[17]

If an infant's smile or cry is equated with the social signals of animals, what kind of assertion is being made? It is hardly a hypothesis. True, we may be led to suggestions about the causes and effects of the behavior; and in fact harmless experiments, of an ethological type, have been done on the stimuli that make infants smile. But to call an infant's smile or cry a social signal is, primarily, to put these actions in a particular ethological *category*; and the main effect may be to determine how we see them. We may even be persuaded to treat the infants themselves as circus animals to be tamed and trained.

As this example shows, many analogical statements are conveniently seen as *classifying*, and their implications should be carefully examined. D.O. Edge quotes a head teacher who talked of children who need to let off steam and who also likened some children to bombs that must be allowed to explode. Hence he implied that the children were dangerous. 'Some fears', writes Edge, 'are appropriate' for we are justified in alarm at an overheated boiler or a bomb, 'but we should be critical of any move to

An infant's smile. A species-typical social signal of Homo sapiens?

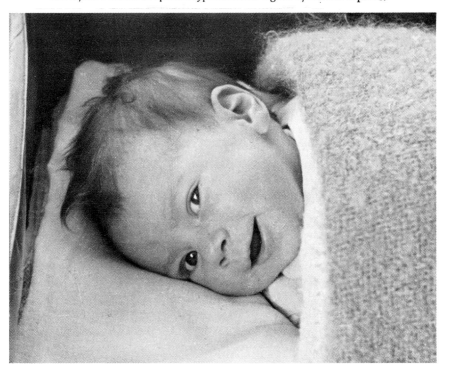

transfer this wariness towards a lively child.' Metaphor can bring about such a transfer unobtrusively.

The changes of attitude, to which Edge refers, include the idea of a human being as a machine or as an animal, both of which are legitimate in some contexts. But a human being, young or old, is primarily a person. I may be justified in saying that a colleague has a screw loose; but that does not authorize me, or his doctor, to treat him merely as a robot with a defective component.

Homology

The temptation to make uncritical transfers appears in the misuse of a special, biological type of analogy, that of homology, in which we meet again the Beast in Man. I give two examples. John Bowlby, a psychiatrist, in an important work on mothers and children, writes: 'whatever behaviour is found in sub-human primates we can be confident is likely to be truly homologous with what obtains in man'.[19] And a zoologist, Paul Leyhausen, has stated that:

> overcrowded human societies reflect the symptoms of overcrowded wolf, cat, goat, mouse, rat or rabbit communities to the last detail . . .; the basic forces of social interaction and of organization are *in principle* identical, and there is true homology between Man and Animal throughout the whole range of vertebrates.[20]

In biology, the word homology usually refers to structures. For instance, the human middle ear has three small bones, but in embryonic development at first it has only one; later, two rudimentary bones move from the embryonic lower jaw into the ear, leaving the jaw with only one bone. But an adult reptile's lower jaw has three bones and its ear, only one. We, like other mammals, are descended from reptiles. Hence two of the minute bones in the human ear are said to be homologous to two much larger structures, with a quite different function, in the jaw of (for example) a crocodile.[21] Many such instances are known.

Organs, then, are said to be homologous when they have similar spatial relationships, especially in embryonic development. The anatomical concept was clearly stated in 1843, before evolution was accepted.[22] Later, these at first sight inexplicable phenomena came to be seen as a result of evolutionary change from a common ancestral condition. The changes can sometimes be traced in fossils. Some fossil reptiles have jaws intermediate in structure between those of typical reptiles and mammals.[23] Homologies are, of course, most interesting when the adult structures concerned differ in function. They are usually contrasted with the functional analogies of organs such as an elephant's trunk and a monkey's hands: these have similar uses, but are anatomically distinct.

Since the concept of homology is anatomical, it seems that it can hardly apply to behavior. Yet, as the quotations above show, it is applied even to human conduct. What then are these writers trying to convey? A common evolutionary origin is implied; but we have no means of confirming or denying such statements. Moreover, even if they could be confirmed they would tell us nothing about the significance of the behavior now. A second implication is that the features referred to (for example, ownership of property) are inevitable, because they are in some way fixed in our genes. This raises a question which is one of the main themes of the present book. I say now only that the existence of a superficial similarity gives no evidence concerning the genetics or the development of behavior.

A third implication is that certain kinds of animal behavior reappear in ourselves *with the same function*. This is a move from what homology usually means to something quite different. As we have seen, homologous organs often differ fundamentally in adult function, as when bones that were once part of the eating apparatus become, during evolution, part of the organs of hearing. Evolutionary speculation cannot, in any case, tell us about the function of behavioral features: for that, we need to observe the behavior as it is today and, if possible, to experiment on it.[24]

The idea of behavioral homology is therefore of questionable use even when animal species are compared. For example, the cats, such as leopards, lions and tigers, all sometimes bare their teeth; so do many species of monkey. For some species we have evidence that this is a 'threat': that is, the grimace sometimes drives away others of the same species. It adds little to say that the bared teeth of a lion are homologous to those of a leopard, but only analogous to those of a monkey. Such statements imply no more than guesses about the evolution of the behavior. We do not know how the ancestors of the cats or monkeys behaved, but only what existing species do. And, most important, we cannot find out the function of an act (such as baring the teeth) in one species, by studying it in another.

As for homologies of human and animal conduct, they may sometimes be thought of as descriptions, but often they emerge as rhetoric: they are an attempt to influence our opinions about the behavior described, and so exemplify the *persuasive* function of analogy.

Political analogy

Persuasion by analogy is particularly prominent in politics and the social sciences, where superficial similarities may be taken for evidence in support of an argument. A.J. Toynbee's famous (and controversial) work, *A Study of History*, contains an account of certain human societies which, Toynbee holds, are examples of arrested development. Among them are those of the Spartans and the Eskimo. The initial turn taken by these societies, he writes, was 'a blind alley'; and he finds '*corroborative evidence*' (my emphasis) in a comparison with the eusocial insects. He discerns in 'an ant-heap and in a bee-hive . . . the same outstanding features that we have now learnt to recognize in all . . . arrested civilizations' – that is, division into castes and specialization of function.[25]

The 'societies' of the eusocial insects have remained stable, generation after generation, perhaps for millions of years. Their behavior is uniform throughout each species, and is adaptable to new conditions only in minor ways. In these respects the insect communities are the opposite of our societies: they are useful in descriptions of human conduct especially if one wishes to *contrast* their conduct with ours. The idea that such remote comparisons provide evidence for a historical theory is, of course, absurd.

As the political scientist, W.J.M. Mackenzie, remarks, biological analogies are nonetheless extensively used in the social sciences, and – he says – give much non-logical satisfaction.[26] For another political scientist, G.D.H. Cole, this is a source of regret. 'Again and again', he writes,

> social theorists, instead of . . . employing a method and a terminology proper to their subject, have attempted to express the facts and values of Society in terms of some other theory or science.[27]

And so society (he continues) has been likened to a mechanism, an organism, or a person. In this passage Cole seems to urge social scientists to surrender all their metaphors, which is perhaps asking too much.

We are, however, justified in complaining when a literary device conceals an attempt to influence our attitudes. A writer refers, say, to the state as a father, and to its citizens as children, and so implies that the state looks after the citizens and that the citizens, in turn, should obey the state authority. (Paternalism is often a euphemism for autocracy.) For the argument of this book, the most prominent of such devices is the use of the prestige of science to shore up a prejudice. As soon as Darwin's theory became scientifically respectable, it was called upon to support at least three political positions, each incompatible with the other two. In a famous letter to Friedrich Engels (1820–1895) Karl Marx (1818–1883), expressed delight at Darwin's discovery, among animals and plants, of the competitiveness and Hobbesian war of all against all which (Marx held) was characteristic of English capitalist society.[28] A later socialist writer, Enrico Ferri, whose work was widely circulated in Italian, French and English, wrote:

> Darwinism has proved that all the mechanism of animal evolution is reduced to the struggle for existence. . . In the same way . . . social evolution has been reduced by Marxian socialism to the law of the *struggle of the classes*. [Original emphasis.]

In a preface to Ferri's book, J.R. Macdonald – Britain's first Labour Prime Minister – stated that 'socialism is naught but Darwinism economised, made definite, become an intellectual policy, applied to the conditions of human society.'[29] Meanwhile, in the opposite camp (as we see further in Chapter 9), the 'Darwinian' concept of the struggle for existence was used to oppose all measures of public welfare, from old age pensions to state education. Lastly, the Russian anarchist, P.A. Kropotkin (1842–1921), in his *Mutual Aid*, assured his readers that natural selection produces an impulse toward cooperation, not conflict.[30]

Imagery inevitable

Yet, despite its many hazards, metaphor is inescapable. It has indeed been justly held to be a model of the ways in which we learn.[31] The transfer of meaning in a metaphor often gives us new understanding of facts or ideas that were at first obscure or baffling. I now, therefore, sum up the ways in which metaphors, especially analogies, can with care be useful even in science. The first is naming. Many names, notably of structures, are based on remote similarities. J.Z. Young, a neuroscientist, has protested at this: 'we should try', he writes, 'to speak of the brain in terms that indicate the operations it performs, rather than by simply giving it names indicating the similarity of its shape to a nut, a rind, an olive, or a pear'.[32] He refers here to the amygdala (walnut), cortex, olive and pyriform. These terms are, however, not misleading: botanists are never consulted about the amygdalae or the superior and inferior olives. A second, related use is in description. The movements of a nectar-bearing honey bee, or of a courting stickleback, are called dances. The similarity to human dances is remote, and again nobody is misled: experimenters on bee behavior do not (as far as I know) play waltzes or reels to these animals.

Third is the persuasive use of analogy. Suppose I wish to convince an audience that certain popular ethological writings are misleading. I may suggest that an ethology derived from them would resemble a historical thesis based on cloak-and-dagger romances ('zounds!' and 'forsooth!'). Similarly, when later I attempt a critical analysis of neo-Darwinism, I could write that the theory of natural selection is a cross between a sacred cow and a white elephant. Wisecracks, however, are preferably accompanied by sequential argument.

Fourth, analogy may be used to display the logical character of an argument. We are told, perhaps, that baboons are closely related to *Homo* and some, like early human beings, live on grassy plains. They are also territorial. Hence – it is held – we must be expected to be compulsively territorial and owners of property in land. I reply that subadult baboons regularly herd young females; and I ask whether we should therefore conclude that kidnapping young women for a harem too represents a universal human instinct.[33] If not, what is the validity of the argument about territory?

Last are aspects of metaphor to which it would be possible to do full justice only at great length. Metaphor plays a crucial part in helping original thinkers (and others) to look in new and rational ways at natural phenomena, or at intellectual problems. Here we have the cognitive uses of metaphor – its role in enabling us to think constructively.

As an instance, one of the landmarks (a metaphor) in the history of science was the discovery by William Harvey (1578–1657) of the circulation of the blood. Owsei Temkin comments on the prominence of metaphor in Harvey's *De Motu Cordis* (1628), in which the discovery is described. The heart is likened to a Prince 'because the heart originates first, because it does not depend on other organs, notably brain and liver, but has the organs of its movements in itself . . . so that all the rest of the body depends upon it.' Harvey also compares the heart with the sun, and this, writes Temkin,

> elucidates Harvey's very concept of circulation. The sun makes vapors rise from the earth; these vapors in turn become condensed and change into rain which moistens the earth . . . Likewise in man, the blood reaches all parts of the body while warm and nutrient, but is itself cooled and worn out so that it returns to the heart . . . in order to be perfected.

In a book full of metaphors, Temkin remarks, those of the sun, monarch and heart present a 'system of physiology which . . . is not modern at all' but is of vast historical importance.[34]

As another instance, of equal historical prominence, D.O. Edge quotes Johannes Kepler (1571–1630) on the universe as a system rather 'similar to a clock'.

> This new, scientific, metaphor was the vehicle for a great cognitive advance. It led inexorably to Newton and to the . . . whole powerful structure of modern physics. *But clocks are things you tinker with.* . . . If the world comes to be perceived as genuinely 'clocklike', . . . other attitudinal changes are likely to follow.[35]

Lastly, we may take the notions of lines of force in the account of magnetism, or of heat as a fluid, used by James Clerk Maxwell (1831–1879) and by William Thomson Kelvin (1824–1907). R.R. Hoffman describes the use of metaphor by these physicists, and by Michael Faraday (1791–1867), the principal founder of modern knowledge of electricity, who 'at almost every turn . . . relied on metaphor'. Faraday proposed a *field* theory; he also wrote of *elastic* fluids.[36]

These examples, among the many possible ones, warn us not to dismiss imagery as merely a literary device. Metaphors do not provide proofs of theories, but they are unavoidable features of our communication and our thinking. In this book we are concerned especially with metaphors of which the impact is strengthened by equating human beings with animals. Animals themselves have a powerful hold on our imagination. Hence, when we find

Homo sapiens represented as an animal, we must be conscious of what kind of statement is being made: whether it is descriptive or persuasive; a literary device or a source of hypotheses; whether it is helpful or misleading. And, always, we should remember Shakespeare's warning about his mistress' eyes.

Part 2

Homo pugnax:
the violent species

See you now:
Your bait of falsehood take this carp of truth:
And thus do we of wisdom and of reach,
With windlasses and with assays of bias,
By indirections find directions out.

Shakespeare: Hamlet

The portrayal of human beings as inherently violent is supported by analogy. We, like other species, are a product of evolution, and are said to 'inherit' enmity from our animal ancestry. Social dominance among animals becomes social status in human communities, and the holding of territories by animals reappears as human ownership of property. These are represented as 'instinctive' or 'innate' and as involving destructive violence, or aggression.

Chapter 4 therefore compares the social lives of animals with those of human beings, and shows up the limitations of the idea of instinct. The argument from animals to humanity then breaks down at every important point: biological analysis cannot cope with either our intelligence or our social life.

Just as our violence has been said to reflect our animal ancestry, so 'primitive man' has been described as especially aggressive. In Chapter 5 we find that this belief is not based on authentic knowledge of our 'Stone Age' ancestors, or of any modern groups. People are often violent when they should be peaceful, and this is held out as natural; but, in that case, so are benignity, helpfulness, moderation and rationality. Violence is of many kinds, and cannot be reduced to a single impulse. Often, it is learned; sometimes, as in an army, it is systematically taught. The opposite kind of training can also be successful.

4 Communication and instinct

*If a man elect to become a judge of these grave questions . . . he will commit
a sin more grievous than most breaches of the Decalogue, unless he avoid a
lazy reliance upon the information that is gathered by prejudice and filtered
through passion, unless he go back to the prime sources of knowledge – the
facts of Nature.*

T.H. Huxley[1]

'Behavioural science', we are told, 'really knows so much about the natural
history of aggression that it does become possible to make statements about
the causes of much of its malfunctioning in man'. Moreover, 'aggression . . . is an instinct like any other', and 'in human social behaviour,
innate species-specific . . . response norms play a far greater role than is
generally assumed'.[2] These are statements from an influential popular work
by K.Z. Lorenz. Similarly, Nikolaas Tinbergen, the principal founder of
the modern science of animal behavior (ethology), states: 'In order to understand what makes us go to war, we have to recognize that man behaves
very much like a group-territorial species.' In saying this he equates his
own species with, for example, baboons, of which some groups occupy
regions of African plains to the exclusion of other groups. Although –
Tinbergen tells us – 'we are forced to speculate', nonetheless 'comparison
of present-day species can give us a deep insight, with a probability closely
approaching certainty, into the evolutionary history of animal species.'[3]
He also states that the same comparative method, that is, the use of analogy,
can help us to understand human conduct. We now therefore ask what are
the relevant facts of animal behavior; and what, if any, are the real likenesses
to our own practices.

Signals
Since social ethology is founded on the interpretation of animal
signals, we begin with them.[4] The peacock's tail is a conspicuous (but little
studied) example. 'The jewelled splendour of the peacock', wrote D'Arcy
Thompson in 1916, 'is ascribed to the unquestioned prevalence of vanity
in the one sex and wantonness in the other'.[5] But ethologists now rarely
express themselves in such terms, and their writings are correspondingly
less enjoyable.

The modern analysis of animal signalling began as a rather mechanistic account in which the traditional concept of communication among persons was discarded. In Latin *munia* are duties; *communicare* was originally to make something common, to share it. Hence the idea of communication had a moral ingredient: it implied intention, and included the principle of social obligation. In contrast, the signals described by ethologists, at least in their simplest form, resemble those that might be used on an automatic railway: green, and the machine goes; red, and it stops. Communication becomes the passage of *information*, as that word is now used by engineers and mathematicians.

The simplest social signals each consist of a single, discrete event such as a squawk from a bird (an alarm call), a flash of light from a mating firefly or a puff of an odorous substance; and in the ideal case each provokes a consistent response, such as taking cover or approaching the signaller. The mindless quality of such interactions is seen in the response of male moths to a substance secreted by a female: put a mature female in a cage in the open, and every male of her species, for several hundred meters down wind, compulsively flies toward her. The odor of a bitch in heat has a similar effect on the dogs in her neighborhood. These are actions of chemical signals, or pheromones.

We now have evidence of human pheromones. Young women who live in college dormitories tend to have synchronized menstrual cycles and, when they are celibate, the cycles are long; but when the women begin to go out with men the cycles shorten. Similar effects observed in experiments on animals depend on chemical signals. The menstrual changes evidently depend on skin secretions.[6]

Probably, human beings of both sexes secrete quite an array of such signals. Nonetheless, human responses to social odors differ greatly from those of animals. To people in many communities, body odors are repellent: often, they are removed by elaborate washing, and even disguised or replaced by perfumes. Here is both a departure from our 'animal nature' and also a great variety of customs and attitudes. Such variety is not found within any one animal species.

There is, however, more to be said, for the visible and audible signals of animals rarely consist of simple on-off signs with mechanically constant effects on the recipient. A monkey in a territorial encounter may chatter, shake a branch and grimace; but, despite its vigor and elaboration, the display does not always drive off an intruder. An engineer would be dissatisfied with an automatic train that stopped at a red light only nine times out of ten. So would the passengers. Yet in some social encounters probabilities are still lower.

As another complication, development of behavior typical of the species may depend on special experience early in life. The songs of birds such as chaffinches, song sparrows and larks are normal only if the birds hear adults of their own species during an early sensitive period. Deprived of this 'training', they sing a different tune.[7] Moreover, distinctive individual sounds do not always depend on abnormal rearing. Each species of gull has a typical repertoire of calls, but there may be subtle variations among individuals. The chicks distinguish the calls of their parents from those of others, hence do not hide when a parent approaches.[8]

Chicks of another sea bird, the guillemot, given a choice between a parent's call and that of a stranger, move toward the former, and this ability appears immediately after hatching. In an earlier period such a response might have been attributed to a mysterious filial instinct, beyond rational analysis. But recorded calls have been played to some of these birds while they were still in the egg, and the newly hatched chicks, given a choice between the familiar and another call, chose the former: they had learned to discriminate before they had hatched.[9]

These and many other, similar findings on how behavior develops illustrate a fundamental principle. Each individual develops in an environment, and environments vary. A pattern of behavior (or any other feature), apparently quite fixed, may need a special environment if it is to develop normally. Change the environment, and the result may be very different from the typical. Sometimes, notably in human development, no typical pattern exists; an example is the language we speak. The fact that we have few standard signals recognizable throughout our species is an aspect of our capacity to adapt our conduct to novel situations.

The preceding paragraph may suggest that after all we can learn something about ourselves by studying animal behavior. Well, so we can. But we learn most when we examine the *contrasts* between ourselves and other species. We can see this more clearly if we look further at the 'meanings' of animal signals. By 'meaning' I refer to the internal state of the signaller (for example, readiness to mate) and also the effect of the signal on the recipient. Most signals or displays convey one of six or so kinds of message, each of which can be roughly rendered in exclamatory English.

First is 'Look out!': alarm calls are common among birds and mammals. Second is 'Go away!': such 'threats' occur at the boundaries of territories, during dominance interactions and when a female weans her young. Third we have many versions of 'I love you!': prolonged and elaborate courtship is widespread. Fourth, an animal may seem to signal 'I'm friendly!', and so avoid a clash. Fifth are distress calls, as when a young bird or mammal is in need of food or warmth. Lastly, a parent may signal 'Come here!' to the young. Despite many elaborations, all such messages are expressive: they indicate what in a human being would be called an emotion or an attitude. They are not descriptions of events occurring in another time or place, nor are they arguments.

There is, however, at least one exceptional case. Consider the following message: there is a new, large source of food, that smells like this, 200 meters to the north-west of here. Such a statement is beyond the powers of any non-human mammal, even a chimpanzee. Yet it is within the powers of a honey bee. When a worker bee returns to the hive, the movements (dances) and sounds it makes can convey such information to others. The messages are fundamentally different from those listed above, for they indicate not merely the state of the signaller but something external, recorded in the past: they are referential. Moreover, when bees swarm, some go out, find cavities suitable for a new hive, return and dance at length, indicating the place they have found. Eventually, the swarm flies off to a site so indicated. Hence, if we include the bees, we find that not all animal signalling is exclamatory.[10]

Human signals

Yet, despite the bees, human languages have little in common with the signals of animals. The principal exceptions are found in the behavior of infants. In the previous chapter I describe how a child, before learning to speak, communicates by sounds and facial expressions. These can be studied by ethological methods. They seem to be universal, that is, typical of *Homo sapiens*. As we know, an infant who has not been fed utters the hunger cry; one that has been hurt gives the quite distinct pain cry.[11] If the analogy with animals were complete, we should be able to state, with fair probability, what kind of response the parent will make. We can do this for other mammals, from mice to monkeys: the young make distinctive sounds that induce instant attention from the mother – such as picking it up, taking it to a nest or refuge, licking it, and allowing it to suck. But human conduct is much less predictable: some mothers take no notice of a baby crying (unless it is the set time for feeding); some rock the baby but do not pick it up; some cuddle it; some promptly put it to the breast. Feeding practices are equally diverse. Some mothers not only feed their babies with a manufactured mixture from a bottle, but regard breast-feeding as improper. The conventional age for weaning from the breast or bottle varies from about six weeks to three years.[12]

Hence, even if infants have standard signals, parents do not treat their babies in standard ways. Moreover, as the gentle reader may have already remarked, a stimulus-response account of the conduct of babies and their parents is absurdly inadequate. Every infant has, from birth, a distinct personality. And most mothers (or foster mothers) treat an infant as a person, not as a circus performer to be trained or conditioned.

Most human communication, unlike that of infants, is by means of words. Some of the features of our speech may be found in the signals of animals; yet human communication is radically different from that of other species. Whether similarities or differences are emphasized may depend on our interests (or prejudices) and on the features chosen for comparison.

To illustrate both similarities and differences, I return to the bees. The sentence above, in which I represent a bee's discovery of food, resembles things we say, not only in referring to external objects or events, but in its potential novelty. A central feature of human language is that it is productive: we daily utter sentences that we have never uttered, or even heard, before. A bee, similarly, can indicate a source of food in any direction and, up to a point, at any distance, whether it has done so before or not. Moreover, not all a bee's 'conversation' is about food: other messages, delivered to a swarm, may point to a new nest site.

Nonetheless, the contrasts with our speech are more significant. Our words usually have an arbitrary relationship with their meaning. (Shriek, grunt and a few others are exceptions.) But the dances of a honey bee are iconic, like a map or a diagram: they are miniature representations of the flight needed to reach the target. Hence they have a very narrowly defined scope: they all convey direction, and many convey distance; when they refer to food, they may also indicate its relative amount; when they report a place for a new hive, they report its suitability. But that is all. Yet a further contrast is in the uniformity, in time and space, of the 'language' of bees. With minor exceptions, the honey bee, wherever it is in the world, uses the same code; and these signals go on, generation after generation, without detectable change. Correspondingly, bees do not, as we do, gradually learn to make their signals; and they have nothing like our changing traditions.[13]

By now the reader may have become impatient with this ponderous comparison. Consider, for example, what I am trying to do as I write these words; or what the reader is doing while reading them. If I am successful, I am conveying elaborate information about a great variety of events distant in time and space from the reader and from me; most of these events I have not myself observed; some are hypothetical. Above all, I am presenting an argument based on logical principles. Animals do not do these things.

Social status

Most animal signals either regulate interactions within groups (including families and mated pairs) or maintain territorial boundaries between groups or individuals. Of interactions within groups, those of dominance have been much used in the attempt to explain human violence ethologically. Unfortunately, nearly all the words we use to describe the social lives of animals are those we use also for our own. The descriptions, therefore, in effect, *presume* a likeness between animals and ourselves, when we should be *asking* whether there is a significant similarity. For instance, I use the word status for the relationships of dominance and subordinacy. In human affairs, status has a number of meanings, of which one is legal: 'the standing or position of a person as determined by . . . membership of some class of persons enjoying certain rights or subject to certain limitations' (*OED*). In sociology, status may be used instead of class: its 'real significance' is then as 'a new and modernizing term for rank'.[14]

The use, in ethology, of words that describe human conduct leads to perennial difficulties: it encourages us unthinkingly to look on animals as if they were human (that is, anthropomorphically), or human beings as if they were animals (zoomorphically). As a result, popular ethological writings have an insidious and widespread influence on current beliefs and attitudes. Here is a passage, not at all exceptional, from a highly literate thriller, set in the Darkest England of spydom. A British agent is talking to somebody whose knowledge of English idiom is imperfect.

> 'I'm here now, sticking my little neck out!' Richardson paused. 'What I mean – '
>
> Freisler raised his hand. 'No. That I do understand. To stick the neck out is a very ancient gesture of trust and submission in the animal kingdom'.[15]

The allusion is to a story, from a book of anecdotes about animals, in which K.Z. Lorenz describes a subordinate wolf presenting its neck to a dominant male in a gesture of submission.[16] The story has reappeared in a scientific journal: in 1964 J.S. Huxley, in a general review of ethology, referred to wolves as deliberately displaying the most vulnerable part of the body in an attitude of appeasement.[17] It is implied that wolves are capable of symbolic gestures, like thrusting forward the hilt of a sword in a declaration of allegiance. This alone should provoke doubt. And, when actual encounters were recorded by careful observers, the story was shown to be wrong: the wolf that presents its flank is the dominant animal.[18] This example illustrates one kind of difficulty for anyone who wants to know about animal behavior: the most amusing and widely distributed writings are often fictions.

Now let us examine status among animals objectively. The early scientific accounts were conveniently simple: one individual dominates all the others, a second dominates all except the first, and so on. The seminal paper on such 'straight-line peck orders' was based on observations of birds, especially the domestic fowl.[19] Later, but before modern ethology was founded, the subject was taken up by American zoologists. Their name for such a system was 'dominance hierarchy' – an expression which, to persons with a classical education, suggests rule by priests.

An obvious question is: just what is meant by dominance, and by its obverse, subordinacy? A possible answer would assign to one individual an array of characteristics: if male, he would be superior in strength and perhaps in intelligence; he would have a number of 'rights', such as priority of access to place, food and females; and he would also have 'duties', such as leading and defending the group. I call this model the Tarzan principle. A group of gorillas has a single dominant male with several of these characteristics. He is the largest member of the group, and may weigh 180 kilograms and reach a height of two metres. The grey hair of his back marks him as older than those with black hair. He can displace any other group member; he determines the direction of march; and, if two females scream and bite in a clash, he moves toward them, grunting, whereupon they stop. His status is maintained without injurious violence.

Unfortunately, the Tarzan principle fits few of the other social systems closely studied. All the 'rights' and 'duties' mentioned have been used to define dominance; but the various social roles are often filled by different individuals. Many species of Primates – apes, monkeys and lemurs – have been observed. Among them an individual is commonly labelled as dominant if it can displace others; but that individual may not have corresponding mating success. It is often stated that males, called dominant because they have pride of place, sire more young than others; but the evidence for this is weak.[20] And the leader of a group, that is, the individual that determines the way the group goes, may not be dominant on any other criterion. There may indeed be no one leader: the direction taken by a large group may depend on the combined influence of several individuals.

To establish such facts, many months of close watching and accurate recording are needed. The interactions most easily observed are the vigorous ones, such as loud screeching, chases and assaults. An example of the importance of looking beyond them is given by Phyllis Jay, in her study of the hanuman langur. In the north of India this attractive (and sacred) monkey lives in small groups. The males of a group form a status system, but only an unobtrusive one. If a male is approached by another, the first one may withdraw; or, if one stares or slaps the ground, again the other male goes away. Even harmless clashes are rarely seen. A casual observer of this species in natural conditions finds it difficult to identify the dominant animal.[21]

Male Primates are usually dominant over females, at least in the sense that they have prior access to food; but there are exceptions. Alison Jolly has described the social behavior of the ring-tailed lemur of Madagascar. These animals are unlike the more familiar Primates in the importance of social odors: during squabbles among males the long, hairy tail is rubbed on skin glands and then waved about in front of the other male.Such 'stink fights' are evidently harmless. So are encounters between the sexes; but in them the female is 'dominant': she may even snatch food from a male, and give him a cuff as she does so. Similarly, female squirrel monkeys are dominant over males in the sense that, at most times, males that approach females and their young are driven off; even the young 'threaten' adult males.[22]

These examples only hint at the diversity of social structures. Some Primates live in 'nuclear families'. The small dusky titi monkey of South America forms pairs of which the members, when still, keep in close contact with tails intertwined; they show no signs of dominance relationships. Another small monkey, the marmoset, has an even closer family life: both parents look after the infants and older juveniles help to carry them. Among the apes, the gibbons also form family groups marked by much play among the members.

In contrast, some macaques, baboons and apes (chimpanzees and gorillas) live in large groups. Each species has its own social system and differs not

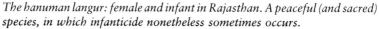

The hanuman langur: female and infant in Rajasthan. A peaceful (and sacred) species, in which infanticide nonetheless sometimes occurs.

only in its dominance relationships but also in other features: sexual patterns range from pairing to complete promiscuity; care of young may fall entirely to the mother or be shared with the father; sometimes adults that are not parents behave in an avuncular way. Combined action against predators may occur but is not universal. Hence a balanced account of a species should include information not only on dominance and intolerant behavior but also on other social roles.

Some popular writings continue to give the impression that the groups of all species, at least of mammals, have an 'overlord' or 'tyrant' at the top and below him an array of others each with a rank. Superior status is said to be achieved by 'aggression'. Now that we have exact knowledge of many species, such stories are no longer plausible. Still less defensible is the argument that our 'animal nature' obliges us always to seek the authority of a Leader.

Territory
There is similar diversity in the interactions between groups. This subject, too, has attracted thriller writers. Here is an author's digression during a frightening incident, when Our Hero meets an opponent in Hong Kong:

> Man is one of the territorial animals and contact between two members of their species gives rise to an immediate issue when one or the other is on his own ground. For both of them – but particularly for the intruder – there is the primitive decision to be made: to fight or run.[23]

And here is a statement from a widely discussed work by a zoologist, in which animal territories are referred to as property-holding:

> In man the primitive systems of property-tenure are essentially the same as those [of animals] already described. Territorial claims are, or were, vested in the nation, the tribe, the family or the individual . . .[24]

In the passage from the thriller, it is taken for granted (by analogy with the supposed conduct of other species) that there are no alternatives to fighting or running; and in the second passage animal territories and human property are equated. Neither assertion corresponds to the facts.

Holding a territory (in the ethological sense) must not be confused with dominance among animals: the latter concerns relationships *within* groups. In the present book I define a territory as a region occupied by an animal or group of animals to the exclusion of other members of the species. Usually, the region is actively defended, but this is not part of the definition.[25] A complete or model territory would be held by a family group at all seasons; it would contain all the resources needed by the group – food, shelter and so on; and it would be defended from neighbors by visible or audible displays. But territories are of several kinds, with different functions, and there are few examples of the model.

The most studied, those of birds, are usually held only in the breeding season, and are defended by visible displays and song. A second kind, also associated with breeding, is the nest territory of a small mammal, such as a squirrel, a mouse, a gerbil or a prairie-dog. A female with young drives off all intruders. A third kind, also seen among small mammals, is held by a group and is permanent; it may consist of a burrow and its surroundings. Sizes of territories vary greatly. A sea lion's territory may be a few square meters of sea shore; that of a vicuña, in the central Andes, averages seventeen hectares. Among the primates, the titi has small territories, the gibbon, large ones; but those of both species are well marked. In contrast, baboons range widely; if they have territories at all they are 'core areas' within the total range occupied by only one group but not defended against the neighbors. Not all closely studied species have territories. The red colobus monkey of African rain forests has no distinct territorial boundaries, and it is doubtful whether rhinoceros or gorillas are territorial.

At first, all this variety may seem to provide parallels to the varied customs of human societies. But among animals the diversity is between species. Each territorial species has its own pattern; there is always some variation, if only in the size or shape of the occupied region, but one can nearly always state, for the whole species, what kinds of region are held, whether by individuals or by groups, whether all year round or seasonally, and so on. And, once again, these patterns are maintained by signals typical of each species as a whole.

Human property and social status

Ever since Darwin, despite the obvious contrasts between our own and other species, we have been repeatedly assured that findings on animal status and territory tell us much about ourselves. In 1924 a leading anthropologist, W.J. Sollas, wrote that for a tribe 'to infringe on the boundaries of its neighbours . . . is to break the most sacred "law of the jungle", and inevitably leads to war'.[26] A zoologist, Walter Heape, in 1929, wrote of 'territorial rights' as a typical feature of civilization which had been evolved not by man but by animals. Another anthropologist of note, Arthur Keith, in a popular work published in 1948, cites Heape and asserts 'the close similarity . . . in the arrangement of bird and human territories'.[27] But in fact, when we turn to property-holding or land-tenure among human beings, and the accompanying status relationships, we find ourselves in another world. Our one species creates immense diversity in any one historical period, and often rapid change in time. Correspondingly, our various kinds of ownership, occupation or status depend, not on species-typical signals, but on local traditions or laws, enforced by custom or by a police and judiciary. These depend in turn on the distinctive features of communication by words. The conduct of animals has none of these features.

Exceptionally, we find tribes with no land of their own. In Niger nomads depend for their lives on their sheep. They drive their herds through the bush, in the areas farmed by Hausas, and pay for the privilege with service or animals.[28]

More interest has been attracted by the few groups that have remained in the gatherer–hunter stage – the condition of all humanity until about 10 000 years ago. Such people, though without agriculture, have an intricate system of land rights. (They also have very elaborate languages.)

The Australian aborigines were divided into groups, each with a fixed region which in a sense belonged to it. Every young man was instructed in the boundaries of the region, in the sources of water and food, and in the sites of 'spirit centers'; the latter were the dwellings of the spirits of dead ancestors, and were related to 'ritual rights and obligations'.[29] The quoted phrases are from an account of the Walbiri of northern Australia. The bands are commonly called kinship groups; but all Walbiri and members of neighboring groups were regarded as relatives, or kin. A band of aborigines that occupied a defined region might be no more than a family, but it often included members of many families. 'Ownership' of land was based on an involved set of rules.

Moreover, there was (and is) conduct that matches no neat system. Meyer Fortes refers to the incorporation of outsiders in communities of aboriginal Australians; when this happens, they are given kinship status.[30] Hence, though not members of any of the local families, the outsiders come to have the same relationship to the land as the primary owners. Similarly, J.P.M. Long describes the life of the Pintubi, in the Australian western desert, some of whom moved widely outside their own clan country. Of one of them, he writes:

> An old man met . . . in the extreme west of the area was said to have lived for many years there, some 100 km from his own country, apparently because he preferred the company of the group there.[31]

A kinship group can adopt as a member a stranger who does not even speak the local language, just as today American and European families accept as kin adopted children from Asian countries.

African groups have customs different from those of Australians, but they display a similar flexibility both in their social relationships and in their attitudes toward the land. In the Kalahari grassland, wells and localities are shared by several bands of Bush people; the bands are associated not by common descent but by being neighbors. The boundaries of the regions are areas of desert. In Zaire, the Pygmies hunt antelopes in tropical forest. Each band has its own region; but there is also, deep in the forest, a 'tabu' area: there the god of the forest is said to live. Maurice Godelier suggests that this area is, in effect, a sanctuary for antelopes and other animals which ensures their continued breeding.[32] Again the bands are not simple family groups: individuals move among them, and hence between regions.

Before the Europeans came, the population density of both Pintubi and Walbiri was about one person to 200 km². Gatherer–hunters in more favorable regions may have reached a density ten times as great. But, during recent millennia, their way of life has been almost completely replaced by farming, and so by new kinds of property relationships in more crowded conditions.

In the Middle East, the first farmers evidently planted wild grasses – the ancestors of our wheat, barley and others – in temporary clearings: they had no attachment to well worked fields or pastures. Today few people practise this itinerant or extensive farming, but relics of it remained even in twentieth century Europe. H.H. Stahl has described how in Romania the historical changes from one kind of agriculture to another left some regions untouched.[33] In the 1930s it was therefore still possible to study living communities of an early type. In some heavily forested uplands in Carpathia villages had no fixed fields: cultivation was itinerant and without private ownership. Each village was managed by an assembly, without a chairman, of which all adults, of both sexes, were members. The whole assembly was the landholder. In other villages, of a more advanced type, land was evenly divided among families.

In most regions, however, these kinds of ownership had, centuries before, been replaced by (to us) more familiar property relationships; the distribution of wealth had become increasingly uneven; and the owners of large estates (local lords or boyars) also controlled the lives of the people (serfs) who worked on the land. When land was sold, the serfs, like the livestock, went with it. Serfdom was formally abolished only in the eighteenth century, when the modern system of large scale private ownership and production for the market became dominant.

According to one definition of feudalism, Romanian communities that included serfs were feudal. But the better known feudal societies of western Europe had an important feature not found in Romania: this was the fee (in Latin, *feudum*), or estate, granted to a free man by a lord; the award was often in return for military service, but landholders also included priests.[34]

The Romanian sequence differs in another important way from that of other European regions. Romania seems to have escaped one of the major elements in ancient, Mediterranean civilization. The Roman and Greek societies, from which we derive much of our law and many of our ideas, were founded on property in people, that is, on slavery; and it was the breakdown of the great slave-owning civilizations that led, eventually, to the feudal order in most of Europe.

The expression, feudal *order*, is entirely appropriate, for feudalism was not merely a means by which a minority held possessions and power: the lords also had duties. Each was required to protect his vassals and, when disputes arose, to see that justice was done. He also relieved severe distress. The concentration of local rule in a single individual, or in a small group, allowed, not only great inequalities of wealth, but also abuse of power. Nonetheless, the system made possible social stability for many centuries. There was even progress, for major advances in agricultural methods were also achieved. The horse collar and the use of iron horseshoes were among the innovations that had a profound effect on agricultural production. In the great civilizations of south and east Asia vast peasant populations were similarly organized on the basis of ownership of land and a form of dependence on the landowners. Indeed, some historians write of Indian and Chinese feudalism.[35]

The wide distribution of such systems has tempted some writers into rash assertions. Anthony Storr, a widely read psychotherapist, writes of 'earlier and more primitive forms of society than Western democracy', in which the 'aggressive drive' may (he holds), have created, as among animals, 'a stable society based on dominance'; and he proposes that this happened in feudal times.[36] Hence a lord or landowner is equated with the 'dominant' male of a monkey troop (but of what species is not stated, nor is dominance defined). Clashes between neighboring barons then represent interactions between troops with adjoining territories, and so on. If such writers made more use of historical or anthropological knowledge, they would find severe difficulties. If feudal societies do significantly resemble those of monkeys, then other, very different human societies, including those of gatherer–hunters and the earliest farmers, do not.

The argument therefore depends on *selecting* human social systems with superficial resemblances to those of some animals. Attitudes and practices concerning wealth, ownership and work provide plenty of variety to choose from. Property, writes R.H. Tawney (1880–1962), is 'the most ambiguous of categories'.[37] P.J. Proudhon (1809–1865), the French anarchist, stated that all private property is theft. An English conservative aristocrat, quoted by Tawney, held that, on the contrary, it was invariably theft for the state to *take* property (especially in land) from its owners. These views were put forward in the same historical period. More important, attitudes change greatly with time. In the European Middle Ages, labor was regarded as 'the common lot of mankind' and was 'necessary and honorable'; trade was 'necessary but perilous to the soul'; and finance was 'at best sordid and at worst disreputable'. Tawney remarks that men had not then [as they have today] 'learned to persuade themselves that greed was enterprise and avarice economy'.[38] The presumption of fixed biological propensities does not help us to understand such variety.

Property in persons

Although usury was condemned in the middle ages, property in human beings was often taken for granted, and hence ignored by moralists. The famous English census of 1086, the Domesday book, recorded ten per cent of the population as slaves. Moreover, slavery continued into the modern period. D.B. Davis writes: 'the Declaration of Independence was written by a slaveholder and . . . Negro slavery was a legal institution in all 13 colonies at the beginning of the [American] Revolution.'[39] Can the study of animal 'societies' help us to understand slavery?

Even the most imaginative writers do not suggest that dominant baboons or gorillas are slave owners. Yet we find slavery described in some zoological works. The reference is then to insects. In this case, however, it is not human beings that are said to resemble animals, but animals that are described as if they were human. Some species of ants are called slavemakers. A book on American eusocial insects gives an account of the 'blackbirding expeditions' of blood-red ants to the nests of a black species. Members of the 'straggling army' (poor discipline?) of red ants enter the target nest, and 'soon the outraged black citizens begin to stream out of their besieged city'. The reds, now referred to as bandits, carry off the larvae of the blacks, and either eat them or leave them in their nest. There they emerge as adults. Colonies of red ants therefore often include large numbers of black workers; these help to rear the red larvae, and so evidently contribute to the survival of the reds.[40]

This orgy of metaphor is obviously misleading. The relationship of red with black ants is between two species. The black ants might be likened to our domestic animals, but there is no close counterpart to them in human affairs. A new name is needed; and so E.O. Wilson calls ant 'slavery' dulosis – a term which, though derived from the Greek word for slave, is unlikely to call up inappropriate associations.[41] To say that ants (of some species) have slaves is one of the many rather extreme anthropomorphisms in our descriptions of eusocial insects: others include the use of the words queen and soldier.

Nonetheless, slavery (in its primary sense) illustrates some of the main themes of this book. A slave may be said to be the property of another person, subject to the person's authority, and obliged by coercion to work for or otherwise to serve that person.[42] A critical reader may remark that, in certain kinds of family, by that definition wives and children (and sometimes other kin) are slaves. We are now concerned, however, with relationships outside the family. The social institutions covered are still very diverse.

In ancient Rome and Greece, and in India and elsewhere, some slaves were household servants, and their lot was not necessarily disagreeable. In Rome, in the first centuries of the present era, it was held to be better to be a slave in a rich family than free but poor. Some slaves had a professional status: they were physicians or tutors, and hardly to be distinguished from free citizens.[43] At the other extreme, and much more recently, there were the Africans rounded up from their villages, packed in barbarous conditions on board ship, and sold into brutally enforced labor in North America.[44] Between these extremes is much variety. Sometimes, for instance in Greece and Rome, slaves were foreigners captured in war. In China, most slaves entered the market as children. Hence the subject of slavery, in itself, illustrates the diversity of human communities.

The lack of uniformity within our species is only one reason why human social life, properly described, seems out of place when set alongside that of animals. We can see the incongruity more clearly if we examine three general questions about slavery. The first concerns its *function*: What use is it to the slave owner? Such a question can be thought of as biological: it is analogous to asking about the survival value of a behavior pattern. And the answer seems obvious: the work of slaves is financially profitable to the owner, or provides the owner with comforts – sometimes with luxury. Hence the owner may be helped to rear his children and even grandchildren. There may even be advantages for the slave: a livelihood and security, in which he or she, too, can rear children. But these statements are very incomplete. We may also ask, secondly, what is the social role of slaves? In many communities, in Africa and elsewhere, slaves did not always produce more than they consumed; nor did they always look after their master's children: they were *status symbols*, like that of (say) a highly trained (and expensive) butler in a rich, modern household.[45] Last is the question of *rights*. Much of our concern about slavery comes from asking whether property in persons is morally justifiable. Today, most people answer no.

The last two questions take us right away from biology: neither ants nor apes go in for status symbols; nor do they debate the rights and wrongs of what they do. These belong solely in the human dimension.

To liken human societies to animal groups is indeed always liable to lead to absurdity. If we observe an animal group or a human community for a brief period, we may find both of them stable and quite peaceful. But the means by which stability is maintained differ. Among animals, as we know, social interactions consist of standard signals common to the whole species. In contrast, the stability of human groups depends on local custom or law; and not even our most rigorous systems of 'law and order' have insured universal acceptance of conventions concerning ownership or control of people, property or land. On the contrary, these sometimes lead to injurious violence. We have no universally accepted signals to mediate our disagreements. Our many kinds of violence toward our fellows are not 'animal' but human.

Our lack of fixed patterns of social behavior allows our social systems to change rapidly. Human history presents many examples of invasion of land, and of one group subjugating or even enslaving another. And, sooner or later, the result has been further violence. Revolts by slaves, peasants and colonial and other oppressed peoples are regular items in our history books. As I write, people in Central and South America, Afghanistan and the Republic of South Africa provide examples. Slavery, protest, revolt and revolution are not represented among the social interactions of animals; nor are the ideas of freedom or of a rationally achieved consensus.

The 'law of the jungle'

Since the territorial activities of animals usually consist of 'keep out!' signs, and most status interactions are similarly bloodless, we may ask why there is still a widespread belief that animals are incessantly violent *to their own kind*. This belief seems to have four principal sources: two concern methods of observing behavior; a third is a matter of bias; the last represents inference from biological theory.

First is the impact, or salience, of a lively encounter. I have spent many hours watching animals interacting socially – or waiting for them to do so. When, at last, something happens, it is the vigorous action that one remembers most clearly, and that one shows to one's colleagues and pupils on film. Second is the fact that the easiest animals to observe are those in man-made environments, where they are liable to be abnormally crowded: if animals are in conditions unlike those of their normal habitat, their behavior is often atypical, and may be exceptionally violent.

Third is the bias exemplified by European travellers' tales. In the nineteenth century, traders, hunters and explorers shot large mammals in vast numbers, and 'natives' too if they were obstructive. They then returned with stories of the savage behavior of the animals (and of the natives). On one interpretation, they were projecting their own hostile feelings on to other species (or races). During the recent historical period, there has been a corresponding readiness to accept Tennyson's image of 'Nature' as 'red in tooth and claw'.

J.L. Mackie has shown how the preference for violent interpretations has distorted our use of the phrase, 'the law of the jungle'. For most of us, as he says, it signifies 'unrestrained and ruthless competition'; yet, when Rudyard Kipling (1860–1938) originated the expression, he was describing and advocating an elaborate system of customs which ensured social stability.[46] Kipling's *Jungle Books* were not contributions to ethology; but, in their emphasis on order among animals, they were nearer the mark than some modern writings on aggression.

Widely read biologists still emphasize the violent aspects of animal life. 'Murder and cannibalism are', we learn from E.O. Wilson, 'commonplace among the vertebrates'.[47] This assertion reflects the fourth reason for thinking of animals as waging continuous war. Wilson, like many others, tries to explain social interactions by the theory of natural selection; and, at first sight, the theory implies that there must be unremitting strife for necessities and mates. Hence we may be led to *expect* lethal violence among the animals we study. And, of course, when we believe we have found it, the behavior is easily described in terms with moral and legal connotations, such as cannibalism and murder. To avoid these pitfalls, we may look once again at the conduct of mammals during conflict.[48] When a mammal meets a rival or adversary, postures that make the body seem larger, and loud noises that advertise the animal's presence, are common. Among our nearest relatives, the Primates, howler monkeys in Central America howl; macaques in India and Japan chatter, grimace and shake branches; gorillas in Africa stand at their full height, roar and beat their breasts. Injurious violence is exceptional.

One kind of exception, formerly believed to be typical, is of animals in captivity or in other association with human beings. Such environments often promote exceptional crowding and frequent violence; but, when the same species are carefully studied in nature, they prove to be quite peaceful. One species of baboon, violent in captivity, forms small social units, each of an adult male, several females, and young. Unattached males that approach are driven off by gestures. During feeding, the family groups assemble in bands which defend food sources from other bands mainly by gestures and noises. The large, powerful adult males, which are a match for predators such as leopards, may relieve a female by carrying her baby on a long march.

Nonetheless, 'murder' – including infanticide – does occur among mammals even in natural conditions. A pride of lions consists of several females with their young and a peripheral group of males. If a pride lacks males, others may take over and kill the cubs.[49] Among the Primates, the sacred langur of northern India is usually quiet and peaceable; but, if the dominant male of a troop dies, a strange male may replace him and kill all the infants; he then, of course, sires some of his own. (According to some accounts, however, such violence occurs only during exceptional crowding.) Similarly, clashes between groups of chimpanzees may result not only in deaths of adults but also in slaughter of infants.[50]

In earlier times such acts would, as a matter of course, have been described in moral terms, as in the Bestiaries mentioned in the previous chapter; and a hankering for morality derived from the beasts still exists. But in modern ethology such conduct is not thought of as an ethical matter. When we see a male langur killing a baby (or a male baboon carrying one), we note the occurrence and try to explain it. In the type of biological explanation most relevant to this book, we ask whether the acts observed are likely to help the actor or its descendants to survive. Thus we attempt an explanation, of a kind to which we return later, in evolutionary terms: infanticide may be presumed to have 'survival value' for the killers, just as carrying a baby is regarded as increasing the Darwinian fitness, or chances of leaving descendants, of the 'altruistic' male who undertakes this burden.

Homicide may similarly contribute to the survival of a human killer; but often it makes likely his or her death or imprisonment. This is because we classify the act in moral or legal terms. But attitudes toward killing vary. Although, in most societies, infanticide is both illegal and held to be morally wrong, in some it has been customary. Jonathan Swift (1667–1775), in his bitter satire, *A Modest Proposal*, on what to do in a famine, proposed killing and eating babies. In doing so he was advocating only an extension of what had been normal in some communities, including those of the Eskimo and Australian Aborigines, and the city states of ancient Greece.[51]

When, in time of siege or famine, people expose their newly born infants to die (or to be adopted), the survival of the parents' other descendants is made more likely. Infanticide is then accepted as a rational act. It is not species-typical conduct, for only a few human groups resort to it. Attitudes to other kinds of homicide also vary. The historical trend is increasingly to accept the commandment, 'Thou shalt not kill'. This moral advance still has a long way to go; but it shows with particular clarity the weakness of the argument from animals to humanity. One animal species may be quite peaceful, another sometimes dangerously violent; but, in either case, all social interactions are typical of the species and depend on standard signals. No species, except our own, has a code of morals or of laws based on what is held to be right or socially necessary; nor do neighboring animal populations differ widely, or debate what code is best.

Crowding and hypothesis

Yet the search for guidance from animals has continued. Among the seekers have been social scientists studying life in crowded cities. Above I mention disruption of social order among animals crowded in exceptional environments. Even 'threats' can be harmful: at the extreme, animals under threat by a member of their own species, though unwounded, may collapse and die. Crowding in itself has therefore been supposed to have ill effects on the social relationships and health of human beings. As an example, Anthony Storr writes: 'in man as in other species, overcrowding is an undoubted breeding ground of hostility'.[52] In contrast, J.L. Freedman, who has done much careful research on human crowding, states: 'there is a great deal of evidence that crowding is not generally harmful'.[53]

The idea of crowding may at first suggest a jostling multitude in which people collide at intervals, like molecules in a gas: all encounters are the same, and the adverse effects of collisions increase in proportion to population density. But of course, the collisions even of animals differ: injury results only from encounters of special kinds, such as those between a territory holder and an intruder.[54] In human societies the complexities are still greater. As an example, R.E. Mitchell reports adverse effects of crowding in Hong Kong, where it is common for three or more people to share a bed. The important factor there is the *kind* of crowding experienced (and described in words) by the inhabitants – for instance, whether two or more families are obliged to share space. The critical question is then not how many people occupy a hectare, but whether people *feel* (in some sense) crowded: if they do, hostility to their neighbors or their own children is more likely.[55] Such effects depend on custom and on what is expected. G.V. Coelho & J.J. Stein write:

> Arab families of eight or ten persons sleep in a . . . hut consisting of one or two rooms; the typical Chinese family of five to seven persons shares a two-room apartment; and many Asian children grow up sleeping together with their parents. Yet these situations are rarely viewed as intolerable.[56]

One reason why high population density seems to induce hostility is that it often goes with a high rate of violent crime. A correlation of this kind

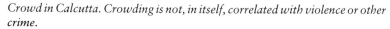

Crowd in Calcutta. Crowding is not, in itself, correlated with violence or other crime.

can mislead us into thinking that we have uncovered a cause; but crime goes also with poverty and other features. Which, if any, is important? Research by Freedman and others has overcome some of the difficulties. In New York City many sub-populations differ not only in crowding and crime rates but also in other features such as income and national origin. When the figures are critically analyzed, violent crime proves to go with poverty, but to be independent of crowding.[57] To test whether poverty is a *cause*, it would be necessary to raise incomes and see what happens.

The presumption that crowding causes 'social stress' can lead to disregard of its opposite, 'social sedation'; yet there is plenty of evidence of a liking for crowds and contact. The San gatherer–hunters of the Kalahari desert represent an extreme case. Patricia Draper gives their population density as one person to between 10 and 26 square kilometers. They live, however, in packed (but peaceful) groups, each of which is separated by great distances. Within a group an individual has about 17 square meters of personal space, or half the minimum recommended for housing in the USA. Draper describes how these people even lean against each other for closer contact.[58] Correspondingly, some studies of urban populations have found density to be *negatively* correlated with 'stress'. Moreover, it is possible to be socially isolated in a modern city. J.D. McCarthy and others have studied the figures of murder and aggravated assault in 171 American cities. The distribution of these crimes suggests that the isolation of persons living alone has more serious effects than crowding.[59]

So the simple hypotheses derived from supposed effects of crowding on animals have not been confirmed. The resulting researches nonetheless illustrate the correct use of the argument from animals to humanity: analogies suggested possibilities which could, up to a point, be tested. I say, up to a point, for two reasons: experiments were, as a rule, impracticable; and the investigators were obliged to use speech and to take into account human feelings and attitudes.

Instinctive knowledge

In the preceding pages the analogies between animal behavior and human action emerge as, at best, superficial; often, we discover contrasts, not resemblances. But, despite the differences in visible behavior, perhaps we share with other species internal impulsions that drive us toward certain goals. A belief in such impulsions is implied in the quotation on the first page of this chapter, in which aggression is referred to as an instinct. The word instinct is disappearing from scientific writings, but the idea of fixed human drives or urges is still important, and is often supported by metaphors concerning stored energies. Each kind of action is supposed to have its own specific sort of energy. In an influential text published in 1963 we find: 'energy with *some* degree of specificity, channeled in some way or another, is fundamental to the modern concept of instinct.'[60] These instincts or drives are sometimes said to be coded in the genes, or preprogrammed by our evolutionary past; but such expressions, though they may sound modern and scientific, have no foundations in genetics or in evolutionary theory.

The writings of all ages contain 'explanations' of mysterious phenomena by words that sound impressive but are empty. A leading medieval biologist, Albert the Great (died 1280), wrote of a *vis formativa* or life force that activates the body; and particular kinds of conduct are attributed to specific forces: the faculty of evaluation is due to a *vis aestimativa*, and a *vis irascibilis* foreshadows the 'drive for aggression' of the present century.[61] Today, the metaphor of accumulated violent energy is sometimes propped up by expressions taken from nerve physiology: an ethology text tells us that 'aggression is a true instinct with its own endogenous excitatory potential'.[62] The last three words are, however, only an additional metaphor, without any physiological basis.

The attractive energy metaphor has led to simple, but at first sight plausible, proposals on how to reduce violent conduct. A philosopher, George Santayana, in 1954, advocated vigorous sports because – he said – they reduce injurious violence, even war.[63] In 1960 an expert on sports medicine wrote:

> Competitive games provide an unusually satisfactory social outlet for the instinctive aggressive drive . . . which constantly seeks expression. Where its direct expression is denied symptoms may develop . . . The most aggressive outlet is seen in those sports in which there is bodily contact.[64]

The vision of vigorous sports as a remedy for social ills has been popularized by more recent writers, notably K.Z. Lorenz, but without discussion of its practicality, or of whether evidence exists in its favor.[65]

The advocates of sport as a cure for animosity often fail to distinguish between watching and doing. The evidence on the effects of *watching* sport has, however, been critically analyzed by J.H. Goldstein.[66] Riots and other kinds of mass violence are a frequent outcome of football and boxing matches, but they are rare during or after less violent performances, such as horse racing. International football matches lead to international clashes. Indeed, on one occasion, even cricket led to a diplomatic incident between Australia and Britain.[67] The belief that national occupation with vigorous sports reduces the danger of war or other conflict between nations has also been tested in a study of 20 nations by R.G. Sipes.[68] His findings give the hypothesis no support. This is not surprising: war is a result of decisions, usually made by elderly men, ostensibly on the basis of rational calculation. The famous aphorism on the subject, that of Carl von Clausewitz (1780–1831), describes war as a continuation of state policy: hence the objective is commonly economic advantage or enlargement of the area ruled by an 'aggressive' government.

Belief in the pacifying effects of *playing* violent games is founded on the notion of catharsis: when one expresses anger or tension by direct action against others, it is held, the feelings subside, and further violence is unlikely. What would happen if presidents, prime ministers and their colleagues took to boxing or football has not, as far as I know, been examined, but on catharsis there is experimental evidence. The findings on children are especially important, for the belief of some people in a need to express pent-up 'aggro' has led them to encourage violent children to be violent. (Recall the metaphor, in the previous chapter, of children as explosive.) Careful studies, however, give overwhelming evidence of an opposite effect: encouraging violence by children either maintains the unwanted behavior or enhances it.[69] Hence children's violence should be deterred – of course, by non-violent means. Children learn much by imitation, and, as we see in the next chapter, they need pacific adult conduct on which to model their own.

Catharsis has also been looked for in adults.[70] J.E. Hokanson and others subjected students to verbal 'aggression' in the form of insult, or to electric shocks. Both led to raised blood pressure and heart rate. It was hypothesized that this state of 'tension' (a metaphor) would be most quickly relieved by an 'aggressive' response. (Such a response could also be called defensive.) Sometimes the findings agreed with the hypothesis; moreover, it proved possible to train initially non-belligerent subjects to make violent responses. On the other hand, some subjects (especially women) made friendly responses to insults; and these, too, were accompanied by lowering of blood pressure and heart rate. In yet other situations, people were trained to give themselves a shock: that is, they were induced to behave masochistically. This again resulted in relief of tension. Evidently, the response to such provocation depends on personality, on the social circumstances and, above all, on attitudes and habits socially acquired.

As usual, therefore, the attempt to interpret human actions and feelings by a simple principle fails in face of our diversity and readiness to change our behavior – our adaptability. We all have the capacity to be angry, and to express our anger; but we can all also feel and express friendship. How we do so, and how such behavior influences our feelings, depend greatly on previous experience, and also on conscious choice. These statements, though trite, are (as we see in the next chapter) also true. They are not compatible with an account of humanity driven, willy-nilly, to a destructive pugnacity.

The limitations of ethology

The present chapter opens by quoting an optimistic statement: that ethology can go far toward explaining human anti-social violence. *Homo sapiens* is a product of evolution and an animal among others; if, then, we know other species, it is held, we also know ourselves. One means of making such equations seem plausible is to use a verbal trick. Animals are first described as though they are human: in accounts of status and territorial relationships, words previously applied to human action are used for very different activities of animals. The descriptions of animals as if they were human are then held out as showing how like human beings are to animals.

At the physiological level our differences from other species are minor; but the fact that we perform such functions as coughing and digesting like other mammals can hardly lead to any profound conclusions on the nature of humanity. In our more complex behavior differences from other species are more prominent: our diverse laws and customs concerning property and landholding are not matched by the territorial patterns of animals; human rank, rule and subjection are not paralleled among animal status systems; human languages are fundamentally different from animal signals. The various concepts of instinct at best only name phenomena that are better described in plain words. The notion that we are driven by mysterious inner forces does not explain but only obscures human motivation.

Hence the idea that human anti-social violence can be usefully interpreted ethologically conceals the real achievements of ethologists and is itself unfounded. The understanding of human combativeness, as of other human action, depends on the study of human beings.

5 The aggression labyrinth

Why this wild strain of imagination found reception so long in polite and learned ages, it is not easy to conceive; but we cannot wonder that while readers could be procured, the authors were willing to continue it; for when a man had by practice gained some fluency of language, he had no further care than to retire to his closet, let loose his invention, and heat his mind with incredibilities; a book was thus produced without fear of criticism, without the toil of study, without knowledge of nature, or acquaintance with life.

Samuel Johnson

'We are all potential assassins', writes P.G. Zimbardo;[1] but, he might have added, few of us are assassins in reality. We may ask why human beings commit appalling acts of violence; the need to answer this question is behind much writing on 'aggression'. But it is equally appropriate to ask why most of us, even in the face of provocation, are peaceable most of the time. The previous chapter disposes of attempts to find answers in ethology, but there are other obstacles to understanding. One is the vast range of phenomena put under the heading of aggression.

Meanings

It is sometimes said that aggression is difficult to define. Such a statement might be accepted if it referred to writings about activities of a single kind with indistinct boundaries. As the lists opposite show, this does not apply to aggression: the term has been used so variously, that sometimes it could be replaced by 'behavior' with no loss of meaning.[2]

Many research reports with aggression in the title are about animals: they may then be concerned with the signals (such as bird song) that tend to separate members of the same species; most of these are non-injurious, and many, defensive. But we are now concerned primarily with human conduct. If we examine the list of individual human actions called aggressive, we find a range from murder to verbal abuse, and from psychosurgery to assertiveness. Moreover, most of the categories in the list can be subdivided. Homicide may be a result of a road accident, a drunken brawl, ritual sacrifice, a public hanging, or neglect of safety precautions; a murderer may be a compulsive killer of young women or the paid or unpaid assassin of a head of state.[3]

A physician, J.R. Lion, and his colleagues have reviewed the clinical management of aggression, by which they mean violent conduct that injures others. Among the many pathological conditions in which violence occurs are: psychotic behavior induced by drugs such as amphetamines; psychosis of which paranoia is a component; depression; and an 'explosive' personality structure. One might add violence induced by treatment in lunatic asylums. The classification of such conditions is controversial, but their diversity is not in doubt.[4]

A much debated question is whether 'insane' conduct should be regarded as culpable. Do mad people *intend* the consequences of what they do? If not, it might be argued, the behavior should not be classified as aggressive; for a commonly used definition of aggression is behavior *intended to injure* another person.

There is therefore much variety not only in the categories of 'aggression' but also in our attitudes toward them. Here are some violent actions that are or have been accepted in some communities, rejected in others: crucifixion, amputation, torture, flogging or the stocks as punishments for crimes;

Some meanings of 'aggression'

===

Animal

Signals that make another of same species withdraw:
    territorial (including those of female with young)
    maintenance of status
    female weaning young

Male controlling female's movements

Violent behavior provoked by pain or prolonged isolation

Injuring or killing member of same species

[Predation; defense against predators]

Human

Individual	*Group*
Self defense by violence	Football hooliganism
Homicide	Repression
Assault	Riot
Rape	Revolution
Insane violence	Counter-revolution
Assault without battery (verbal abuse . . .)	Annexation
	War
Humiliating behavior	
Dominating behavior	
Assertiveness	
Punishment	
Psychosurgery	
Destructive feelings	
[Hunting other species]	[Hunting other species]

===

hanging or flogging to maintain military discipline; wife-beating; thrashing school children; boxing. There are many more. The various kinds of group action called aggressive add to the diversity. And this diversity has political implications. An earlier version of the list on page 57 included riot and revolution, but not repression or counter-revolution. The added terms cover the violent actions of authority, by means of police, soldiers, clandestine murder squads and others, against dissenters. Sometimes, such 'aggression' is much more prominent than that from the other side; but it is often not remarked, because the most prominent pronouncements about violence come from those who rule.[5] Whether a police action during civil disturbance is called aggressive or defensive must often depend on where the speaker stands, that is, on which side of the barricades.

Repression and riot may be analyzed at a variety of levels, including the psychological as well as the political. The same applies to war. Indeed, one inquiry on population and proneness to war of European nations begins with a reference to the supposed effects of crowding on animals. The authors, however, find no relation between population density (or change in density) and war-making.[6]

Attempts to describe war psychologically are sometimes based on the idea (unsatisfactory, as we know from the previous chapter) of a drive for aggression. There are many such attempts to explain war (and other kinds of human violence) by reduction to an apparently simple concept, such as animal territory, overcrowding, a drive for reproduction – or original sin.[7] They seem to reflect a craving for easily understood explanations of complex (and frightening) phenomena. They also disregard the most obvious explanation: that war is a result of calculation. The decision to go to war may be both stupid and wicked, but it may still be rational. In a study of modern European history, governments were found to have decided on war one to five years before war began.[8] 'War fever' among the victims came afterwards. Similarly, when a person behaves in a hostile way, the primary intention may be not to inflict harm but to coerce: the objective may be to cause or to prevent some action. Preoccupation with 'aggression' may disguise the importance of the desire for power. The resulting confusion would be reduced if catch-phrases were avoided: as J.T. Tedeschi and his colleagues remark, the concept of aggression has so many meanings that it interferes with understanding, and 'perhaps should be avoided in scientific discourse'.[9] If this advice were taken, we should be able to see the features of *Homo pugnax* more clearly.

The beast in man

These features appear in western writings of the past three millennia; but we are now concerned with the period since Darwin. For more than a century, a prominent idea has been that of the beast in man. 'From the point of the view of the moralist', writes T.H. Huxley, the animal world is on about the same level as a gladiator's show . . . the strongest, the swiftest, the cunningest live to fight another day. . . . [Similarly] among primitive men, the weakest and stupidest went to the wall, while the toughest and shrewdest . . . but not the best in any other sense, survived. Life was a continual free fight and . . . the Hobbesian war of all against all was the normal state of existence.[10]

It is implied that the differences between the strong and the weak are genetically determined ('inherited'); hence the strong have more children than the weak, and these children are themselves strong. But Huxley adds:

The history of civilization – that is, of society – on the other hand, is the record of the attempts which the human race has made to escape from this position.[10]

And so we have the double image of our species which Alexander Pope had summed up more than a century before:

A being darkly wise and rudely great . . .
He hangs between; in doubt to act or rest;
In doubt to deem himself a god or beast.

The view of humanity as dual was helped by the belief that our evolutionary past is accurately recorded in the stages of our development from the fertilized egg. For several decades, recapitulation (the biogenetic law) was part of conventional biological teaching; and, like other biological half-truths, it appears outside biology. S.J. Gould gives many examples.[11] Among them are some famous lines of Kipling:

Take up the White Man's Burden –
Send forth the best ye breed –
Go, bind your sons to exile
To serve your captives' need:
To wait in heavy harness,
On fluttered folk and wild –
Your new-caught, sullen peoples,
Half-devil and half-child.

The 'primitive races' were supposed to represent an earlier, more embryonic stage in the evolution of that pinnacle of Nature, the White Man. More generally, we were held to 'climb our family tree' as we develop. Our adult form, on this view, is a stage stacked on top of earlier stages: the late fetus then represents the ape, and so on back to the egg, which corresponds to a single-celled protozoan (like *Ameba*). But in fact evolutionary change takes place at every stage of development, not merely the last. The newly fertilized human egg is just as much *Homo sapiens* as an adult, though less obviously.

Of twentieth-century writers on aggression, Sigmund Freud (1856–1939) has had the most widespread and profound influence on our thinking. Freud took recapitulation for granted.[12] He held not only that we pass through ancestral stages as we grow, but also that, when we are adult, we retain mental processes that belong to earlier, 'atavistic' conditions; neurosis, he held, results from such retention. To this he added an account of human motivation in terms of drives (*Triebe*) of which the stored energies have to be expressed in some form.[13]

Freud's vast, poetic interpretation of the dark side of human impulse, fear and desire led him to write that human beings are 'creatures among whose instinctual endowments is . . . a powerful share of aggressiveness'; and he believed that 'holding back aggressiveness is in general unhealthy and leads to illness'.[13] Here we see again the influence of the metaphor of stored energies which are supposed to drive us to action (pages 52–4). Freud adumbrated the picture, derived later from misunderstood ethology, of human beings endowed with destructive and self-destructive impulses which are controlled, if at all, only with difficulty; such impulses are said to be 'inherited' from primitive ancestors.

So we now turn to the conduct of our pre-human and human predecessors, and of groups that still remain in the stage of gathering and hunting. The crude image of *Homo pugnax* suggests that such people lead lives of incessant and compulsive strife.

Confusion about predation

First, a red herring about hunting must be disposed of. One popular version of our evolution emphasizes a change from (supposedly) pacific plant eaters to (supposedly) aggressive hunters. Hunting other species is held to result in violence among the hunters themselves. This story has been put out especially in popular, dramatically written books by Robert Ardrey (1908–1982).[14]

The hunting hypothesis depends on three presumptions. First, predators should be violent, not only to their prey, but also to their own kind, while plant eaters should be more pacific. Second, the behavior of other species should tell us what to expect of ourselves – or, at least, of our primitive ancestors. Third, these ancestors should at some stage have depended largely on hunting.

The first presumption equates a relationship *between* species with social interactions, that is, with interactions of members of the *same* species. As I define the word social, if I milk a cow that is not social behavior, but cannibalism is. If I eat the cow, that is predation. In predation, an animal attacks and eats another of a different kind: herring eat small crustaceans, and we eat herring. Among familiar predators are wrens, dolphins and tigers. These are neither more nor less violent to members of their own species than are, say, parakeets, elephants and gorillas, all of which are exclusively plant eaters.

The second presumption, that we can know ourselves by studying animals, confuses metaphor with evidence. The analogy of animal predation with human hunting could indeed be more reasonably used in the opposite sense: animals such as wild dogs, wolves and lions combine in groups to run down their prey; hence it could be argued that hunting goes with cooperation, not conflict.

Lastly, there is the question whether our ancestors, as they evolved into human beings, did become to an important extent hunters. As we see below, there are reasons for doubting this long-held belief.

'Primitive' violence

Even if the hunting hypothesis is not valid, early human beings, or their forebears, may still have been violent. In 1959 Raymond Dart, an anthropologist, published a popular account of fossil *Homo* and prehuman Primates. The latter include *Australopithecus*, the 'man-apes' of southern Africa; and Dart gives a lurid account of these hominids as violent, murderous and cannibalistic. In an earlier work he had described *Australopithecus* as

> carnivorous creatures, that seized living quarries by violence, battered them to death, tore apart their broken bodies, dismembered them limb from limb, slaking their ravenous thirst with the hot blood of victims and greedily devouring livid writhing flesh.[15]

The beings named *Australopithecus* lived between about two and four million years ago.[16] They are marked by a patchwork of human and non-human features. The body resembled ours, but was often small like that of a present-day pygmy: the posture was usually erect; correspondingly, the skeleton – skull excepted – is remarkably similar to that of a human being. The ape-men walked and ran not in forests but in open grasslands. Their thumbs were like ours and unlike those of apes: the hands were efficient grasping organs and seem to have been used in similar ways; associated with some of their bones are stone tools of simple design. *Australopithecus* was evidently a tool-maker.

Yet the australopithecine brain, like that of a gorilla or chimpanzee, was less than half the size of our own. The skull and teeth were of ape-like proportions, and the jaws were massive and protruding. Not long ago there was much talk of the 'missing link'; these fossils, with their combination of human-type limbs and partly ape-like skull, have seemed to represent a link that is no longer missing. *Australopithecus* has therefore been held to be a stage in our evolution from primitive apes.

Dart's picture of his and our savage ancestors is based on the condition of a number of skulls, including some of baboons: these, according to Dart, had been lethally fractured by blows from blunt instruments; others had had their teeth knocked out. The weapons most favored by the australopithecines are said to be the long bones of antelopes. (There is an echo of this belief in the science fiction film, *2001*, when a pre-man catches the bone of a slaughtered animal, and it turns into a club for bashing other pre-men.)

There is, however, no good ground for attributing the damage described by Dart to blows from weapons. Fossil skulls are often badly damaged, and fossil jaws rarely have all their teeth. It is quite usual for fossil skeletons, of any species, to be crushed, after petrifaction, by the rocks in which they are preserved. Some fossil bones nevertheless still have marks made before or shortly after death. C.K. Brain has described 15 years of research on the ways in which the various kinds of marks could have been made.[17] The bones found in limestone caves include those of baboons, as well as australopithecines. These seem to have used the caves for shelter in cold weather, as baboons do today. Judging by the way they are marked, the bones were accumulated by carnivores, including very large members of the cat family that are now extinct. If so, our ancestors (if they were our ancestors) were not predators, but prey.

There is quite a tradition of imaginative writing on human fossils; but Dart's narrative is more than a case of private eccentricity, for it provided inspiration for Ardrey, whose works still turn up on bookstalls and in university reading lists. Ardrey was an extreme proponent of *Homo pugnax*. Human history, he wrote, must be interpreted as an account of an 'armed predator'; and our species 'emerged from the anthropoid background for one reason only: because he was a killer'.[18] K.Z. Lorenz, in a popular and influential book, took up a similar theme soon after Ardrey. 'There is evidence that the first inventors of pebble tools, the African Australopithecines, promptly used their new weapon to kill not only game, but fellow members of their species as well'. On our own genus, *Homo*, he adds: 'Peking Man, the Prometheus who learned to preserve fire, used it to roast his brothers.'[19]

Similar myths, with some further distortions, were quickly propagated by journalists. Nicholas Tomalin, in a left-wing English weekly, presents our ancestors as 'jingoistic brutes with the meanest of property instincts'; hence 'any idea of progress in politics which ignores these ape-like qualities [or] our aggressive impulses' is bound to fail. 'The old adage "you can't change human nature" becomes true once more.' And another English journalist of distinction, Katherine Whitehorn, writes:

> The desire to have and to hold, to screech at the neighbours and say 'Mine, all mine' is in our nature . . . Ardrey and his allies have let us off perfection and I for one feel a lot better for it.[20]

No doubt such passages are dashed off, with little thought, to meet a printer's deadline, but they still both reflect, and also help to create, an outlook based on wrong information.

Popular fiction, too, has (again) made use of these ideas. A leading film director, Sam Peckinpah, who was responsible for *Straw Dogs*, a celebration of violence, in an interview described how people have 'instincts that go back millions of years' and that explain the 'violence in every human being'. A critic comments that in Peckinpah's films 'a good man is incapable of doing good, and a noble action ends by destroying its maker.' Peckinpah supported his image of humanity by distributing copies of Ardrey's books, and naively referred to one of them as a 'scientific statement'. (He also confessed failure to convert his daughter, aged 20, to his views.) Stanley Kubrick, director of another successfully brutal film, *A Clockwork Orange*, has described 'man' as 'an ignoble savage'; and he too acknowledged Ardrey as his source.[21]

While these stories were being spread around in the media, serious research was going on. One method of finding out about the diet of extinct forms is to examine their teeth microscopically, and to compare them with those of existing primates whose diet is known. The microwear of molars from one group of australopithecines proves to resemble that of mandrills, orang utans and chimpanzees. These live principally on fruit and other plant food. (Chimpanzees also eat insects and occasionally meat.) There is no sign that these australopithecines ever gnawed bone; on this evidence, they were fruit eaters.[22] Others may have had a more varied diet that included small animals, such as lizards. Whether they were occasional eaters of small mammals cannot be decided. But the concept of australopithecines as bloodthirsty cannibals is fantasy.

Nonetheless, at some stage some human or pre-human beings took to hunting, at least, some of the time; they also adopted a diet of seeds as well as fruit.[23] But the picture of our ancestors as predominantly seed-eaters unfortunately lacks drama, and so has hardly appeared in the media.

Whatever they ate, our predecessors became increasingly dependent on the use of stone tools. The latter began as pebbles with a crude cutting edge. Later came fully shaped hand-axes and, later still, elegantly chipped blades. Some of the last seem to have been as much works of art as implements.[24] If our ancestors were indeed australopithecines there was, during the same period of about two million years, an exceptionally rapid increase in brain size.[25] There are only guesses on how this happened. Perhaps there was selection for a good memory and the ability to communicate by speech. An additional mystery is that the enlargement seems to have occurred not gradually but during three separate periods of rapid change.

A million years ago, or less, between the australopithecines and ourselves, is *Homo erectus*. These beings had brains nearly as big as ours, but more massive skulls and teeth. In the grasslands of eastern and southern Africa *H. erectus* lived, no doubt, a wandering life; but we now find evidence of home bases with accumulations of stone tools.[26] Chunks of rock must have been carried considerable distances to these sites. The debris includes bones;

hence the diet included meat, but we do not know in what proportion. Gatherer–hunters of our own time, such as the Australian aborigines or the people of the Kalahari, eat more plant than animal food. The plants are commonly gathered by women.[27]

It is also uncertain to what extent the meat was obtained by hunting. The bones that remain are usually from large mammals, not from the small or medium species that are eaten by the pre-agricultural peoples of today. Large mammals could be obtained by scavenging, and by killing weak or young animals. This is not merely speculation. Two intrepid research workers, G.B. Schaller and G.R. Lowther, took a walk of 160 kilometers across east African grassland. In less than two days, they reckon, they could have picked up 35 kilograms of meat – plenty, not only for themselves, but for their wives and children at a (hypothetical) home base. The main source was gazelles, including dead adults and newly born young. On another occasion, during a dry season, they walked 95 kilometers. Their finds included animals killed by lions and a dead (but nutritious) buffalo.[28]

Hence if one should find oneself isolated on an African plain, time should not be wasted making a bow and arrows, or a sling. In addition to edible plants, dead or dying meat is the thing to look for. One should, of course, (like *H. erectus*) have the means of making fire. One would also be well advised to co-operate with one's companions, not to eat them.

Woman as gatherer: collecting berries in the Kalahari.

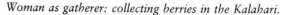

Exactly how protohumanity co-operated or competed, in all these activities, we do not know. As G.L. Isaac has said, the conduct of our ancestors of nearly a million years ago may have been quite different from anything we know today.[29] The evidence we have suggests a peaceable existence; but whether a writer emphasizes hunting or carrying, egoism or sharing, co-operation or conflict, tools as implements or tools as weapons, often depends on personal choice rather than the strength of the evidence.

Social diversity

After the romantic stories of strife and violence among our 'Stone-Age' ancestors, such a non-committal conclusion may be disappointing. Perhaps we can do better by looking at the lives of existing (or recently existing) gatherer–hunters. There is a tradition of attributing violence to present-day people who live (in part) by hunting or trapping animals.

Statements on this theme are often confusing because, as usual, very different phenomena are brought together under the one heading of aggression. In 1958 a prominent anthropologist, S.L. Washburn, writing on the evolution of human behavior, referred to the pleasure man takes in hunting. 'Unless . . . training has hidden the natural drives, men enjoy the chase and the kill. In most cultures torture and suffering are made public spectacles for the enjoyment of all.' In another article Washburn refers to 'man' as 'naturally aggressive', and states that hunting has 'exerted a profound effect on human psychology'. He also equates war (until recently) with hunting.[30] No evidence is given on the incidence of enjoyment of torture. In some communites, including those of ancient Greece and our own, enactments of tragedy can arouse terror and pity, not malicious satisfaction.

And it is not at all clear that hunting and sadistic pleasure should be put in the same category. The pleasure derived from hunting (successfully), apart from the provision of food, may often be due to the exercise of skill. But, as Erich Fromm remarks, the psychology of hunting has been little studied.[31]

A zoologist, Irenaus Eibl-Eibesfeldt, provides another instance of confusion, when he criticizes reports that groups, such as the Eskimo and the Kwakiutl of North America, and the San of the Kalahari, have no (or few) aggressive tendencies. On the Eskimo he cites 'face-slapping rituals or song duels' as examples of aggression. Aggression among the Kwakiutl is said to be expressed in the potlach custom, under which the richest individuals, the chiefs, acquire prestige by destroying their own property. (Compare the slaves as status symbols mentioned on page 47.) As for the San, the children are described as performing 'aggressive acts, such as hitting with the fist . . . jostling, biting, kicking.'[32] The obvious comment is that, if 'aggression' consisted only of face-slapping, competitive duetting, wasteful showing off and childish aggro, there would be no problem of aggression:

if such actions are usually the worst things that happen in these communities, then the Eskimo and the others do indeed qualify as notably non-violent.

Washburn, in the passages quoted above, represents those who (like Dart) take primitive violence for granted; Eibl-Eibesfeldt exemplifies those who (like Ardrey) wish to make a case for aggressiveness as a human instinct. Advocacy of the opposite case is represented by Fromm. We now ask first, what are the facts and, second, to what extent are they important?

Some of the most detailed information comes from Australia, where all the aborigines were gatherer–hunters until recently. They were present in sparse populations (page 44) throughout the mainland and Tasmania. The aboriginal Tasmanians, who became extinct as an identifiable group before the end of the nineteenth century, were evidently a peaceful people. An Englishman, Robert Thirkell, who came to Tasmania in 1820, lived in an isolated house among them for many years. He describes encounters in the bush and visits by aboriginals to his home, and says they were never a source of danger. They did, however, attack white men who treated aboriginal women badly; and, when they were driven from their usual living areas, they tried to resist. Hence they were described by the invaders as 'treacherous, aggressive, ungrateful and cruel', and exterminated.[33]

On the mainland, when European immigrants encountered aboriginals, some thought of the Australians as friends who, though inferior, 'could be taught by patience and kindness' to perform useful services; but another belief was that they were filthy, thieving, murderous and idle. And, sometimes, so they were. Eloquent accounts of the calamitous impact of white on black have been given by C.M.H. Clark and by Henry Reynolds.[34] If one wants evidence of human violence from the nineteenth century, one should examine first the conduct of the immigrants, who came from the technically most advanced and civilized nations.

In Africa, two closely observed pre-agricultural groups are the forest Pygmies of Central Africa and the Bushmen or San people of the Kalahari desert. Whether one describes them as pacific or aggressive again depends on the kinds of behavior one has in mind. C.M. Turnbull, after three years among the Pygmies, describes them as noisy and quarrelsome; but assault rarely goes beyond slapping, and quarrels are soon settled.[35] Similarly, the San people evidently indulge in frequent bickering, but seldom resort to dangerous violence. R.B. Lee gives exact information. In a population of about 1500, twenty-two killings occurred in 50 years. This is about one-third the homicide rate of the USA; but American victims of violence often survive owing to skilled medical care, hence the actual difference is even greater than the figures imply. On the other hand, homicide rates in Britain, France and Germany are *lower* than those of the San. These gatherer–hunters therefore emerge as mostly peaceful, but not as saints. They keep their killing low without police, judiciary or prisons. Lee concludes:

> A truly communal life is often dismissed as a utopian ideal . . . But the evidence [from] foraging peoples tells us otherwise. A sharing way of life is not only possible, but has actually existed in many parts of the world.[36]

In the modern accounts of the lives of gatherer–hunters, by anthropologists aware of the dangers of misinterpreting a strange culture, the people studied prove to be neither 'noble savages' nor ferocious, cannibalistic killers, neither angels nor fiends. The accounts of their 'aggression' or lack of it, by Europeans who first encountered them, turn out to be of two main kinds. First are those by invaders who occupied their land, often took the women, and tried to make them give up their customs and religion in favor of some form of Christianity; the resulting picture was usually adverse. Second are the detailed descriptions by individuals who went to live as visitors in communities of gatherer–hunters; these commonly give us portraits of peoples whose prominent features include charm and friendliness.

This is an agreeable conclusion, and perhaps will not evoke much dissent. But to suppose it relevant to questions on antisocial violence in modern, advanced communities it is necessary to presume some form of (social?) recapitulation, and to treat present-day pre-agricultural people as representing an early stage in our social evolution; *Homo sapiens* would then be held out as primarily or naturally friendly and co-operative – *Homo amans*.

Unfortunately, this has little more validity than the argument that leads to *H. pugnax*. We learn nothing of why human conduct often departs from what is obviously desirable, or of the causes of the variety we find among human communities. This variety is well illustrated among proto-agricultural groups – people at a technical stage resembling that of our ancestors not long after agriculture was invented. By biassed selection of examples, it would be easy to present the 'primitive tribes' at this level as fierce and disagreeable. In the high country of New Guinea, for instance, the Dugum Dani make 'warfare' a way of life. Battles are rather like football matches between teams from neighboring cities; but they are more dangerous, and a much higher proportion of the opposed populations take an active part.

Battlefields, like football grounds, are set aside for fighting. Men get themselves up, in rare feathers and elaborate head-dresses, to look their best. Battles are a result of shouted challenges, and do not begin until both sides are ready. They have a formal pattern of advance, challenge and brief but lethal clashes between men armed with bows and arrows or spears. They end in time for everyone to get home before dark. The victors gain no ground, goods or women – only the satisfaction of having fought and won. Boys are systematically and energetically trained in the arts of this kind of war.[37]

In barely accessible regions of southern Venezuela and northern Brazil there are the Yanomamö, among whom N.A. Chagnon spent 19 uncomfortable months. These people are energetic beggars and tricksters, and also make regular use of hallucinogenic drugs, extracted from plants. Perhaps this practise helps them to put up with their continually repeated resort to violence, and its consequences. Clashes between men who are neighbors are frequent, but they are not what we should call fights. One combatant punches or pounds at the chest of the other; and then the other takes his turn. Or, if the chests are too sore, there is resort to side-slapping. These rituals, though painful and injurious, may end with the opponents as friends. Even when machetes are carried, as a rule only the flat of the blade is used. Fights in which clubs are used produce worse injuries: many men have intricately scarred scalps of which they are proud. (There is an analogy here with the duelling scars which were popular in Germany as recently as the 1930s.) Such fighting is usually over women, but may be provoked by theft of food. Rarely, spears are used, by arrangement.

Clashes also occur between villages. The objective is then to kill one of the enemy, and to get away undiscovered. But the violence, though frequent and severe, is not lawless. Chagnon writes:

> Much of Yanomamö fighting is kept innocuous by ... rules so that the concerned parties do not have to resort to drastic means to resolve their grievances.

Moreover, the Yanomamö 'aggression' has not infected neighboring groups. Chagnon remarks on the quite different, pacific conduct of the nearby Makiritare.[38]

Other such 'unaggressive' cultivators have been closely observed. Tahiti (in the Society Islands) has been visited and described by Europeans since the eighteenth century. All have been impressed by the gentle conduct of the Polynesians who live there. During the period 1928 to 1962, only two homicides are recorded. R.I. Levy has described in detail the attitude of these people to violence and anger. The former is systematically discouraged at all ages; and, if a person is angry, he or she should express the feeling in words, not in injurious action. Timidity, among the Tahitians, is a virtue. Among children, quarrels are commonly settled by harmless rituals: a child who has been badly treated chases the offender but does not catch him; soon he stops the pursuit, and throws a piece of coconut – but misses.[39]

Another group with a simple agriculture, the Semai of Malaysia, is of special interest. They are notably non-combative, and dislike killing even animals. Yet some were recruited into the British army and, as soldiers, were appropriately violent. R.K. Dentan describes their military experience as shut off in a compartment separate from their normal existence. When former soldiers returned to their villages, they were no less gentle and afraid of violence than before.[40]

A still more extreme case of pacific conduct was discovered in 1971. In Mindanao, in the Philippines, the Tasaday live, cut off from the rest of humankind, in a densely forested mountain region. John Nance describes the adults of this tiny group as completely harmonious, without conflict. The children, however, squabble normally. Like the people of Tahiti, the Tasaday discourage youthful clashes and praise friendly and helpful conduct. They greet strangers with affection.[41]

For a last example I turn to a quite strange and much discussed people. The Ik until recently lived a nomadic life in eastern Uganda, but were moved, by government action, to an infertile upland region which they were required to farm. C.M. Turnbull describes their conduct during a drought, when many were starving: those unable to care for themselves, owing to weakness, age or illness, were treated with indifference or brutality, and were denied access to what food was available; children of three years were left to look after themselves; the misfortunes of others were greeted with laughter. It is a horrible story.[42] Turnbull's study has a special interest, because it was used in the media to support the well publicised picture of humanity as primarily and essentially beastly. Since the Ik had been ruthlessly driven from their home region, it would be reasonable to look for the sources of the 'aggression' inflicted on them by authority. The causes of their own aberrant conduct are obvious enough. To suggest that, in their desperate state, they stand for human normality, is itself a strange aberration, and one quite contrary to Turnbull's own conclusion.

We have now seen something of the variety of ways of life among peoples who live by gathering and hunting or by subsistence agriculture. Especially we have examined the ways in which they express violence and its opposite. There are no general, simple rules that enable us to understand all their many ways of life. And there is similar diversity in other communities. Paul Bohannan has compared murders in African tribal societies with those that led to the death sentence in England during 1949–1954. In England the commonest type of murder was of a mistress or girl-friend, and the next commonest, killing while committing another crime such as burglary. In the African groups, both were rare or absent.[43] There are also differences in the forms of violence within single communities. Among them are the effects of age: in 'western' cultures violent assaults are most likely among men in their twenties; and there are effects of class: homicide is commoner, and suicide rarer, among the poor than among the rich.[44]

To try to explain all friendship and enmity, conciliation and quarreling, peace and violence, by reduction to biological drives is to ignore characteristic features of human social life. Among these features – to which we return repeatedly in this book – is the (usually unexplained) diversity of our societies. Two other features should have prominent positions in any account of humanity: our capacity to adapt intelligently and thoughtfully to circumstances, and the transmission of knowledge, wisdom and moral principles, by teaching and imitation. When human beings defend themselves, even by violence, against intrusion, such action may be a result, not of succumbing to an ungovernable drive or instinct, but of calculating the consequences of alternative actions: it may seem the only sensible thing to do. Similarly, when people behave peaceably, this may be an outcome of rational debate: co-operation and friendliness may be held out not only as agreeable and virtuous but also as sensible. Second, though teaching has been neglected by anthropologists, its importance has been acknowledged throughout history. 'Train up a child in the way he should go; and when he is old he will not depart from it.'[45] In all communities it is taken for granted that whether a person is violent or tranquil, criminal or law-abiding, domineering or co-operative, depends greatly on upbringing.

Learning enmity

How then do children become violent or restrained, loving or hating? In our century, the ways in which personality traits develop have been systematically studied.[46] When questions are asked about the training of children, we are likely to think first of the uses of rewards, or even punishments. But children are trained largely by example: they model their behavior and attitudes on those of parents, older children, teachers and others. And even when one speaks of a child being rewarded for a peaceable (or a violent) action, the reward is often not an object, such as a sweet, but an expression of approval: the child's conduct is accepted socially, and may be a result of what the child has been urged to do, or has seen others doing. Hence copying a model and reward (in this sense) often go together. As for punishment, it is notoriously an unsatisfactory way of regulating conduct. In extreme cases, children may be violently punished for violent behavior. Such children are often found to be *more* violent in their daily activities than others who have been treated gently.[47] Perhaps this is because a child who has been beaten does not 'learn a lesson' of the kind intended, but instead copies the adults who treat him violently.

Experiments have confirmed that children imitate violent conduct after seeing it. They have been shown people ('models') shouting and bashing at a large doll. They were then given things to play with, and watched to see whether they behaved more destructively ('aggressively') than before. There were of course control groups not exposed to violent behavior. (The experimental details are much more elaborate than I describe here.) In other experiments, children saw 'aggressive' actions, but sometimes these actions were punished or verbally condemned. The immediate effects were usually those expected: children (especially boys) tended to copy the 'aggression' they had seen, and also to display other forms of violent behavior chosen by themselves; but, if they observed punishment or disapproval of violent acts, the tendency to be violent declined. Correspondingly, seeing violence on television increases 'aggressive' conduct, again especially among boys. Such influences may last for months, but are least in families in which pacific and friendly attitudes are encouraged. Similar findings come from studies of adults. As we know, even watching sports such as football increases the readiness to behave violently. But only some people are so affected, and only sometimes: J.H. Goldstein remarks, reassuringly, that the findings 'do not mean that anyone who watches an aggressive sport will go home and beat his wife'.[48]

Some kinds of violence are systematically encouraged, but the encouragement may then have to be violent to be effective. To produce soldiers who readily kill opponents at close quarters, young men are relentlessly trained in the ruthless use of weapons and made confident in the rightness of their cause. Some armies have been exceptionally brutal. The Nazi *Schutzstaffeln* (SS), or protection squads, were set up by Adolf Hitler in the 1920s. Later, some of the SS were organized in Death's Head units (*Totenkopfverbände*, or TKV). During the Second World War the TKV provided guards for concentration camps, and detachments for the slaughter of the inhabitants of countries occupied by German armies.

H.V. Dicks, a psychiatrist, interviewed 138 German prisoners of war, and published a detailed study of eight who had been convicted of the mass murder of defenseless people. He describes the training of the TKV:

> To break them in, . . . recruits had to undergo the worst excesses of barrack-square bashing as well as insult and humiliation . . . Next they would be paraded to see 'official' flogging and torture of prisoners, and were watched for signs of compassion or revulsion . . .[49]

For these young men, expressing moral scruples was a breach of discipline to be severely punished. The TKV came from a people with a tradition of high artistic, scholarly and scientific achievement: they were not 'savages'; and, as Dicks shows, the mass murderers he describes were not insane: of the eight interviewed at length, only one was without signs of a conscience – a psychopath. 'And', writes Dicks,

> – their SS roles ended or interrupted – these same 'fiends incarnate' in various ways disappeared quietly into civilian life, in some instances resumed orderly and normal careers.[49]

This does not signify that anybody can be trained for ruthless murder. In the German male populations studied, only about eleven per cent were wholehearted supporters of the Nazi ideology. These men had in common a number of features which together qualified them as 'high-F' (for Fascist). They enjoyed inflicting pain, that is, had strong sadistic tendencies; but they were paranoid, that is, excessively ready to suppose themselves persecuted or the victims of misfortune; they were also often irrationally anxious. They were without religion, but strongly disposed to become loyal members of an institution such as the Nazi party or a regiment. They were hostile to people of other races or nations. They had been brought up in households dominated by the father, who held his wife's views to be of little account; in such families expressions of tenderness were frowned upon. They had some homosexual tendencies; women were accepted, if at all, as sexual partners and breeders with a status inferior to that of men. Lastly, they had authoritarian attitudes: that is, they were insensitive, punishing and arrogant toward subordinates, sycophantic toward superiors.

A similar personality structure was independently identified among Americans studied by T.W. Adorno and others. This controversial work began as research on anti-semitism, but came to have a wider scope and was published as *The Authoritarian Personality*. Middle-class American men assessed as high in racial prejudice also scored high on a measure (resembling the 'F' scale used by Dicks) of 'pre-fascistic tendencies': among the features isolated by Adorno and his colleagues were conventional attitudes, submission to authority, domineering conduct to inferiors, 'toughness', destructiveness and cynicism; and they refer to the tabu on tenderness also emphasized by Dicks.[50] Such findings point to the family environment as critical: if a boy is deprived of gentleness and mothering in early life, he is in danger of developing at least an authoritarian personality; he is promising material for membership of a 'paternalistic' organization, and perhaps for training in violence.

The studies of Dicks, Adorno and their successors provide a description of personalities. When we say that we understand something or somebody, we often mean that the object or person has been put in a familiar or at least comprehensible category. These studies also show how certain kinds of person satisfy their needs by action. The needs of the people with whom we are now concerned are by no means only for violence: subjection to an organization or leader, with its resulting feeling of security, may be much more important. People who succumb to violent and 'inhuman' political programs may not be marked by some notable psychological peculiarity, any more than they are distinguished by a characteristic color or shape.

Of the leading Nazi criminals, Adolf Eichmann (1906–1962) had one of the worst records: he supervized, with the utmost efficiency and dedication, the murder of millions, killed only because they were Jews. Yet, as Hannah Arendt's celebrated study shows, Eichmann was personally agreeable and unassertive, a joiner of clubs. The subtitle of her book, *A Report on the Banality of Evil*, shocked many readers. The history of this extraordinary yet ordinary man shows how a person can organize appalling acts of violence without being obviously strange or 'inhuman'.[51]

Correspondingly, the original portrait of the authoritarian personality has been modified. A later finding is of the large numbers of 'decent ordinary people', in any population, who can be induced to 'tolerate, serve and creatively abet astoundingly vicious policies.'[52] Much depends on the climate of opinion in the community. In Hitler's Germany, brutal and paranoid attitudes were conventional; hence people with strongly conformist personalities were likely to adopt such positions. The same process is found among South African Whites today.

But emphasis on the social environment does not relieve us of responsibility. We remain capable of making choices, and the choice can be the opposite of vicious. This other face of 'human nature' may appear even in unfavorable conditions. John Buchan's massive history of the First World War describes incidents of the early weeks of trench warfare. In 1914, during the 'extraordinary truce of Christmas Day', men of the opposed armies met in joint celebrations. In one place 'a famous clown in the German trenches occasionally went through performances amid the applause of both sides.' In others, the truce included games of football and visits to the trenches on the other side. As Buchan remarks, 'outposts have always fraternized to some extent – they did it in the Peninsula and in the Crimea'.[53] In committing those gross breaches of discipline, men succumbed to impulses of friendship and tolerance that are at least as much features of humanity as are violence and hatred. No doubt, by 1918, nearly all were dead or seriously injured. Yet their most significant contribution to history was not military: it was the brief, peaceful interlude in which they took part – provided that we remember it.

Political violence and power

The analyses, described above, of violence by followers of Hitler, are concerned with individuals. But Nazism was also a political phenomenon; and current writings on 'aggression' include much on violence related to 'the form, organization, and administration of a state' (*OED*). The quoted phrase is part of a definition of politics. For the present argument, political violence is important in three ways. First, it illustrates once again the variousness of human enmity. Second, it exemplifies the rationality of some violent conduct. Third, it shows how preoccupation with 'aggression' can distract attention from vital questions.

Political violence within one country has at least five sources.[54] First is religious or racial conflict: Muslims clash with Hindus, Catholics with Protestants, Malaysians with Chinese. Second, separatist movements may be violent, as in Kashmir, and Bangladesh before it became a state. Third, we have revolution and counter-revolution: in this century the leading examples of the former have been in the USSR and China; counter-revolution was notably successful in Germany after the First World War, and in Spain before the Second. Fourth, there are coups on behalf of single individuals or small groups – common events in South America and Africa. Lastly, narrower political issues sometimes lead to violence – for example, those concerning university education and the treatment of students.

For a balanced account we must recognize also the opposite situations: even in the present century, many communities have remained calm. Kerala, for example, in the south-west of India, has probably more religions to the hectare than anywhere else; but Kerala, like most European countries, is usually free from sectarian conflict. Similarly, in this century of revolution, only a minority of countries have experienced revolutionary upheaval. It remains possible that most readers of this book will have experienced political violence only at secondhand, on television.

In analyzing all this diversity, we are not helped by studies of baboons or chimpanzees: we must use the methods of political science and sociology. Whether a government, or a religious, racial or national group resorts to violence, or refrains from it, may depend on a decision reached after debate: it need not be an outcome of some ungovernable impulse. Similarly, in some countries, such as Britain, the police rarely carry firearms; but this is not because the British have some special, inherent drive for non-violence: it is the result of the (now changing) policies of governments; and these in turn are influenced by historical, economic and other factors without counterparts in the 'societies' of other species.

Governments and political groups generally use violence not for its own sake but as a means of maintaining or acquiring political or economic power. Since the collapse of Hitler's Third Reich, governments have increasingly turned to violence of kinds more subtle than that of jackbooted stormtroopers. As an example, psychosurgery originated as a means of treating severe mental illness. It has, however, had other uses: the indications for destroying parts of a patient's brain have included 'aggressive' conduct. Medical men, enthusiasts for this kind of treatment, have therefore proposed it for rioters.[55] As a result, they have received endowments from the United States government. There are also non-surgical methods of violence against the brain. A Canadian, D.E. Cameron, while head of a Psychiatric Institute in Montreal and first president of the World Psychiatric Association, used a form of electro-convulsive therapy (ECT). ECT originated as a treatment for severe depression. Cameron applied it to people with a greater variety of conditions, some very mild; and he used a higher voltage and longer and more frequent shocks than those recommended. The result was permanent psychological injury. Cameron's work, too, was secretly supported by an agency of the United States government.[56]

In political affairs, however, completely non-violent means are often more important; and preoccupation with 'aggression' can be dangerous if it makes us ignore them. Here is a comment by a master in the art of politics:

> Through clever and constant application of propaganda, people can be made to see paradise as hell, and also the other way around to consider the most wretched sort of life as paradise.

This cynical, but by no means wholly false, statement was not written by a statesman in one of the democracies: it is from Hitler's *Mein Kampf*.[57] It is quoted by an American political scientist, Bertram Gross, in an account of the ways in which governments and other ruling agencies, in ostensibly democratic countries, can override individual rights without any parade of armies, police or security guards and without even the use of clandestine violence. 'The open brutality of the old-fashioned fascist regimes [he writes] has misled many observers into equating fascism with official violence'; but this obscures the rewards often provided by authoritarian regimes. The rewards include membership of an organisation or Nation and adherence to a stimulating idea.

Such rewards do not, of course, satisfy everyone; but, when they fail, a government has other non-violent sources of power. Even in the world's most powerful and most influential would-be democracy, democratic aspirations clash repeatedly with the desires of authority. Gross, indeed, suggests that under the Nixon administration, the USA 'escaped by a hairsbreadth from becoming a police state'. Today, in a technically advanced country, a police state is likely to be a computer state: each person's health, finances, education, politics, eccentricities and much else may be recorded and made accessible to government. There is never any guarantee either that the information is reliable, or that it will be used benignly. Another American political scientist, F.J. Donner, in 1980, gives a fully documented review of surveillance in which he expresses the anxieties of informed people. Americans, he says, 'have betrayed the principles by which we have agreed to live as a people.' And he writes:

> No aspect of our common life has been so battered by misconduct and betrayal as our commitment to the fullest measure of political freedom. And none of the manifold excesses of the past can compare in scope and intensity with the secret war waged . . . against all shades of dissenting politics by the domestic intelligence community . . . dominated by the FBI.[60]

Even the role of secret police may be diminishing. While warnings, such as those I cite above, were being published, others were appearing on the changing character of the news. Even in countries with substantial freedom of thought and expression, authentic information on what is going on is, for most people, hard to come by: as television becomes dominant, matters of life and death are reduced to the kinds of visual stimuli calculated to promote the sale of soft drinks.[61] To acquire information on matters of public concern often requires prolonged exertion and access to large libraries and to people with special knowledge. Such difficulties for the public make it easier for those who wish for arbitrary rule. The hoodwinking process is not complete, but it is real and growing. Hence, while aggression, in the sense of violence against persons or property, is still of vast importance, in political affairs it is only part of a larger picture, that of state power.

Preoccupation with aggression as a biological phenomenon has led to several kinds of misdirection. At the beginning of this chapter I show how diverse activities are lumped together as aggressive, when they ought to be treated separately. This disguises the real achievements of those who attempt systematic study of human violence. Worse still, it distracts attention from coercive actions, often hidden, that do not include violence. We have two large areas of relevant knowledge. One concerns the conduct of individuals, from infancy: we know much that can help our children to lead independent lives and yet contribute to the common good. In the other area, that of politics, despite heavy and repeated reverses, we are slowly making progress. I defer further support for that optimistic statement to the final chapter.

Part 3

Homo egoisticus:
the selfish species

Man, in a word, has no nature; what he has . . . is history.

Ortega y Gasset

The image of humanity as intractably selfish includes some features of *H. pugnax*, but also uses the modern theory of evolution. We are told that even our social lives and attitudes can be fully explained by the action of natural selection on our ancestors. Just what the new Darwinism does and – equally important – does not tell us comes, with some unavoidable technicalities, in the next chapter. Many questions remain; in particular, no theory tells us how the human species acquired its special features. The explanatory value of natural selection has often been overrated, partly because, for some influential writers, it is more than a scientific theory: it has become a substitute for the workings of divine providence.

Evolutionary theory rests on genetics, and so does human biology: it is impossible to make sense of biological accounts of humanity unless one understands the interaction of heredity and environment. This, as we find in Chapter 7, is a continuing source of confusion. It is still usual to describe characteristics as inherited or genetically coded, and to write of genes *for* traits such as (for example) intelligence or mental deficiency. As a result, our genes come to be thought of as imposing an unchangeable destiny. The error in such expressions arises from disregard of the environment – one, moreover, that in part we create by our own choice.

In Chapter 8 we come to the shifting ground of human evolution: who were our ancestors, and what did they do? Early speculation was based on a few fossil fragments. We now have much more evidence, and much more cautious

78

interpretations. But imaginative inventions are still current. In their most extreme form they imply that all our moral principles are spurious; that human existence is a condition of unremitting strife; that the strife extends to the relationships of the sexes and of parents with their children; and that women must always be subordinate to men.

These, or similar, less extreme conclusions, do not follow from any scientific findings: their importance, as we see in Chapter 9, comes from their political impact. For more than a century, biological arguments have been used to deny the possibility of improving the health and educational level of the poor, and to assert the inferiority of certain races and of women. The customs of a particular time and community have been held out as representing an unalterable human nature. The social role of *Homo egoisticus* is to resist social change at a time when changes are occurring as never before.

6 Evolution and natural selection

Vast chain of being! which from God began,
Nature's aetherial, human, angel, man,
Beast, bird, fish, insect, what no eye can see,
No glass can reach . . .
From Nature's chain whatever link you strike,
Tenth, or ten thousandth, breaks the chain alike.

Alexander Pope

'One hundred years ago', a zoologist tells us, 'Charles Darwin bequeathed to us a theory which is, in principle, able to explain every characteristic [of] every species.'[1] There are believed to be 3 to 30 million species of organisms, of which fewer than two million have been named and described.[2] Each species has features that match it closely with a particular range of habitats or ways of living. If our descendants have the desire (and retain the resources), they will be able to spend millennia working on the systematics, distribution, life histories and physiology of known and as yet undescribed forms.

The all-embracing theory of Charles Darwin is that of natural selection; and it is an attempt to state *how* the diverse species of today have evolved. It must be distinguished from the quite separate proposition that organic evolution has indeed taken place: this states only that existing forms have arisen from very different forms in the past and that others will continue to do so in the future; it does not say anything about the agency, or agencies, of these vast changes.

Evolution as fact

Today, arguing against the fact of evolution is like holding that the earth is flat. Yet the argument is still heard, and is even influential in the USA.[3] I therefore say something about the evidence.

It is commonly held, with reason, that the principal ground for believing in organic evolution is the existence of fossils; but in fact, even if we had no fossils at all, we should still conclude that animal and plant forms have relationships analogous to those shown in a family tree. We should infer this from the facts of structure, especially the changes undergone by embryos.

In earlier chapters I mention the concept of recapitulation. Although complex organisms do not 'climb their family tree' during development, their development does have features which are intelligible only if we assume evolution. A flatfish, such as a plaice or flounder, lies on its side concealed in sand, with both eyes poking above the surface. The two eyes are on the same side of the body. But the young flatfish has a structure like that of the young of a normally rounded species: the eyes are symmetrically arranged. Later, one 'migrates' to join its fellow. We say that this peculiar sequence leading to the adult form reflects a common ancestry with fish species of the usual shape.

The development of a human being carries a similar message. At about four weeks the gentle reader was about 6mm long and looked much like an embryo fish: there was a substantial tail, with the rudiments of powerful muscles; the heart was operational but merely pumped the blood forward (a fish-like arrangement); there were pouches that looked as though they were destined to become gill clefts; there were rudimentary kidneys of the piscine type; and so on. Many of these facts were known before Darwin, but they were not satisfactorily explained. Today we say that they reflect the ancestry we share with other vertebrates: the fish, amphibians, reptiles, birds and mammals all have early stages with the same features; and all, we conclude, must have had a common precursor at a remote period.

The findings and conclusions from comparative anatomy and embryology are matched by those from geology. Our understanding of fossils is founded on our knowledge of the rocks that contain them. For hundreds of millions of years ice, wind, rain and tides have been eroding the earth's surface, while in other regions the seas and rivers have been depositing layers of rock made largely of the eroded particles. Sedimentary rocks are slowly laid down, one layer on top of another, so that, ideally, the most recent are nearest the surface. Earthquakes and volcanic and other upheavals cause massive distortions but they do not upset the principle. Today, analysis of radioactivity often enables us to date rocks and fossils with some precision. Fossils are the petrified remains of animals, plants and other organisms, embedded in rock.

At about the time Darwin was born, it was observed that fossils occur in rocks always in the same order. Rocks of about 300 million years ago contain fossil fish, but no land vertebrates. We can trace how, later, some early fish gradually developed the capacity to live on land, and so gave rise to the land vertebrates. The earliest were amphibians (now represented by frogs and newts); but for much of the time after this the most prominent were reptiles, including the notorious dinosaurs and pterodactyls. One branch became birds.

Meanwhile, reptiles of a separate lineage developed complex teeth instead of pegs or spikes; and some became 'warm-blooded' and acquired the capacity to bear young in an active state (instead of laying eggs). The mammals have been the main land vertebrates for 60 million years. We can trace in great detail the evolution of some modern hoofed animals, and of large carnivores such as tigers, from small, undistinguished creatures. Other groups have left fewer fossil remains, evidently because they did not live in surroundings which favored fossilization. Among these less well recorded orders is unfortunately that of the Primates, the one that includes the monkeys, apes and human beings.

Nonetheless it is possible to give an account of the evolution of mankind from ape-like creatures. In the Miocene rocks of about 20 million years ago are the skulls and a few limb bones of apes that lived mainly on the ground. These dryopithecines had spread over Asia, Africa and Europe, and seem to represent an ancestral stage common to modern apes and ourselves. Later come the australopithecines, described in the previous chapter, with their tools and their patchwork of human and non-human features; and, later still, fire-making *Homo erectus*, clearly human but with a distinctively massive skull. There is plenty of room for debate about details; but the fossil record which precedes modern *Homo sapiens* is, by itself, good evidence for the fact of our evolution.

The preceding paragraphs are, however, misleading in one respect. They describe sequences of fossils which we readily describe as advances. It is quite usual for textbooks to refer to evolutionary progress. And R.A. Fisher writes: 'Evolution is progressive adaptation and . . . nothing else.'[4] But vast numbers of types are as dead as the dodo. The most celebrated are the giant reptiles of around 100 million years ago, about which many modern children first learn at their mother's knee. Many other giant forms too are, alas, extinct. They include dragonflies with a wing span of 70 centimeters – the largest insects that we know of, and millipedes 2 meters long. There were also completely unfamiliar animals. One, called *Hallucigenia*, had seven pairs of limbs and seven tentacles, each of which seems to have ended in a mouth.

In the nineteenth century, the idea of evolution, and especially of the 'descent [or ascent] of man' from primitive apes, caused quite an uproar. In Europe, North America and elsewhere people had been reared on the Jewish myths, in Genesis, of a creation achieved in six days. This was believed to have happened about 6000 years ago. On that foundation was built the idea of the Great Chain of Being, celebrated by Pope in the epigraph of this chapter. A.D. Lovejoy describes the widespread belief in, and use of, this idea in western thought. It was a

> conception of the universe as an infinite number of links ranging in hierarchical order from the meagrest kind of existents through 'every possible' grade up to the highest possible kind of creature.[5]

At the top was a Creator. The adaptedness of living things was held to be evidence of a transcendental design. To accept evolution, people had to turn the picture upside down. As Gregory Bateson puts it, the traditional account had Mind at the top: Mind was an explanation and indeed the source of all the rest. But now Mind 'became that which had to be explained'.[6]

Some Christian fundamentalists still demand that the creation myths in Genesis should be presented, in the schools, as an alternative to the account of evolution in textbooks of biology. But these are not alternatives. The fundamentalists acquire their beliefs by reading and interpreting ancient scriptures. Biologists derive their conclusions by observation of nature and logical argument. There are, however, true alternatives to *Genesis*. Bateson gives an example. 'The extraordinary achievement', he writes,

> of the writers of the first chapter of Genesis was their perception of the problem: *Where does order come from?* They observed that the land and the water were . . . separate and that species were separate.

He refers to a ruling (made in California) that

> students be told of other attempts to solve this ancient problem. I myself collected one of these among the Stone Age head-hunters of the Iatmul tribe in New Guinea. They, too, note that the land and the water are separate even in their swampy region. They say that in the beginning there was a vast crocodile, Kavwokmali, who paddled with his front legs and paddled with his back legs, and thereby kept the mud in suspension. The culture hero, Kevembuangga, speared the crocodile, who then ceased to paddle, causing the mud and water to separate. The result was dry land upon which Kevembuangga stamped his foot in triumph. We might say he verified that 'it was good'.[7]

Accepting 'Darwinism' a century ago entailed a double upheaval. There was first no longer a cosy picture of the past of humanity and of the world: it was replaced by a vaster canvas, less clear and partly (like all science) provisional. Second, there was the implication that our view of the world should no longer be laid down by authority (of the church or of the scriptures): instead, it should be founded on evidence critically examined, vigorously debated and always open to question.

Nonetheless, people quickly absorbed the new ideas. There was no interruption of historical processes attributable to theoretical biology. Even Christian doctrine has, in most places, come to terms with the concept of evolution. And so we may ask whether, for our understanding of ourselves, the idea is now of great significance. As early as 1883 Jane Welsh Carlyle did not think so. She wrote:

> But even when Darwin, in a book that all the scientific world is in ecstasy over, proved the other day that we are all come from shell-

fish, it didn't move me to the slightest curiosity whether we are or not. I did not feel that the slightest light would be thrown on my practical life for me, by having it ever so logically made out that my first ancestor, millions of millions of ages back, had been, or even had not been, an oyster. It remained a plain fact that I was no oyster, nor had any grandfather an oyster within my knowledge; and for the rest, there was nothing to be gained, for this world, or the next, by going into the oyster-question, till all more pressing questions were exhausted![8]

Today, some may readily accept Jane Carlyle's cheerful philistinism. We face plenty of pressing questions to which our evolutionary past gives no answers. But we not only accept our origin from an animal lineage; we also sometimes presume that we know how this evolution has been brought about; and we are thereby led to important issues – even political issues – that concern not our past but our future.

So I now turn to ideas on the causes of evolution, especially those relevant to *Homo egoisticus*. This entails bringing out, not only what we know about evolution, but also the limitations of our knowledge. Much of this chapter is therefore unavoidably technical. Readers who are willing to take evolutionary theory on trust may skip to page 95 or 102.

Natural selection: stability and change

One of the best argued of modern works on evolution is, the author tells us, 'based on the assumption that the laws of physical science plus natural selection can furnish a complete explanation for any biological phenomenon'.[9] A reader might suppose that, to fill such a role, 'natural selection' must represent an elaborate and difficult set of theories. Yet in an authoritative text published in 1977 we learn that, like other 'really great ideas in science', it is 'remarkably simple'; the writer continues:

> The idea is basically that healthy and vigorous individuals have better chances of surviving and leaving progeny than ailing and frail ones.[10]

But the appearance of simplicity is misleading. There is still debate on how the theory of natural selection should be rigorously stated; and the most ruthless scientific critics even question whether the expression refers to a theory at all. If it is a theory, it should have explanatory power. Can it explain adaptation?

The phrase, natural selection, is itself metaphorical (page 21). Darwin made an explicit analogy between selection of seed or stud males, by farmers and stockbreeders, and what happens in nature. Cereal crops vary in their ability to withstand unpredictable weather, and in their yield of grain; cattle in their resistance to disease, and in the beef and milk they yield;

cats in their readiness to catch mice; and so on. Human beings have made use of such variation to breed types adapted to their needs. Yet, when we look at wild plants and animals in nature, we may be more impressed by the uniformity of each species than by variation. Unusual forms do occur on occasion. A black species of bird, or a brown species of mammal, may each produce the rare white individual, just as there are albino people. Usually, such abnormal forms have shorter lives, or are less fertile, than normals: they are 'weeded out'. Often, the difference from the norm is genetically determined. We may then say that the difference in survival or fertility ensures both a uniform appearance and also uniformity in genetical make-up.

This is stabilizing selection. Human beings are exceptionally variable, but we have evidence of such selection even in human populations. In every population, some babies die at or soon after birth. The greatest number of such deaths occurs among the smallest and largest babies.[11] Differences in birth weight are, we assume, partly of genetical origin. (They are, of course, also influenced by environmental effects, such as the mother's nutritional state.) Different death rates in the first weeks after birth therefore tend to favor the genetically typical or average, at the expense of extremes. Similar evidence has been found for adult stature and for the length of the menstrual cycle.[12] Hence, it seems, natural selection *prevents* change.

In the preceding sentence 'natural selection' is a concise way of saying something like 'differences, of genetical origin, in survival or fertility'. Darwin's greatest contribution was to see natural selection, in this sense, as a source of gradual change which could transform species. Moreover, he did so without direct evidence of such change in natural populations.

Today we have such evidence. Much of it is a result of human activity, and is familiar. We try to rid ourselves of insect pests with poisons, and they develop resistance. Forms of disease-causing bacteria arise that thrive despite the use of antibiotics. A new toxin is introduced to deal with rats and mice, and populations of these animals appear which are not harmed by it.[13] Sometimes, the difference of the resistant types from the others is genetically determined. Some such changes are only incidental results of human action. In parts of Europe and Britain many trees have a coating of pale-colored lichens; and some insects, especially moths, have a speckled appearance that camouflages them marvellously against such a background. But industrial pollution in some regions has killed the lichens, and left the tree trunks bare and dark. Against them, a speckled moth stands out 'like a bar of soap in a coal scuttle'; in these regions black (melanic) varieties (formerly uncommon) were less preyed upon by birds, survived for longer, and at one time went some way toward replacing the speckled forms. Later, pollution was reduced, lichens returned, and speckled moths became more numerous again. In at least one species, the peppered moth, the difference between black and speckled is genetically determined in a quite simple way.[14]

In human genetics, the standard example is sickle-cell anemia, a disease which in most populations is very rare, and which kills at an early age. It is a recessive condition: that is, to have the disease one must have acquired a certain gene from each parent. Clearly, there must be strong selection against such a gene. Yet in certain regions, notably of west Africa, where sub-tertian malaria occurs, the sickle-cell gene is common. Possessing *one* sickle-cell gene, plus a normal gene from the other parent, allows a full life span and also confers resistance to malaria. Hence in those regions selection favors the gene, despite its lethal effect in the double dose. But among the descendants of African slaves in the USA, which has no malaria, the incidence of the gene is lower than that in the parent African populations. There the gene confers no advantage, and its incidence is steadily declining.[15] Hence genetically different types may have different chances of survival; advantage may depend on where the individuals are living; and natural populations have both stability and the capacity to change.

Questions on variation
The form taken by evolutionary change depends, as far as we know, always on just what genetical variation is already available. When our ancestors bred cattle for milk or docility, or rice for grain yield, they were making use of such variation; and Darwin saw that the results of domestication suggest the existence of plenty of variation in all animal and plant populations. Since then, artificial selection in the laboratory has revealed it in many species, including plants such as maize, insects such as fruit flies and mammals such as mice.[16] A few generations of breeding may produce changes in structural features, such as size; in resistance to disease; in behavioral features, such as an altered pattern of courtship; and in much else.

Of course, when maize or mice are subjected to artificial selection, they remain maize or mice. Similarly, domestication does not originate completely new species. Compared to the vast transformations in fossil sequences, the modifications achieved by human selection are minuscule. When we observe the formation of new species today, it is usually a result of a structural change in the chromosomes, and therefore involves many genes.[17] Yet the accumulation of small genetical changes, such as those observed during artificial selection, is usually held to show us the causes of evolution.[18] This is necessarily a presumption, for we have no time machine in which to go back for a look; and some evolutionists question it.[19] In the absence of direct evidence, the dispute is unlikely to be settled quickly.

In most experiments, after a time selection ceases to have any effect: all the genetical variation seems to have been used up. D.S. Falconer selected a stock of mice for large size on one line and smallness on another. Initially the mice averaged about 20 grams, but the first group went up to 30 and the second, down to 13. After 23 generations, however, change had almost stopped.[20] Hence for small variations to be responsible for evolution, a continuing source of 'new' genes is needed. This is provided by mutation. Yet, when a mutant gene occurs in a sperm or an egg, a common result is death or impaired survival or fertility. In human genetics, one of the best known examples is hemophilia, in which the blood fails to clot normally. Until very recently, this condition was always fatal quite early in life. About one in 300 000 males is hemophilic. This represents a rather high mutation rate.[21]

If all mutants were disadvantageous in all circumstances, they clearly could not be responsible for evolution. The presumption is that a few have favorable effects. Such an outcome is especially likely when a population (such as that of the moths in polluted regions) is exposed to a change of environment: what was formerly a handicap may then become beneficial. Moreover, in the evolution of a complex structure, such as a sense organ, each step probably requires the coming together of several favorable mutant genes – a very rare occurrence.

As an additional complication, selection for a particular feature may entail disadvantageous changes. The human species has evolved from four-footed ancestors. We are now completely upright and our hands are fully released for carrying and manipulation. The change is supposed to have aided the survival of our ancestors; but the resulting narrowing of the pelvis has increased the hazards of childbirth, and trouble with the vertebrae and attached structures causes much backache. Presumably, however, on balance uprightness was worth it.

The variation on which evolution depends is said to be random. The idea of randomness is a difficult one;[22] but we do know that mutation is *undirected*: its effects are unrelated to environmental demands.[23] Exposure of a population to a new disease or other hazard does not induce new and appropriate mutations. An opposite belief, named after the French natural philosopher, J.B. de Lamarck (1744–1829), was generally accepted almost until the present century. Lamarck wrote a treatise on the theme that 'acquired characters' are inherited. According to him, use and disuse of organs and abilities during an individual's lifetime have a corresponding influence on the individual's offspring. Lamarck also believed that the will of an organism had a similar effect. Unlike Darwinism, the Lamarckian theory requires no pre-existing variation: altered forms arise in response to the direct, current demands of the environment. Hence no selection among different types is needed. We have, however, no good evidence for this notion, and much against it.[24] The modern position conforms with everyday experience. My grandfather, an Englishman, spoke French like a

Frenchman, but his children and grandchildren still had to learn French laboriously and, alas, imperfectly. Children benefit from their parents' learning not by altered genes but through imitation and teaching.

When people acquire skills by such social means, they do so as a result of intention: a child emulates a parent, a parent hopes to transmit abilities to the child; each has a plan with an end in view. We sometimes write casually of 'adaptations' as though they too were the result of conscious design: a zoologist may seem to follow Lamarck and write that a concealing pattern was evolved by an insect *in order to* deceive a predator.[25] But this is misleading: our account of genetical variation and its consequences has no room for intention or for a designer. On the contrary, much importance is sometimes given to 'blind' chance.[26]

In human genetics chance effects are most clearly illustrated by disadvantageous states. Porphyria is a serious disease transmitted as a dominant condition. In most populations it is very rare; but among South Africans descended from Netherlanders the incidence is about one percent. The original small group from the Netherlands evidently included one porphyriac who, despite the handicap of the disease, has had many descendants. If so, this is an example of the founder effect – a consequence of the action of chance in a small group. Huntington's chorea, a progressive disease of the nervous system, is another dominant condition; it often develops only in middle age, and so allows production of children. The many sufferers from this disease in the USA are believed to be descended from a single immigrant. Many such instances are known.[27]

A more general term for this kind of phenomenon is drift: if small groups are sufficiently isolated, otherwise rare genes may become common in some of them even if the genes do not increase fitness. The result is then unusual forms and exceptional populations. Some hold that evolution has depended, to an important extent, on such events. In that case, the role of natural selection would be no more than 'weeding' or 'winnowing': novel forms would be a result of unpredictable changes like the movements of flotsam in a restless sea. We have no firm basis for a conclusion on these questions. In particular, the effects of selection and chance in human evolution are not known.

Adaptation: an enigma

Evolutionary change is usually regarded as adaptive. We can see that this apparently obvious notion contains difficulties if we examine the expression, the survival of the fittest.[28] In biology, the fitness of an individual or of a type is defined as some measure of its contribution to later generations. Athleticism is not necessary, but fitness in this sense is related to fertility and longevity. For clarity, therefore, we speak of Darwinian fitness.

Darwinian fitness is neutral in relation to human values, and its consequences can be disconcerting. Among the Hopi American Indians albinos are surprisingly numerous. Albinism is usually a recessive condition: that is, both parents may be normal but, to have an albino child both must transmit to the child a particular rare gene. It is also a handicap: tolerance of bright light is low, and the skin has to be protected from the sun; hence albinos do not live as long as normal people. On the face of it, therefore, albinos are less fit (in every sense) than non-albinos, and the albino gene should tend to be eliminated whenever it appears. How then did the Hopi albinos do so well? Evidently, while the healthy men were in the fields, the albino men stayed in the villages with the women. As a result, the albinos were biologically fitter than the others.[29]

Albinism is only one example of a general difficulty: we cannot expect to identify a feature as biologically advantageous or disadvantageous merely by looking at at. Here are two further conditions which would usually be taken for granted as always maladaptive: rheumatic heart disease, and severe near-sightedness (myopia). (Recall the reference to 'ailing and frail' individuals on page 84.) But in fact, in some conditions, each – like albinism – could increase fitness: in time of war they could lead to exclusion from military service and so improve the chances of survival and of the opportunity to produce children. Variation in these conditions is partly of genetical origin: hence selection in their favor is possible. The principle has been expressed more generally by Santiago Genovés: 'if the strong were always fighting, the weak would inherit, if not the earth, at least the wives of the strong.'[30]

We have now reached the notorious logical impasse of evolutionary theory: the survival of the fittest becomes the survival of those that survive. Much authoritative writing on the subject contains such tail-chasing utterances. A distinguished evolutionist provides an extreme example:

> Natural selection . . . states that the fittest individuals in a population (defined as those that leave the most offspring) will leave the most offspring.[31]

It is easy to see why the theory of natural selection is sometimes said to be a tautology – 'the repetition of the same word or phrase, or of the same idea or statement in other words' (*OED*). If, however, the theory is nothing more, it must appear as merely a grandiose confidence trick.

It is not quite that. A more valid comment is that the theory of natural selection is tautologous in the same sense as are mathematical proofs. Pure mathematics is derived from a set of axioms, not from observation of the external world. Mathematical statements are essential for many investigations of actual events; but, as we know, they are of a kind very different from those derived from, say, timing actual runners in a race (Chapter 2). Yet natural selection is usually regarded not as a self-contained set of equations but as a theory that tells us about real organisms: it is held out as a theory about *causes*.[32] The obvious comparison is with genetics, which allows us to predict the outcome of many breeding experiments with confidence. So we may ask whether modern evolutionary theory has similar predictive power.

I mention above changes in populations of bacteria, moths, mice and human beings. No known theory, of natural selection or anything else, would have enabled us to *predict* them. We cannot even take for granted that an environmental change, such as air pollution, will be followed by *some* corresponding adjustment in the populations affected. Genetical adaptation always depends on the genetical variation already present. The speckled moths occasionally produced black individuals *before* the lichens were killed. The theory of natural selection says nothing about the origin or nature of such variation. The moths might have lacked the necessary rare genes and so have become extinct; but that would not have contradicted any theoretical principle. Other outcomes can be imagined. Some insects are poisonous or distasteful and also conspicuous: they are then said to have a warning or aposematic appearance. Mutant forms of the speckled species might have arisen, protected by brilliant color and toxicity, and replaced the others. They would then have been quite justifiably greeted as a striking example of natural selection in action. As Niels Bohr once remarked, prediction is very difficult, especially if it refers to the future.[33]

Darwinian fitness

Evolutionary accounts of changes in natural populations are descriptions made after the event. Once the observations have been made, the story is told in Darwinian terms. In such a context, natural selection is not a theory but a *name* for differences in the survival rates of types or of genes. The descriptions use the idea of Darwinian fitness. Sometimes, especially in the laboratory, certain narrowly defined measurements can be made, for instance the number of young produced in a given period or in a lifetime; these may be called measures of fitness. Another kind of index is the rate at which a gene spreads or declines in a population. If the gene is compared with an alternative, we say that its relative fitness has been recorded. The decline of the sickle-cell gene among Africans in North America is an example.

We have no single definition of fitness for every occasion: if the term is used, it has to be separately defined for each situation studied. That is why it is often stated, paradoxically, that fitness cannot be measured.[34] Moreover, the findings from such researches, being primarily descriptions, cannot test a general theory, of natural selection or anything else.

Where then do we find *general* statements on evolution *that can be tested*? One possibility is to propose that any clearly defined characteristic, such as color or a behavior pattern or a chemical property of the blood, is the best possible, or optimum: all departures from it would impair survival or fertility. Each species today is an outcome of eons of evolution, and so may be expected to be well equipped for the conditions in which it lives. Here is an example.[35]

Certain birds collect food and carry it to the nest. If they took any edible fragment indiscriminately, some would be too small to be worth the effort (that is, the energy expended). On the other hand, if only the largest possible fragments were accepted, they might be too few to justify the work of finding them. So the hypothesis is of a 'strategy' of optimal foraging; and a calculation may be made of just what a bird should do to achieve it. Birds of the species concerned are now observed; and they do not behave quite as expected: they take more large fragments than – according to the hypothesis, or model – they should.

So the hypothesis is wrong – or, at least, not quite right. But perhaps something has been left out. If the nestlings are left alone, they are liable to be attacked by predators. So the foraging 'strategy' may represent a compromise between optimizing energy balance and ensuring security for the young. Unfortunately, however, this additional hypothesis is difficult to test.

To say that a feature typical of the species is optimal is not a finding from the facts of nature but a working hypothesis; and a peculiarity of the procedure is exposed when what is expected is not found, as in the example above. The discrepancy can always, with sufficient ingenuity, be explained away by means of supplementary hypothesis – in the example, the effects of predation.

The axiom that all features of all organisms are the best possible has been wittily criticized by S.J. Gould and R.C. Lewontin. They call it the Panglossian fallacy.[36] I venture to soften their verdict by calling it the Panglossian *principle*. In Voltaire's novel, *Candide*, the philosopher, Dr Pangloss, holds that everything is made, not only for a purpose, but for the best purpose: noses, for instance, are made for the wearing of spectacles. As it happens, satire apart, the human nose does present a problem for adaptationism. It also illustrates the scope and limitations of Dr Pangloss's doctrine. The noses of Europeans are narrow, those of Negroes, broad. The difference is genetically determined. How is it explained? One may always speculate: a narrow nose is perhaps advantageous in a cold climate, for it warms the air better; in a hot climate, a narrow nose confers no advantage, and a broad one is less liable to be obstructed. Human anatomy provides other such instances.[37] What, for instance, should we make of the coiled growth of a Negro's hair and the straight growth of Chinese hair? The difference has no obvious neo-Darwinian explanation, but we cannot *prove* that it has never had an effect on fitness. Hence we may either presume that the difference has some selective significance, *or* ask for a demonstration that the difference adapts the two populations to their respective environments. Either hypothesis can be a framework for research. The research, however, must be on existing populations; and it could be done without making any guesses about our evolution.

What we need, and do not possess, is a general criterion of adaptation which *does not use the test of contribution to later generations*. I illustrate the difficulty above, by showing how conditions, such as albinism, that we regard as weakening and undesirable, can enhance Darwinian fitness: that is, they sometimes improve the chances of leaving descendants.

It may be argued that albinism and the rest are not advantageous in the long run. If, for instance, a community had more than a certain proportion of albino men, it would be so handicapped that it would die out; or perhaps 'selection' would be reversed, to favor non-albinos. But in fact this argument exposes another dilemma. How long must the run be before we are willing to call a characteristic truly adaptive? How many generations or how much time must be covered? There is, as before, no general answer to such questions.

Once again, we see that each research program on Darwinian fitness has to be considered separately: for each enquiry a decision must be made on the number of generations to be studied. And, since the duration of human life is short, and of research grants even shorter, the scope of most investigations is narrow: experimentally, we study only ultramicroevolution.

The contribution of sociobiology

There is, however, an alternative to experimenting. In 1930, R.A. Fisher quoted the astronomer, A.S. Eddington, on the value, in natural science, of contemplation 'of a wider domain than the actual'. He comments: 'For a mathematician the statement is almost a truism'; and he advocates study of 'possibilities infinitely wider than the actual'.[38] Half a century later, his advice is being taken by the members of a flourishing school of mathematical biology.[39]

Their method is to design on paper a simplified model of what might happen when populations of organisms change by selection. One of their special interests is in how certain kinds of social interaction could have evolved. They assume that a population contains two or more genetically distinct types that differ in breeding success. Calculations are then carried out to discover what would happen if such a population actually existed.

Suppose we wish to speculate on how threat and attack ('aggressive behavior') have evolved.[40] During an encounter a hypothetical animal may adopt any one of several 'strategies': for example, it may take up a distinctive posture ('threaten'); it may begin by threatening but retaliate if attacked; it may attack but flee if wounded; and so on. A computer program simulates a population containing the several types; and the outcome is that one type of individual (that is, of strategy) does better than any of the others: the one that threatens at first and becomes violent only if the opponent becomes violent.

In this instance, neo-Darwinian calculations seem to fit the facts of nature. Both territorial contests and those between group members of different status usually consist of harmless signals (Chapter 4). But John Maynard Smith, the originator of the model, in an article entitled *Evolutionary Games*, has pointed to a flaw. In some species that have been studied in detail, differences in strength or size decide the outcome of a clash; and these differences are indicated by signals, such as the loud sounds uttered by deer and toads. The weaker or smaller (hence less noisy) individual, after a period of display by both parties, withdraws. In such a situation it would therefore pay to conceal weakness from an opponent: deceptive conduct or appearance should have evolved; small individuals should *appear* larger than they are. But, if deception became the norm, it would pay to ignore an opponent's signals and to attack regardless of appearances. And, if that strategy were adopted, it would be a waste of effort to give signals at all. We are led to infer that the elaborate systems of social signals observed among animals during hostile encounters should not occur. As Maynard Smith remarks: "The conflict between theory and observation is thus clear."[41] Zeno the Eleatic (page 8) would have enjoyed Maynard Smith's argument for, like his paradoxes, it implies that what we know to be true is impossible.

Should we then conclude that the theory of natural selection has been refuted? In practice, biologists do not draw this conclusion. How they avoid it can be illustrated from another case. Many birds and mammals make distinctive sounds at the approach of a predator such as a hawk or a fox. Others of the same species then take cover or become still. We ask again how such behavior has evolved, for it seems to endanger the alarmist by attracting the predator's attention. Consider then another imaginary population with two genetically different types: the helpers give alarm calls; the others, the unhelpful, do not, but do respond to the calls. Numerical values are given to the effects of calling and of responding. The computer concludes that the proportion of helpers in the population will diminish and so, correspondingly, will the genes which make them differ from the others. Eventually, few or no helpers will remain. The inference from this model is that alarm calling cannot occur as a species-typical act.

Yet in fact such co-operative interactions are common. More generally, many animals seem to sacrifice themselves for the sake of others of their species. The problem is to show how such behavior could after all evolve by natural selection. The most important answer emphasizes kinship, that is, close genetical relationship. Care of parents for offspring is to be expected because an offspring has half of each parent's genes. A similar argument applies to helping other close relatives, and can be expressed quantitatively in terms of cost-benefit. Brothers and sisters, like parents and offspring, have half their genes in common; the corresponding figure (or coefficient of relationship) for uncle and nephew or grandparent and grandchild is one quarter. Nepotism, then, becomes a law of nature. Hence follows J.B.S. Haldane's much quoted epigram, that he was prepared to lay down his life for two brothers or eight cousins.[42]

As a result we have the concepts of inclusive fitness and kin selection.[43] Inclusive fitness is a measure of the frequency of an individual's genes in future generations. It provides a framework for investigating real organisms, hence its role resembles that of the optimality model. And, as before, the findings often fail to match predictions based on a simple concept of kin selection: the model has therefore to be made more elaborate.

One useful addition is reciprocation. A pride of lions, for instance, consists of adult females and their young, and usually an associated group of adult males. Prides have been watched for several years in Tanzania. The males co-operate, and share access to the females. Their reproductive success is enhanced by living in a pride; and the larger the pride, the better. The concept of kin selection implies that the males should be closely related. Yet members of male coalitions are often unrelated, and the degree of relatedness does not affect the amount of conflict.[44] Once again, it seems that such conduct could not have evolved by natural selection: an animal that exerts itself helping non-kin should be handicapped out of the race.

Among many other examples of apparently unDarwinian conduct, adoption or helping with the care of young occurs in more than 120 species of mammals.[45] Sometimes help is given to nearly related individuals, but often it seems indiscriminate. It is here that reciprocation rescues neo-Darwinism. If an animal expends energy or endangers itself, and so improves the chances of survival of another, the actor's own chances would also be enhanced if the beneficiary later responded in kind. The animals must be able to recognize each other as individuals – an ability possessed by both mammals and birds.[44]

Yet all this elaboration still fails to match reality. W.D. Hamilton, a leader in this field, has written of 'the world of our model organisms, whose behavior is determined strictly by genotype'.[47] His remark points to one of the severest limitations of sociobiological models. The most elaborate mathematical programs rarely allow for individuality. Yet even lowly mammals have distinctive personalities. They develop from helpless sucklings through a number of stages, and during development they learn from the particular conditions they experience. This is one source of individuality. Models may allow for several distinct behavioral types in a simulated population, but these types, as Hamilton shows, are treated as genetically fixed. Hence a gulf remains between Fisher's 'possibilities infinitely wider than the actual' and the actual itself. And the gulf is widest when we are concerned with humanity.

What do we know?

I now try to sum up what we know, in statements which (I believe) can be readily accepted. First, all organisms possess features closely adapted to the conditions in which they live. Second, in all species individuals vary, and some of the variation is genetically determined. Third, alterations in the genetical structure and visible (phenotypic) characteristics of populations can sometimes be observed, and can even be related to their effects on survival and reproduction. These statements are all descriptive. Fourth, it may be hypothesized (as a corollary of the first statement) that all the features of an organism are optimal for its survival (or reproduction) in its usual environment; and this statement can, in principle, be usefully tested for some features of some organisms. Departures from optimality must, however, always be expected. Fifth, changes in gene frequencies provide a means by which evolutionary change can come about; such changes may sometimes be a matter of chance, that is, they may be unrelated to environmental demand. Lastly, differential survival of genetically different types is the only means we yet know, by which evolution could have occurred.

But, as we have seen, no theory enables us to predict what will happen to particular populations or species. R.H. Brady writes: 'Neither Darwinism nor neo-Darwinism contains a theoretical reduction of an organism to a determinate system.'[48] This correct statement has special interest because many current assertions on human evolution imply that our 'nature' is firmly determined by the past action of natural selection. Brady also asks what sort of finding would lead to rejection of neo-Darwinism by biologists. The reader may think of 'the inheritance of acquired characters'; but in fact even a convincing demonstration of 'Lamarckian' transmission would not do so: it would be treated as an exception. Darwin himself accepted that such transmission could occur.

The concept of natural selection provides an inescapable (though often misused[49]) frame of reference for biology. It rests on one natural phenomenon, the existence of genetically determined variation in fitness: that is its empirical content. But it derives its strength from its tautological component: that is, from statements on the necessary consequences of such variation, given certain presumptions. This component cannot be rejected, unless we refuse to accept the conventional, algebraical logic on which it is based. Moreover, the formulae are useful in scientific practice, for they encourage us to look for certain kinds of regularity in nature. When, as usually happens, nature proves to be more complicated than the formulae, we have still found out something about the real world.

Natural selection as transcendental

Readers may find these summary statements lacking in respect. So perhaps I should, like Marjorie Grene, hasten to say, 'Now naturally . . . I too cross myself when speaking of natural selection.'[50] Her irony was no doubt intended to acknowledge the importance of the concept; but it could also allude to the treatment, by prominent biologists, of natural selection as a transcendental principle. Charles Darwin himself wrote of nature as a (feminine) deity and as the agent of selection; sometimes she is called a power.

> Nature, if I may be allowed to personify the natural preservation or survival of the fittest, . . . can act on every internal organ, on every shade of constitutional difference, on the whole machinery of life. Man selects only for his own good: Nature only for that of the being which she tends.

But he also writes:

> It has been said that I speak of natural selection as an active power or Deity; [but] every one knows what is . . . implied by such metaphorical expressions. . . . I mean by Nature, only the aggregate action and product of many natural laws. . . . [51]

Others have been less cautious. Among the most celebrated was Herbert Spencer (1820–1903), whose voluminous sociological writings had a powerful influence in their day. An editor of Spencer's work emphasizes his originality in presenting evolution as a 'cosmic' process.[52] For Spencer the survival of the fittest (to use his own phrase) appears as the source of our moral beliefs and as the agent of human progress toward higher things. (We return to Spencer in Chapter 9.)

In the mid-twentieth century modern genetics has given new strength to evolutionary theory. (I sketch some of the results above.) With this has come a revival of 'cosmic' attitudes to evolution, and especially to natural selection. These attitudes are personal: they reflect no consensus among scientists, or even among biologists. All have in common a metaphysical component: that is, they invoke ideas or principles which fall outside science. An extreme example is contained in the writings of C.D. Darlington (1903–1981), a botanist who made a lifetime study of chromosomes – the 'machinery' of heredity. It is commonly assumed that genes have no knowledge of the future. Darlington seems to disagree. In *The Evolution of Genetic Systems* he criticizes the conventional notion of natural selection as 'blind and automatic'.

> What we observe of the evolution of genetic systems is hardly consistent with these views. Mechanisms of heredity never benefit the individuals who first manifest them. . . . The species is prepared by its genetic system for what we may call unexpected events.

Natural selection has provided it with a system which although automatic is not properly described as blind. On the contrary it has been endowed with an unparalleled gift, an automatic property of foresight.[53]

Darlington concentrates his admiration on the 'genetic system' he studied, rather than natural selection itself. J.S. Huxley (1887–1975), a famous popularizer and public figure as well as a zoologist, represents a more conventional attitude. The appearance of design or purpose in organic evolution is accepted as misleading: 'natural selection converts accident into apparent design', and so produces 'biological improvement'.[54] Evolution is thus a continuous advance, culminating (so far) in the human species:

Living substance demonstrates its improvement during evolution by doing old things in new and better ways . . . by increasing its efficiency and enlarging its variety.[55]

Moreover, for Huxley the evolutionary process has moral implications and could even become 'the germ of a new religion'. 'We can obtain', he writes,

from the past history of the evolution of life not only reassurance as to the basis of some of our ethical beliefs, but also ethical guidance for the future.

Huxley held that even organic evolution was coming under human control, but that human destiny had been made clear by evolutionary biology: it was to be

the agent of the world process of evolution, the sole agent capable of leading it to new heights, and enabling it to realize new possibilities.[56]

This is a terrible muddle. Huxley asserts the dominance of humanity over nature, yet also seeks some moral agency outside humanity. His argument is therefore circular: he decides that evolution is good by conventional moral standards; then he judges those very standards by evolution.

C.H. Waddington (1905–1975), a notable and versatile experimenter, an evolutionist and polymath, tried to avoid such circular arguing. He expounded a belief in a general evolutionary trend and held that moral principles should be judged by whether they helped, or at least conformed with, the evolutionary process.[57] Hence, like his predecessors, he seems to say, with Alexander Pope,

All nature is but art unknown to thee,
All chance, direction which thou canst not see;
All discord, harmony not understood;
All partial evil, universal good,
And, spite of pride, in erring reason's spite,
One truth is clear, Whatever is, is right.

More recent writers, undaunted by criticism, have continued in the same vein. In 1966 another renowned populariser, K.Z. Lorenz, whom we know as a prominent exponent of *Homo pugnax*, wrote this:

> The great constructors of evolution will solve the problems of political strife and warfare but they will not do so by entirely eliminating aggression . . . This would not be in keeping with their proven methods.[58]

Elsewhere, the Great Constructors have initial capitals. But, unlike the authors quoted above, Lorenz says nothing of substance on evolutionary theory. Instead, he resorts to metaphor: Darwin's 'struggle for existence', he tells us, 'drives evolution forward'; and evolutionary progress is due to 'the profitable "invention" that falls by chance to one or a few . . . in the everlasting gamble of hereditary change'.[59] So here we have another paradox – the Great Constructor as a gambler, or perhaps a croupier. Even G.C. Williams, in his distinguished monograph on natural selection, gives 'Nature' an initial capital[60]; and W.H. Durham, in a discussion of war and Darwinian fitness, treats 'Selection' in the same way.[61]

Design or tinkering?

The views of the evolutionists cited above recall a typically cynical remark by Edward Gibbon (1737–1794): 'So urgent . . . is the necessity of believing, that the fall of any system of mythology will most probably be succeeded by the introduction of some other mode of superstition'.[62] Of those who, in a period of religious decay, became absorbed in the grand spectacle of organic evolution, Antony Flew similarly remarks, 'for many the process of evolution by natural selection becomes a secular surrogate for divine providence'.[63] J.C. Greene calls the works I have quoted the Bridgewater Treatises of the twentieth century. He is referring to an endowment, made in 1825, for works 'On the Power, Wisdom, and Goodness of God, as manifested in the Creation'.[64]

In the previous century William Paley (1743–1805) had written of the maternal behavior of a sitting bird:

> I never see a bird in that situation, but I recognize an invisible hand, detaining the contented prisoner from her fields and groves.[65]

This appears in a chapter on instincts. Paley is celebrated for finding, in the apparently purposeful design of living things, evidence of a divine designer. The examination of the eye, he held, is a cure for atheism.

Man, it has been said, makes gods in his own image. Both Paley and some evolutionists seem to think of the Designer (or Constructor) as an advanced engineer. But intimate knowledge of organisms makes them appear as very unlike the artifices of human manufacture. François Jacob, a distinguished biochemist, describes the action of natural selection as tinkering: it 'uses' whatever turns up; a wing evolves from a leg, a part of an ear from a jaw bone. Such changes take a long time. The result, from a human designer's point of view, is absurd complexity. Jacob takes three biochemical examples: blood clotting; the inflammation that goes with wound healing; and the immunological processes that protect us from infection. In each, about ten proteins are involved in an intricate chain of reactions.[66] Our capacity for developing resistance to disease illustrates particularly well, as P.B. Medawar remarks with excusable overstatement, that evolution 'is a story of waste, makeshift, compromise, and blunder.'[67] Our lives depend on this capacity, yet it often goes wrong. Allergic reactions are the commonest examples. Others are the autoimmune diseases in which the body attacks itself as if it were defending itself against infection.

The most familiar cases of bizarre elaboration, already mentioned in previous chapters, are those of embryonic development: we begin as if we were about to be a primitive marine organism; but cells that arise in one region migrate in various directions to other regions, and we change rapidly into what looks like an embryo fish; structures, such as the tail, become prominent only to disappear ... It is as if an engineer began to make a sailing ship but ended by producing an enormously complex space craft. Astonishingly, it works. As Jacob says, there is nothing in common between human design and organic systems.

Nonetheless, even biologists often write and speak of the 'adaptations' of organisms as *for* some end. In doing so, they imply (usually inadvertently) that organisms design themselves (or are designed by an agency) with an end in view. R.I.M. Dunbar, in a critical discussion, writes that adaptation, 'strictly speaking ... refers to some "problem set by nature"' that the organism overcomes efficiently; and the 'term "efficient" here conveys the notion of "design for a purpose".'[68] Such a teleological component in biological thought is, however, usually rejected when brought into the open, for human beings are the only agents known that can describe a plan and then carry it out. Intention in this sense belongs to humanity alone.

Rigorous accounts of evolution today are in terms not of a plan but of chance and probabilities. Jacques Monod (1910–1976), like his colleague, Jacob, a famous biochemist, has expressed the prominence of chance in biology in an extreme form.

> Pure chance, absolutely free but blind, at the very root of the stupendous edifice of evolution: this central concept of modern biology is ... today the *sole* conceivable hypothesis, the only one compatible with observed and tested fact; and nothing warrants the supposition ... that conceptions about this should, or ever could, be revised ... The universe was not pregnant with life nor the biosphere with men. Our number came up in the Monte Carlo game.

It is surprising that a scientist of distinction should assume such finality in current theorizing. Perhaps Monod is allowing himself to be carried away on wings of metaphor. In the same work Monod puts forward two opposed notions without showing how they can be reconciled. First, he points to special phenomena in living matter which must be added to those that belong to the physical sciences; but he then asserts that these very phenomena can be understood only through biochemical analysis, and only 'our present-day ignorance' prevents us from explaining them by applying the laws of physics.[69]

Apparent inconsistencies apart, Monod's position is reductionist: explanations at the level of molecules, atoms or ultimate particles (such as electrons) are held to be more fundamental, or superior, to those at other levels. This is not physics but metaphysics: it is a presumption, widely accepted by scientists, to which I return in Chapter 12. Here I cite only comments by a physicist, David Bohm, made in a discussion of evolutionary theory. He writes:

> it may well be that the whole structure of physics is inadequate and misleading when extended too far into the processes of living matter.[70]

Monod, in the passage quoted above, presents 'pure chance' as a fundamental principle. Bohm makes a cooler assessment of the idea of randomness. We say that two or more kinds of event are random when, like the fall of dice, we believe them to occur with equal probability. Hence to refer to *laws* of probability is misleading: use of the word law implies an 'iron-bound' necessity for disorder, indeed for lawlessness. But the mathematical theory of probability does not state a natural law: it provides only a *description* of natural phenomena. To be dogmatic on such matters can make us fail to investigate, or even to see, serious problems: 'it is generally assumed', writes Bohm, 'that evolution is due entirely to purely random mutations of the genes. The real question here is that of finding whether this process is . . . purely random or not'.[71]

Our picture of the evolutionary process is therefore provisional and certainly incomplete. Here is Darwin again:

> Why should a Thrush always line her nest with mud, and the Blackbird always with fibrous roots? No answer can be given. . .
> I am more and more convinced that variety, mere variety, must be admitted to be an object and an aim in Nature; and that neither any reason of utility nor any physical cause can always be assigned for the variations of instincts.[22]

We still have no answer.

Ethical naturalism

To look for moral guidance from evolution, in the manner of Huxley, Waddington or Lorenz, is a particularly strange form of ethical naturalism. This doctrine states that the fixed laws of nature tell us not only what is the case but also what is right and good.[73]

The primary objection to naturalism is the confusion of statements about phenomena in nature with the rules we make to regulate our actions. We speak of laws of nature by analogy with legal (and moral) codes, but this is misleading. When we say we have uncovered a natural law, we are speaking of a regularity which is independent of what we do. We may dislike, say, gravitation, for it does stand for something that holds us down when we want to fly; but we cannot rescind it. If, however, an enactment allows (for instance) property in persons, that can be changed.

It is of course open to a person to declare that for him or her what is natural is also good; but this leads to difficulties. We must first decide what, for our species, is natural. In all ages and circumstances, people have said that what is familiar to them is both natural and inevitable. In some societies today the total subjection of women to their husbands is still held to represent a natural law; but in others this norm is rejected. Another possibility is to equate the natural with the primitive. If we do this, we should presumably classify wearing clothes, using fire, making tools and perhaps even speech as unnatural. And if we then also identify the natural with the good, we should be led to a generally unacceptable conclusion. In K.R. Popper's words,

> The choice of conformity with 'nature' as the supreme standard leads ultimately to consequences which few will be prepared to face; it does not lead to a more natural form of civilization, but to beastliness.[74]

Of course, the evolutionists who drift into ethical naturalism do not intend any such conclusion: they merely fail to examine the implications of what they are saying.

The human species is a product of evolution. In the preceding pages I therefore try to give some notion of the difficulties we face when we try to understand evolution, and of the achievements of evolutionists in overcoming them. It has also been necessary to make clear the limitations of our knowledge. The theory of evolution is principally concerned with genetically determined differences between individuals or populations. But much variation is environmentally determined. To assess *Homo egoisticus* we also need to know what modern genetics can tell us about the interactions of 'nature and nurture'.

7 Environment and heredity

Can you, Socrates, tell me, is human excellence something teachable? Or, if not teachable, is it something to be acquired by training? Or, if it cannot be acquired by training or by learning, does it accrue to men at birth or in some other way?

Plato: Meno

The interaction of heredity and environment is crucial for our ideas about humanity and for social policy. It also presents severe difficulties, and not only for non-biologists. The reader should beware of anyone who says that argument on this topic is now settled, and can therefore be brushed aside.

One difficulty arises from our language: it is legitimate to say that C.P.E. Bach inherited his harpsichord, his surname and his musical ability from his father, J.S. Bach. First, the (hypothetical) harpsichord was itself handed over; we have here the primary, legal meaning of inherited. Second, the statement about the name, in contrast, represents a social convention: a man may not be required by law or custom to use his father's name. (In some societies, Bach's sons would not be called Bach.) The third statement refers neither to bodily transfer nor to convention but (by implication) to the development of C.P.E. Bach from a fertilized egg: that is, to his ontogeny. This was inevitably influenced by genes derived from both his parents; but it also took place in a notably musical environment.

This leads to a truism of the utmost importance: C.P.E. Bach's musical ability, like all his (and our) other characteristics, was influenced by both his heredity and his environment. We assume that he differed genetically from other people, and that this contributed to his difference from others in musicality. But we also assume that, had he been reared in another environment, his abilities would have taken a different form: if he had been adopted at birth into an Indian family, he might have become a composer for the sitar – or not have revealed any special musical ability.

As a corollary, we come to a fundamental principle that most people find difficult to accept. *No characteristic* (of a person or other organism) may properly be said to be inherited (or genetically determined): to state that musicality, or stature or temper or any other feature is inherited, is to confuse a biological process with a form of legal transfer of property. Equally, *no characteristic* may, strictly, be said to be environmentally determined. Only *differences* may be described as genetical *or* environmental. To support these assertions, we need examples.

Individual differences

Consider first a child that grows up an imbecile and incapable even of normal speech. Biochemical enquiry shows that the child's developing brain has been injured by a substance in food. The substance, therefore, is toxic, and giving food that does not contain it could prevent the damage. Clearly, we have here an environmental effect. But the preceding sentences describe in fact a condition, phenylketonuria, that appears in all the lists of 'inherited' diseases: it is a recessive condition; two normal parents, each with one copy of the same rare gene, may have the misfortune to have a phenylketonuric child.[1]

When the genetical basis of this condition was first identified, it was taken for granted that the imbecility was inevitable: a child with a double dose of the gene (a homozygote) was doomed. In technical terms, phenylketonuria had at that time 100 percent heritability. But biochemical research uncovered the toxic substance responsible. This, phenylalanine, is present in milk and in all ordinary diets. Further research showed how a child with this genotype could be identified at birth and given a non-toxic diet. With this environmental change, the heritability diminished: in a rich community, with advanced medical services, there is no need to have any phenylketonurics at all. Hence even a genetically simple condition should not be written off as exclusively a product of heredity: to say that it is inherited is not only misleading but is also liable to prevent effective action.

We have here a particularly clear example of the need to distinguish between genotype and phenotype. The former is the assemblage of genes derived from the parents; the phenotype is all the characteristics of an organism, and is influenced by both genes and environment.

The principle can be reinforced by reference to another toxic substance in food. In this case the effect is on adults, and is a digestive disorder which may include diarrhea and severe flatulence. The substance is lactose, a sugar present in milk. To most people in the world the idea of milk as mildly toxic to adults is not strange; but among northern Europeans lactose intolerance is rare. Genetically, the condition is similar to phenylketonuria. And, practically, it is the same in one obvious and important respect. When intolerance occurs, the remedy is environmental action: avoid milk.[2]

Most socially important characteristics have more complex origins. Tuberculosis, the 'white plague', can develop only in a person infected with tubercle bacillus. Yet it is not a simple infection: its history shows how the disease is influenced in a complex way by the environment.

Youth grows pale, and spectre thin, and dies,

wrote Keats in the spring of 1819. This was an accurate description of what was happening at the time, not only to poets and their friends, but to the enfeebled, 'consumptive' young men and women of the new manufacturing towns.[3] Since then, bad working conditions (as in mines), crowding at home and circumstances that tend to induce neurosis have all appeared as increasing the incidence of tubercle.

But there is also a genetical component in susceptibility to the disease, revealed only after exacting researches. Tuberculosis has a 'familial' distribution: in a particular population it is liable to afflict several members of some families, while others remain free. This leads to suspicion of a genetical effect. But members of a family usually have similar environments. The problem is to distinguish between environmental and genetical influences. To do this one observes the incidence of the disease among people of different degrees of relationship, especially twins. The members of a twin pair may be hardly more alike, except in age, than ordinary brothers and sisters; they are then presumed to be binovular, that is, to have arisen from two eggs separately fertilized. The other kind of twin results from division of a single fertilized egg or early embryo; such uniovular twins are presumed to be genetically identical. Among hospital patients with tuberculosis, a few are twins, and one inquires whether, in each case, the other twin is concordant, that is, tuberculous. The higher rate of concordance then found among uniovular twins is evidence of a genetical influence. (The environmental causes, of course, remain for practical purposes far more important.)[4]

Another disorder, schizophrenia, makes a curious counterpart – almost a mirror image – of tuberculosis. A person diagnosed as schizophrenic is withdrawn from reality, and often believes that he or she is the victim of persecution or threatened by mysterious, hostile agencies. (This condition has nothing in common with 'hysterical' dissociation, or dual personality, with which it is often confused.) In most populations, nearly one percent are schizophrenic for part of their lives.[5]

The condition is familial: among people who have a schizophrenic near relative the incidence is slightly higher than in the whole population. It is often said that schizophrenia is strongly 'inherited'; but it has proved to be exceedingly difficult to find out to what extent (if any) the familial distribution is due to genetical similarity. One surprising finding is that, even when both parents are diagnosed as schizophrenic, only 40 per cent of the children are similarly diagnosed. More surprising still, the concordance among uniovular twins, found in the most careful studies, is only 25 per cent or lower. This, however, is higher than that among binovular twins.

The findings on twins indicate some genetical variation in the liability to become schizophrenic, but there is clearly also a substantial non-genetical contribution.[6] The problem is to identify the environmental factors. Much attention has been given to the influence of the family: one school of psychiatry regards certain kinds of upbringing as critically important; but reliable evidence is sparse. An answer has been looked for in the fates of adopted children with schizophrenic parents but normal foster parents, or the obverse, but no firm conclusion has emerged. One such study suggests that developing a schizophrenic condition (widely defined) depends entirely on conditions of rearing.

Although we may suspect that a certain kind of family may drive a child mad, another environmental effect is possible. Even unrelated people are slightly more likely to become schizophrenic if they have lived or worked close to a schizophrene. Perhaps, then, infection is involved.[7] If so, it will still be uncertain whether infection is necessary for the disease, or only a contributory factor.

In their lack of Mendelian simplicity, the last two histories are typical of human genetics. Other difficulties include the deceptive concept of heritability. For instance, in a population such as that of Cape Town or New York City, most of the variation, from black to very pale, reflects genetical differences: the heritability of skin color is high. But now compare Whites who stay in New York during winter with their rich relatives who resort to the sunshine of Miami: the color differences are of environmental origin; heritability in this case is near zero.[8] Heritability measures the extent to which genetical differences contribute to observed variation. The other source of variation is the environment. As the environmental contribution rises, the heritability declines. The heritability of a trait such as skin color may vary with the population studied and alter when the environment changes.

In contrast with color, variation in the hair length of adult Europeans obviously depends on the environment – the actions of hairdressers. Yet, during most of the twentieth century in Europe, North America and Australasia, hair length has been precisely related to genotype: a person with two X-chromosomes had long hair; but a person with only one X-chromosome, being a man, had short hair. Exceptions, until recently, were rare. One could therefore reliably predict a child's adult hair length from its genotype. Yet to say that hair length was genetically determined would have sounded ridiculous: the significant causes of variation were in the social environment.

Hair style is not in itself important; employment is. Suppose a being from space, with an advanced knowledge of genetics but unable to distinguish black from white, visits the Republic of South Africa. The visitor observes a correlation between possessing certain genes and unemployment. In colloquial terms, being out of work (and hence poor) is evidently an inherited character. (We see later, in Chapter 9, that something like this was recently believed by educated people.) The incidence of unemployment among South African Blacks has indeed long been high; but the relationship of genotype with unemployment depends on the social environment – one that may soon change.

A fundamental principle follows: a trait may have a high heritability, but be drastically alterable by environmental means. Of all complex human traits, intelligence or, more precisely, intelligence quotient (IQ) has been most subjected to supposedly genetical analysis. But a statement on the heritability of IQ has no meaning unless the population and the time are specified: an environmental change, for instance in schooling or nutrition, could rapidly alter the figure.[9]

Moreover, no limit can be put on the variety of environments in which children can be reared. Sometimes, novel action produces startling results. The ability to play complex, classical music on the violin is uncommon, and is usually regarded as requiring special qualities which mark musicians off from the rest of us. Yet Shinichi Suzuki has developed a method by which thousands of young Japanese children, not selected for special ability, have learned to play the violin well.[10] The children begin at a very early age (even in their third year), and learn to play by ear, not by reading from a score. A critic may say that nevertheless the children must have had some 'innate' musical ability. But such a statement is empty: it could be said of any developed characteristic whatever. If children are reared in a northern city on a diet mainly of cereals and potatoes, they are likely to grow up stunted, with pigeon chests, bow legs and bulging foreheads. So we may say that these children had an 'innate' capacity to develop rickets. But rickets can be completely prevented by proper diet, that is, by environmental action. We know nothing precise about genetical variation in musical ability. What we do know is that Suzuki, by providing an unusual environment for his pupils, produced results which most people would have supposed impossible. (The extent to which such early training is *desirable* is a separate question.) Perhaps, however, Suzuki's achievement is less surprising than that of modern societies in which nearly everyone can read, write and do simple arithmetic. Not long ago, these magical arts were confined to a small, privileged minority. Widespread literacy and numeracy have resulted, not from any genetical change, but from environmental action – providing schools for everyone.

The preceding pages illustrate principles which are so fundamental that they are often left unstated. They arise from our diversity. We all have some capacity for self-expression; but each of us has distinctive abilities, needs and desires which appear, usually without warning, as we develop. The diversity has three sources. First, a person's genetical makeup (genotype) is not predictable: even sex is an outcome of an accident at conception. Second, complete knowledge of the genotype would not, by itself, enable us to predict a child's development: that depends on the environment, and no limit can be put on the scope of environmental variation. Third, we can *choose* what kind of future environment we should try to bequeath to later generations.

The error of believing in genetically fixed traits goes with support for privilege, usually for members of particular families. In the past, a landowner's children have had better opportunities than a serf's or a laborer's, and a merchant's better than a clerk's. Yet familial privilege would be defensible on genetical grounds only if spouses usually resembled each other in their whole range of abilities, and if children regularly and closely resembled their parents: a caste system like that of Hinduism might then seem inevitable. In fact, however, children are often disconcertingly unlike their parents: unexpected abilities and disabilities can turn up in any family. Each individual is genetically unique (uniovular twins excepted), and environments are infinitely variable. Moreover, in a time of rapid social change, children are as much products of their own times as of preceding generations. (To their parents, they must often seem almost as changelings.)

Given such variation, natural justice demands, for each person, access to a wide range of opportunities, regardless of parentage. Granting such access is not only a matter of communal altruism, for it confers a general good: the greater the growth of individual abilities, the greater are the benefits to the community from the contributions of people whose development would otherwise have been stunted.

Racial differences

Privilege commonly goes with membership of a distinct group, and is supported by beliefs on biological differences between groups, especially those between races. By race I mean a population genetically different and geographically separated from others; the separation may have been in the past. To illustrate some of the principles of studying racial differences, I resort to a parable.[11]

Consider two countries, Ruritania and Kukuanaland (names taken from celebrated works of fiction). The Kukuanas are 'black' (have much melanin in their skins); the Ruritanians are 'white' (have rather little). We are now concerned with a measurable character called zing, highly valued in both communities. Possible scores are from 0 to 100. Zing is not correlated, either positively or negatively, with skin pigmentation in either population; hence skin color, though often referred to in discussions of zing, is irrelevant. The estimated mean value of zing for the whole population of Kukualand is 61: the corresponding figure for Ruritania is 58. The difference, though small, is statistically highly 'significant'; that is, we can confidently presume that it is not due to chance. Certain Kukuanas assert that the figures indicate their genetical superiority to Ruritanians. They point to the fact that investigations of zing in their own population indicate a high heritability. Their views are, however, disputed. It is argued that the environments of the two populations are very different: Kukuanas live in fertile uplands of central Africa, the Ruritanians in a mountainous region of central Europe. Moreover, the two cultures, including their schools, differ greatly. Hence calculations of the heritability of zing in one population have no necessary significance for its heritability in another.

If some Kukuana and Ruritanian children were reared in identical environments, it might be possible to say something decisive about the differences between them. Only a few families of each group live in the country of the other, and attend the local schools. But, of course, the children are brought up by their own parents, and so the family environment is still different for the two races. Nonetheless, when such children have been compared, their mean zing scores have proved to be about the same. But those who emphasize genetical differences refuse to accept this as useful evidence: they say, for example, that the immigrants are not representative of their respective populations.

Most geneticists and zingologists hold that no valid conclusions are possible on the relative contributions of environment and heredity in the distribution of zing within either population; the same applies even more strongly to the difference between Kukuanas and Ruritanians. Those interested in social action also point out that, although there is a 'statistically significant' difference between the populations, there is a very big overlap; correspondingly, both populations have a small but important minority of people with very high zing (and, of course, a corresponding group of people who are almost zingless). Hence, for practical purposes, the two populations are hardly different. More important, there is much evidence that improving zing training in the schools leads to higher zing scores in later life, regardless of parentage.

A curious complexity arises when the sexes are compared. Men in Kukuanaland score rather higher than women; but the reverse is true of the Ruritanians. Some Kukuana women assert that their lower mean score results from social discrimination against their sex in Kukuanaland. If so, it reflects an environmental difference from the situation in Ruritania. The debate continues.

The validity of environmentalism

During real controversies of this kind, general statements are often made on the relative importance of environment and heredity, and it may be said that the two sources of variation are of equal weight: in individual development, the genes are always present and influencing developmental processes; and so is the environment. These statements are correct; and they warn us against absurdities such as disregarding the environment *or* taking literally that all human beings are 'created equal'; but they are not otherwise very helpful. It is more useful to have information on the causes of particular traits. When we have it, we may ask, for each trait, whether genetical or environmental knowledge gives us more guidance for action.

Information on genetics is most obviously helpful when it concerns genetically simple defects.[12] The simplest case is that of dominance. Split hand or lobster claw (ectrodactyly) is a rare, usually dominant abnormality: that is, it is passed on by an affected person to half his or her children, on average. It is transmitted only by persons with the condition, and never skips a generation. It is a severe defect, though it may be accompanied by excellent health in other respects. Persons with this condition may decide not to have children. If all did so, there would be a rapid decline almost to zero in the number of ectrodactylics; the condition would then reappear only as a result of mutation.

It follows that even dominant conditions cannot be wholly eliminated by controlled breeding, for mutation, though rare, always recurs. As another example, achondroplasia, a form of dwarfism in which growth is gravely abnormal, usually prevents reproduction. (Our knowledge of its genetics comes from exceptions who have children.) Since most such persons are naturally barren, the continuing appearance of this condition must be due almost solely to mutation. Other such conditions are known.

Another dominant condition presents a different kind of problem. Huntington's chorea is a severe disease of the nervous system in which there are involuntary movements of progressive violence. The difficulty arises because the disease often becomes evident only in middle or old age. The affected person may already have had children. Diagnosis is then too late for prevention.

Some defects are recessive. I mention the example of phenylketonuria above. The parents of people with these conditions are usually normal: a child with 'congenital' deaf-mutism may have parents both with unimpaired hearing: each parent has then passed on one copy of a particular rare gene to the child. Such a misfortune can happen to any couple, out of the blue. Another recessive condition, albinism, illustrates the limitations of preventive eugenics. Even if all albinos stopped having children, albinos would continue to be produced by normal parents; and it would take about 60 generations or, say, 1500 years, to halve the number of albinos in the population.[13]

Eugenic action has even less application to more important features. These may be unwelcome – for instance, schizophrenia or mental deficiency; or they may be valued – for instance, highly developed skills. We may guess that variation in such characteristics is influenced by many genes and many environmental agencies; but usually, in the present state of knowledge, we cannot proceed beyond guessing.

As an example, mental deficiency, or retardation, has loomed large in eugenic propaganda. Before human genetics had been rigorously studied, it was sometimes treated as a single 'inherited' condition; and it was supposed to threaten the intellectual level of whole populations owing to the excessive breeding of defectives. But the expression, mental deficiency, does not refer to a clearly defined trait, 'handed on' from parent to offspring. The children who attend special schools for subnormality include a great variety of types, each owing the deficiency to a different set of causes. Some can be treated successfully, by psychotherapy, hormone therapy or other means. As knowledge grows, the proportion that can be benefited increases. As for genetics, out of every hundred children with one or both parents retarded, about seven or eight are themselves retarded; a larger proportion are backward, but most are normal. And, just as retarded parents can have normal children, so can normal parents have retarded children. Indeed, most retarded children have normal parents. In this situation, eugenic action has again little scope.[14]

Nor can genetics help when we wish to enhance normal intelligence. It is easy, on meeting children or adults who seem helplessly stupid, to presume that they and their relatives must be 'innately' ineducable. The danger of thus underestimating environmental influences emerges with special clarity if we return to 'racial' differences. Europeans, on meeting tribal people, have often commented on their strange ways of thinking. Some have concluded that members of 'primitive' groups can never be educated as Europeans can. And, certainly, the concepts of the formal teaching to which, no doubt, all readers of this book have been exposed, are utterly foreign to unlettered people who live by subsistence agriculture.

A Russian psychologist, A.R. Luria, has described the impact of such education on peasants in a remote region of central Asia. Some were still living in traditional villages; others were working in collective farms and had had some years of schooling. He applied a number of tests of the ability to form abstract categories. In one, four pictures were shown, of a saw, an axe, a spade and a log, and the farmers were asked to choose the three that belonged together. Most of the educated farmers, but none of the others, chose the tools. The uneducated farmers responded in terms of actual occasions in which the objects could be used: the axe, the saw and the log would be grouped because an axe is used to fell a tree, and a saw to make logs from it.

Luria also tested for the ability to use a formal, logical principle, regardless of actual experience. One question asked: if it took six hours to go from here to Fergana on foot, but a bicycle went at only half that speed, how long would it take on a bicycle? The unlettered could not accept the absurd premise, and replied from experience, not from logic; but the educated farmers had no difficulty.[15]

Sylvia Scribner and Michael Cole describe similar experiences with African peasants: the children given 'western' schooling thought like Boston children, not like their unschooled cousins.[16] In an earlier work, *Culture and Experience*, A.J. Hallowell tells of an unlettered group whose topographical knowledge of their region was highly developed. They quickly grasped the idea of a map of the area, and they became accustomed to studying photographs of places they knew. But they had difficulty with Hallowell's account of his own country. One asked for a photograph of the United States. The learning of people in these cultures is informal and is based on imitation and co-operation with elders, rather than verbal instruction; it concerns particular actions and methods, such as those of cultivation and stockbreeding: it is not abstract.[17] (The difficulties such people have with logical principles that we take for granted are perhaps analogous to those that we ourselves have with the interaction of nurture and nature.)

Intellectual development, however measured, is not a function only of schooling. Practical action, especially on behalf of the severely deprived, requires attention to health as well as education. Harrison McKay and others describe a project in the city of Cali in Columbia. Young, undernourished children from poor families were given extra food and health care, the principles of nutrition were explained to their parents, and the children received some education. Favorable effects were evident in physical development, language and other features. Intellectually, they approached children of wealthy families in the same city. Improvement was greatest when the 'treatment' began early. Every change is thought to have contributed, but the procedure was not intended as a well designed experiment leading to precise conclusions on causes.[18]

For rigorous research, children adopted early in life are of special interest. If some foster parents differ sharply from the true parents, either in economic class or in educational status, evidence on the effects of upbringing can be obtained. A path-breaking study of one hundred American children, published by Marie Skodak and H.M. Skeals, is notable both for its scale and for widespread misunderstanding of the findings. The intelligence quotients of the children were more closely correlated with those of their natural mothers than with those of the adoptive parents. That is, those that scored near the top of the children's range also, on the whole, had true parents near the top of the parental range, and so on. This suggests a strong genetical influence on variation in IQ; but it disregards the actual scores. The IQ of the natural mothers averaged 85 – well below the population mean of 100, but that of the foster parents was over 100. And the adopted children also averaged over 100, in fact, 21 points above their natural mothers. This finding suggests a marked beneficial effect of improving the family environment.[19]

What would have happened to these children if they had not been adopted? We do not know. In other words, we need a control group of similar children who remained with their own families. Such a group is, however, included in a study of French children, from poor families, who had been abandoned by their mothers at birth and raised from about four months by rich adoptive parents. The children were therefore affected by the whole range of differences between rich and poor households. Thirty-five such children were compared with 20 others, all of whom were children of mothers of one of the 35 adopted children, but had been reared by their own mothers. At 13 years the intelligence quotient of the adopted children averaged 109; that of all French upper middle-class children was 110. The children who had remained in poor households scored 95, which is identical with that of the children of all French unskilled workers. The differences in IQ were paralleled by performance in school.[20]

The same principles, and similar conclusions, apply in comparisons of children of different economic classes and of different racial types. Black American children reared by their own parents often score lower in tests of ability than white children of the same economic class. In the past, such differences have been assumed to be genetical. Barbara Tizard describes a study which, designed to reveal environmental effects, also gave findings on racial differences. The children were all healthy but, as infants, had been put in residential nurseries of high quality. Some were white, some black, and some had mixed parentage. Since all were reared in similar conditions from an early age, genetical differences might be expected to show up clearly. But tests, given after a period in the nurseries, with one exception indicated no differences between the types. The exception was a non-verbal test of 'intelligence' on which the non-white children scored higher. Such a finding does not show that the black children were genetically superior (in any sense): it merely contradicts the hypothesis that people with dark skins are genetically inferior in the kind of ability tested. As Tizard remarks, differences in test scores 'could result from differences in the microenvironment of the family, such as linguistic practices and attitudes to achievement.'[21]

The emphasis on the role of the family, and on the experiences of children before they go to school, is based on studies of children in rich countries with schools for all. A survey of 29 countries, in four continents, has provided a more balanced picture. In poor countries, differences in the intellectual development of children are found to reflect, in the main, differences in the quality of their schooling. In most of the world, the numbers and quality of schools are rising, and so the trend is toward a higher educational standard. Disconcertingly, the leading exception is the USA, where, in the late twentieth century, even basic literacy has been declining. Nobody, as far as I know, has suggested a sudden increase in illiteracy genes: on the contrary, American schools have been severely criticized, and the influence of television condemned.[22]

Chapter 5 gives some of the evidence for environmental influences on anti-social violence. As we now see, the findings on intellectual development, too, are what we should expect. To quote Irving Lazar, 'we find ourselves in the peculiar position of asserting the commonsense notion that children will benefit from good experiences.'[23] The consequences of environmental action are so great, and of so many kinds, that for practical purposes attention to the environment is the only sensible policy. Today, the most important assessments of ability are those that compare black with white, rich with poor, women with men; and they especially concern educability, and qualifications for training in professions, crafts and trades. But the debates on such questions are misconceived, for two reasons. First, as usual, environmental effects are ignored. These are clearly visible if we only glance at history. In the early Middle Ages, the Arabs of the Middle East and North Africa were far ahead of most Europeans in mathematics and other branches of scholarship. Correspondingly, they regarded the large, 'white' men of the north as dim-witted oafs incapable of learning. (Contrast pages 105–108.) At that time, the environment provided by the Arab states was educationally much superior to that in Europe.

Second, when population averages differ, there is always a substantial overlap. Hence, whether we are selecting managers, mathematicians or matrons of hospitals, the objective should be to choose on merit. When we do this, differences between populations are irrelevant. Suppose it could be proved that there is a higher proportion of first class mathematicians among black women than among white men. A white man might still become a professor of mathematics: the question for an electoral committee would be whether he was the best among the applicants for that post. Hence statements on average differences between groups have a practical significance only if there is prejudice and consequent inequality of opportunity. This is a matter not of biology but of natural justice and social efficiency.

Not in our stars

Nothing in modern genetics justifies us in thinking of ourselves as bound to an inexorable genetical destiny, any more than our fates are, as astrologers assert, fastened to 'the yoke of inauspicious stars'. This conclusion corresponds to that on instinct in Chapter 4. The concept of instinct includes the idea of a compulsion to behave in certain fixed ways or to achieve certain ends. The supposedly fixed behavior has been called innate or even genetically determined, and it has been sharply contrasted with other activities said to be acquired or learned. A simple classification of actions in two kinds is easy to grasp; but in this case convenience leads to error, for it disregards the development we undergo from conception onward: what is given in the fertilized egg is a set of genes, not of characters – instincts or anything else.

Modern genetics was made possible because its founders discovered variation which, in a restricted environment, had a simple relationship with genotype. (I give human examples above: phenylketonuria, ectrodactyly and others.) Hence genetics is often presented as though each gene had a single effect. Even biologists are then led to speak of genes 'for' characters. William Etkin, in an analysis of sociobiology, finds

> a gene for altruism, a gene for the detection of subtle forms of cheating . . . , a gene for the emotion of gratitude for altruistic acts, a gene for sympathy, a gene for guilt, a gene for establishing reciprocal relationships, a gene for learning about establishing reciprocal relationships, a gene for learning about altruistic and cheating tendencies in others.[24]

This does not exhaust the list.

The expression 'a gene for' is sometimes defended as shorthand for 'a gene that produces a certain effect in a given environment'; but, as it is used in practice, it does not convey this. And when we are discussing human affairs it is always inappropriate. All species, but especially the human species, live in a range of environments. From the moment of fertilization, the course of development is influenced by an inconstant environment which differs from one individual to another. Each developmental change is influenced by the interaction of genes and environment, and also affects the next stage of this interaction. Nor do the complexities end with the development of the individual. The environment of a human being is first that of the uterus; later it includes the mother's milk and other aspects of parental care. But these in turn are influenced by the interaction of the parental genotypes with the parents' environments . . . Such interactions are very difficult to analyze even when one studies mice of which the breeding and the environment can be strictly controlled. (I have tried it.[25])

It may be objected that beneath and behind human variation and adaptability, there is still a fixed human nature. Such a statement is true in the sense that the species, *Homo sapiens*, is distinct from all other species: human beings have unique features, of which the most easily identified and described are structural. *Homo sapiens* 'naturally' walks upright (with all that that entails in skeletal and muscular features) and has a very large brain, a nose and chin, and so on. These characteristics are very stable in embryonic development: it is rare for a human being to be without them. But it is not unknown. The effects of certain drugs and infections early in pregnancy remind us that all features develop in an environment, and that environments vary: a human being can be born without arms or legs.

We are mainly concerned here not with anatomy but with behavior, attitudes and rationality. These are much less stable in development than are hands with five fingers. We face again, as in Chapters 4 and 5, the unpredictability of custom and belief. Other species have their typical signals and social systems; we have diversity in space and, especially today, rapid changes with time.

Those who talk of an unchangeable human nature are often over-impressed by the customs and outlook of their own familiar culture; and they may be anxious to find reasons why no attempt should be made to alter them. The idea of a genetically fixed nature is then helpful, and may be supported by presumptions on our evolution. Natural selection is held out as having produced a species (which I call *Homo egoisticus*) with a uniform set of characteristics that fits it for a single mode of living. Equipped with modern evolutionary theory and genetics, we are now in a position to dissect this creature.

8 Stories of human evolution

Yea, ye would cast lots upon the fatherless,
And make merchandise of your friend.

Job: 6:27.

The picture of our species, framed in evolutionary biology, that emphasizes aggressiveness (especially among men), egoism, conflict between the sexes, nepotism and deceit, is – we are told – designed to demystify human conduct.[1] But when we try to assess this portrait we find that it conceals nearly everything important about us.

Comparisons

We begin with humanity before agriculture. Rudyard Kipling, with questionable poetic licence, tells how

In the Neolithic Age, savage warfare did I wage,
For food and fame and woolly horses' pelt.

The cartoon figure of the club-wielding cave dweller, with women and children only in the background, is one form of *Homo egoisticus*, and it reflects not facts but the outlook of male anthropologists and other writers.[2] Statements on the ancestral roles of the sexes, and on ancient social organizations, are still surmise; but the importance of women as food gatherers in pre-agricultural groups has been recorded in detail: the idea of women as stay-at-homes has been given up; even in late pregnancy a woman can cope with heavy burdens and often does so. To what extent women have, in the past, been hunters is unknown; but in some modern groups women hunt or hunted regularly.[3] In the north of Australia Tiwi women caught bandicoots, possums and a variety of reptiles, including iguanas. In Luzon, in the Philippines, the Agta eat much animal protein. The women hunt game and fish in the rivers; they also barter with the Filipinos in the lowlands for goods and services. In other respects, their societies are typical of gatherer–hunters: for instance, there is no system of formal rule or management of the group.

Genuine knowledge of human social life without agriculture consists of descriptions, such as those above, of people as they are today. We also have accurate profiles of existing animal species. The latter are not our ancestors, but we can usefully compare their structure, physiology and even behavior with ours. We now continue the story begun in Chapter 4.

Zoologically, *Homo sapiens* may be described as an erect, bipedal ape, with an arched foot, exceptional manual grasp and dexterity, the ability to make complex tools, a very large brain, small teeth, and elaborate systems of audible signals made possible by a vocal tract with a unique structure. How these features evolved is mysterious.[4] Walking on two legs instead of four is, at first sight, disadvantageous: it entails a slower pace, severe stresses on the spinal column and difficult childbirth. As usual, it is possible to make guesses. The upright posture frees the hands to carry tools and food, or to throw rocks or spears during hunting; perhaps there was selection for uprightness because of advantages conferred in these ways. But alternatives have been proposed. One is that our ancestors went through a stage when they lived much of the time partly submerged in water. In such conditions, a hominid on all fours would be more often out of its depth than one on two legs. This idea at least illustrates the scope of the scientific imagination.

The human hand has a distinctive structure: the thumb, with its large muscles, allows a powerful grip, and may be a result of selection of our ancestors for the ability to make and to use tools. The earliest known stone tools are more than two million years old. Microscopic examination reveals their principal use to have been for cutting up carcasses; but their owners also whittled wood and cut up plant tissues. On this evidence, early tools were precursors of kitchen and garden equipment. An important later aspect of paleolithic technology was the making and control of fire. We know of the use of fire for only half a million years. Judged by modern gatherer–hunters, its primary application was in cooking.

The ideas of a tool contained in a chunk of rock and of controlling fire represent not only technical advances but also novel intellectual abilities. It is easy, but entirely speculative, to relate them to the accompanying steep increase in brain size. G.L. Isaac calls this the 'central puzzle' of human evolution.[5] It entails additional demands on the ability of a mother to nourish her child, a longer period of dependency in early life and a number of structural changes to cope with the swollen head. A late change, of the past half-million years, was a lightening of the bones, especially of the skull – a loss of robusticity. At some unidentifiable time, during the millions of paleolithic years, language developed or was invented; and – as we see in Chapter 13 – this too entailed structural novelties in the brain.

In addition, we have distinctive patterns of reproduction, and these loom especially large in current biological images of humanity. Here is a summary of facts about present-day human reproduction, derived largely from the work of a physiologist, R.V. Short, who uses sociobiological concepts.[6]

A woman usually bears only one child at a time, and the newly born infant is exceptionally helpless. (To avoid such helplessness, the human gestation period would have to be nearly two years – an unattractive prospect.) Much of the growth of the head takes place after the child is born, during the long period of dependence. Moreover, among gatherer–hunters and many agriculturalists, children are fed at the breast for some years; during this period, women do not menstruate and are infertile. An

Breast-feeding may continue until the child is three years old. A woman of the Sudan.

additional cause of a low birth rate is a high incidence of miscarriage.[7] Births occur every three or four years.

If this were the only reproductive pattern of *Homo sapiens*, the typical birth interval of human beings would resemble that of apes. But of course human reproduction is often at a much higher rate; and when it is low today, that is due not to breast feeding but to other methods of contraception.

Nonetheless, copulation is frequent. The rule among mammals is that the female allows coitus, and is attractive to males, only around the time of ovulation. But ovulation in a woman is normally undetectable or, at least, not detected; and human beings may copulate at any time during the menstrual cycle and even during pregnancy. Moreover, at about 50 years of age, the menstrual cycle stops (the menopause); but this does not prevent coitus. The mammary glands make another conspicuous difference from the apes: a woman's breasts enlarge at puberty, whereas those of other species wait until late pregnancy, just before milk is needed.

A man, like a male gorilla, is larger than the female of his species (though not to the same extent); but he differs from a male ape in his exceptionally large penis. (This makes possible a great variety of positions during coitus.) Yet the human testis is, on ape standards, quite small, and has only a modest reserve of sperm: coitus with a high sperm count is possible, at best, no more than once a day. R.V. Short remarks that we are adapted for a low level of continuous sexual activity. Our pattern is quite unlike that of male chimpanzees, which can copulate more than once an hour and can produce an ejaculate with plenty of sperm at least four times a day.

Some of the comparisons with other species have practical implications. Short points out that in rich countries, even when breast-feeding is practised, it no longer has its usual contraceptive effect because women 'do not breast-feed for long enough, or enthusiastically enough': alternative foods are easily available; feeding at night is inconvenient; babies spend much time in a cot or pram, not on mother's back or arm or in her bed as in primitive conditions; and there is a tabu on breast-feeding in public. 'It may well be', writes Short,

> that Western civilizations have lost forever the contraceptive effects of breast-feeding. [But] there will always be . . . women who . . . are attracted by the idea of 'natural' contraception, and may prefer to use breast-feeding to space their births adequately and safely. The best advice we can give them . . . is that they should breast-feed as frequently as possible, and certainly more than five times a day, and that they should maintain a night-time feed . . . [8]

This passage may understate the prospects for breast-feeding. There is now a movement in rich countries toward feeding infants on demand, based on the several other advantages of human milk; but it seems unlikely that women will soon revert to allowing their children to suckle until they are three years old. In poor countries, however, interference with prolonged breast-feeding can be disastrous for both mothers and children.

Short also points to possible implications concerning the type of contraceptive pill that should be used. At present, oral contraception is designed to allow for a regular monthly period. But 'enthusiastic' breast feeding induces total absence of menstruation (lactational amenorrhea). Perhaps it would be better if contraceptive pills imitated this aspect of normal physiology.

Whatever conclusion one reaches, the argument so far can be recognised as founded on observation and as actuated by concern for human welfare. It does not depend on any presumptions about our evolution. Yet Short asks, 'What is the normal pattern of human social and sexual behaviour?' And he writes:

> There can be no aspect of human existence that has been so prone to cultural influences as our reproduction. This social overlay has obscured what might be termed the natural history of our species . . . genetically speaking we are still primitive hunter–gatherers.[9]

He also asks 'whether we are *by nature* a monogamous or polygynous species' (emphasis added).[10] We do not, however, know what genetical changes have taken place, during the past (say) 10 000 years, in human populations: human 'nature', in the sense of the range of genotypes in human populations, may have changed greatly in that time. As for the 'social overlay', this – the variety of our social lives – is a central feature of *Homo sapiens*: if we have a 'natural history', social diversity is one of its most prominent aspects: there is no good ground for regarding it as a veneer loosely stuck on to something more fundamental.

Each species of ape has its typical (or 'natural') mating system. Orang utan males maintain territories in which several females also live (polygyny); coitus is a rare event. Gorillas, too, are polygynous and infrequent copulators, but move about in groups. Chimpanzees are promiscuous, and the females are polyandrous. Among other Primates we find examples of permanently mated pairs (monogamy), single-male groups, multi-male groups; male 'dominance', female 'dominance' (page 39); and a great variety of rearing methods. But, with minor qualifications, each species has its own single mode. Only human beings exemplify, within one species, polygyny, polyandry (rare), monogamy and promiscuity, each culturally sanctioned, together with uncounted rearing practises.

We may guess that some of our ancestors were polygynous, because men are larger than women; the males of polygynous Primate species are always larger than the females. But it does not follow that we are by nature or by instinct polygynous: a man who wants a harem should not justify his demands by saying that they are natural and therefore inevitable.

Short also follows sociobiological custom when he writes this:

> the human breast is always there and reveals nothing . . . about when the woman is likely to ovulate . . . Breast development can therefore be seen as yet another female stratagem for reinforcing the strength of the pair bond.

And on a man's beard and other distinctive features:

> These facial characteristics have presumably served both in the ritualization of threat and aggression between men and also in the enhancement of the attractiveness of the man to the woman.[11]

These are very complicated assertions. Some are plain facts (the breast is always there), but they are not clearly separated from two other kinds of statement. One concerns present-day function. We can accept that a woman's breasts are often attractive to men; but there is still the question of cultural differences (which are great), and even individual differences, in standards of feminine beauty. The same applies to standards of male beauty: if the beard is so attractive to women, what are we to make of the habit of shaving? And sometimes our natural features are artificially embellished. One of Kipling's characters complains that being kissed by a man with an unwaxed moustache is like eating an egg without salt. This grievance, like the assertion on the attractiveness of breasts, could no doubt lead to an experimentally testable hypothesis; but it seems not to have been tested yet. It is easy to propose casual, and superficially plausible, ideas on 'function'; but it is much more difficult to make well founded statements, such as those made by Short on structure and physiology.

The other kind of assertion made by Short refers, by implication, to the evolutionary origin of human features. The mention of 'ritualization' is based on an ethological concept. It is believed that some of the signals of animals have been derived, during evolution, from behavior patterns that originally had a different function: a posture, say, adopted during fighting becomes an attitude taken up merely as a harmless 'threat' or deterrent signal (page 49). Some examples are quite convincing but, as usual, we have no way of testing whether they are correct.[12]

The description of the breast as a stratagem (an artifice, a trick for deceiving the enemy) is, of course, a metaphor. It is supposed that in ancestral populations breasts varied in prominence and that — as a result of the evolution also of a liking for breasts among males — women with large mammaries reared more young than did their flat-chested rivals. S.G.H. Cant has provided physiological support for this notion. A woman's breasts and buttocks are, he suggests, signals of her nutritional state: the

122

bigger they are, the better she is able, at least in primitive conditions, to rear her young; for successful breeding it is therefore advantageous for a man to choose a big-breasted and markedly gluteal spouse.[13] It seems to follow that beauty contests (judged by men) should always be won by extremely steatopygous women from the Kalahari or the Andamans.

Since this hypothesis is not confirmed by experience, it is open to us to propose an alternative. For instance, the hair of a female monkey or ape provides something to cling to, and is a source of security to an infant. Human beings have little body hair; so perhaps the human breast is a similar source of comfort to our own infants. The logical status of such proposals is no different from that of other suggestions seriously debated. As an instance, the absence, in our species, of the cycle of female receptiveness has been energetically explained. It is often referred to as loss of estrus, but could be described as the acquisition of permanent estrus. At first sight, coitus when there is no egg to be fertilized seems merely wasteful. How could it contribute to Darwinian fitness?

The ideal sociobiological woman? Steatopygia illustrated by a San Bushperson of the Kalahari.

The obvious answer is that it does so by cementing the 'pair bond' – a biological counterpart of the sanctity of marriage. This is called the fidelity–maintenance theory. But of course we cannot know that such a 'scenario' represents what actually went on among our paleolithic forebears. Two alternatives have been proposed. One of them requires questionable presumptions about early gatherer–hunters: that they depended heavily on meat, and that the men hunted while the women stayed at home. If so, it would hardly do if a woman could get a hunk of meat only when she was ovulating – and never during pregnancy. She therefore remains attractive to men all the time. This is the fornication scenario. The other – more subtle – suggestion is that it increases a woman's inclusive fitness to have genetically diverse children by several fathers. This is called the female philandering scenario.[14]

The debate has the merit of entertainment value; but none of the proposals can be tested: all or none could be correct. There are yet other examples. Donald Symons discusses the menopause.[15] On simple Darwinian presumptions, a woman ought to remain fertile throughout her life. To explain the 20 or more barren years late in life it has been suggested that it pays for a woman to spend a substantial period looking after her youngest children or her grandchildren, finding them mates, and advising her daughters or daughters-in-law. One version of the theory assumes group selection: the elders of the tribe (of both sexes) are a source of wisdom, and contribute mightily to its survival.

It has, however, been doubted whether our gatherer–hunter ancestors often lived for longer than 40 years. If so, there can hardly have been natural selection for or against the menopause. The ending of the fertile period could then be a mere by-product of a physiology presumably 'selected' for its effects during the first four decades of life.

We have no means of choosing decisively between these alternatives; indeed, both might be wrong. What we know is that today most women in rich countries experience the menopause. It would, however, be possible to investigate whether, in modern populations, this confers an advantage on their relatives. But, as far as I know, nobody has yet published research on the present-day survival value of having a mother-in-law. (I suspect that it would be great.)

There is also still plenty of scope for more speculation. Janice and John Baldwin recommend giving students exercises in which they are required to suggest how natural selection could have produced apparently useless or unfavorable features. My contribution to this program concerns the pains that some people feel in wet weather. How could they help one to survive or to breed? One answer is fairly obvious. The Baldwin's own suggestion is to take acne as a model: this affliction could enhance grooming by others, increase the chances of coitus, and so have a positive effect on Darwinian fitness.[16]

After the matter-of-fact beginning on human physiology, a reader may feel that this account has collapsed into levity. If so, it is because much writing in this field sounds like self-parody. We now ask whether the confusion extends to neo-Darwinian accounts of our willingness to help others.

Our neighbors and our kin

In Chapter 6 I discuss the conundrum of how helping others of the same species could be an outcome of natural selection. The question is especially acute for our own species: we not only co-operate elaborately but have moralities which urge us to do so.

Traditionally, moral codes have been regarded as products of human reason, but some prefer to look for rules of conduct fixed in us by our evolution. 'It seems clear', writes R.D. Alexander, that

> *if we are to carry out any grand analysis of human social behavior, it will have to be in evolutionary terms, and we shall have to focus our attention almost entirely upon the precise manner in which both nepotistic and reciprocal transactions are conducted in the usual environments in which humans have evolved their social patterns.*   [Original emphasis.]

He also tells us that situations in which social learning occurs must be analyzed 'explicitly in the light of their effects on inclusive-fitness-maximizing'.[17]

Love, of course, in any ordinary sense, does not appear in such an account: the notion that one should love one's neighbor as oneself is quite unbiological. Natural selection is an amoral process, and Alexander and others apply this amorality steadfastly to human, as to animal, conduct. E.O. Wilson, the leading popularizer of sociobiology, attributes 'hate, love, guilt and fear' to processes in 'emotional control centers' in the brain. These centers 'evolved by natural selection. That simple biological statement must be pursued to explain ethics and ethical philosophers . . . at all depths.' He also states that modern evolutionary theory 'has taken most of the good will out of altruism'.[18] And Richard Dawkins, in a popular work, asserts that

> we and all other animals are machines created by our genes. A predominant quality to be expected in our genes is ruthless selfishness . . . We are born selfish.[19]

In the history of Europe, during the past two millennia, statements such as that of Dawkins have usually come under the heading of theology, not natural science. In his letters to the Romans, Paul of Tarsus (d. about AD

65), the principal founder of Christian doctrine, 'proved' of both Jews and Gentiles 'that they are all under sin';

> There is none righteous . . .
> There is none that doeth good . . .
> With their tongues they have used deceit; . . .
> Their feet are swift to shed blood:
> Destruction and misery are in their ways:
> And the way of peace have they not known.[20]

Later came the doctrine of original sin, owed especially to St Augustine (354–430), which has influenced attitudes almost to our own day.[21]

But attempts to prove scientifically that we 'are all under sin' are hardly helped by genetics: we know nothing useful about the genetics of morality even in existing populations. The sociobiological method therefore is not to use actual knowledge of heredity or of what people do, but to propose conclusions about our social lives from suggestions on how natural selection has operated in the past.

As we know, giving indiscriminate help to others is difficult to reconcile with simple theories of evolution; but helping close relatives, and others who are likely to reciprocate, may be expected. The presumption is that there is, or has been, genetical variation affecting helping behavior, and that, during our evolution, people who did not sufficiently restrict themselves to favoring kin and to reciprocation failed to breed at the rate achieved by others. That people help their kindred is, however , not a new finding: it is well established that blood is thicker than water. As for reciprocation, people combine for mutual benefit both in the simplest communities and in those with an elaborate law of contract. Hence so far neo-Darwinism applied to the human species seems to reveal only what is already perfectly familiar.

It is, however, possible to investigate helping behavior by the methods of the social sciences. S.M. Essock-Vitale and M.T. McGuire have asked whether patterns of helping seen in diverse human cultures are 'consistent with predictions following from the sociobiological theories of kin selection and reciprocation'; and they remark that 'speculation concerning the evolution of human social behavior has far outstripped what would be justified' by the evidence. They therefore critically examined 13 reports, on societies ranging from those of middle-class Whites in Connecticut to villagers in Thailand and Venezuela; the reports are selected because they give *measures* of reciprocation and kinship. The principal conclusion is 'that kin are more likely than non-kin to be the recipients of assistance for which no reciprocation is expected'; but, the authors add, 'to say this is far short of saying that [the findings] are sufficient to accept the theory.'[22]

They might also remark that their research concerns what goes on today, not what has happened in the past: it tells us nothing about evolution. For this exceptionally careful and critical study faces the same difficulties as those outlined in Chapter 6. First, it implies genetical variation in the relevant aspects of behavior and enough time for corresponding changes in gene frequencies; but this is always a presumption without direct supporting evidence. Second, if what is expected is not found, alternative proposals can always be made to account for discrepancies. For example, people often clearly endanger themselves, or sacrifice wealth, on behalf of unrelated persons who could not possibly reciprocate. The answer might be that such conduct is an incidental by-product of practises which fit sociobiological presumptions: it is, in effect, an error – a result of imperfections in the behavioral rules imposed by natural selection. A further and more ingenious suggestion is that self-sacrifice is so much admired that the relatives of a person who displays it are favored: if they are children, they are so well cared for that their fitness is enhanced. These stories are, like the algebra of natural selection, only on paper; they are unsupported by evidence, and they hardly match the complexities of real human relationships.

Among the complexities are those associated with marriage, a subject which has been given a neo-Darwinian analysis by Pierre van den Berghe. 'Although much sex occurs outside marriage', he says, marriage is 'the cultural codification of a biological program'.

> Marriage sanctions both kin selection for the maximization of inclusive fitness, and complex systems of reciprocity between spouses and their respective kin groups.[23]

But this states only that, in many kinds of marriage, parents look after their children, children sometimes look after their parents, and people related by marriage often help each other.

Marriage in many societies has been an economic transaction. A bride's parents may pay a dowry to the husband. Sociobiologically, this makes sense: the chances of propagating the parents' genes are enhanced. But, in other societies, a man may buy a wife. In yet others, such customs are regarded as absurd or immoral. Yet sociobiological statements about human 'pair-bonding', expressed in economic terms, are held out as stating universal features of human nature.

'Pair-bonding', according to van den Berghe, 'is a mechanism favoring paternal investment . . . [and] the reproductive game is . . . quite different for men and women.' A male's 'investment' when he inseminates a female is small: the time, materials and energy required for courtship and coitus are trifling, in relation to total expenditure on all activities. A woman 'invests' proportionally much more in each offspring, for she provides all the material for the growth of her young during a substantial period both before and after birth. For a male to maximize his fitness, he should therefore have as large a harem as he can cope with, and he should also inseminate as many other females as possible. But a female may be supposed to do best (that is, to maximize her fitness) if she does not have to rear her young single-handed: a female's fitness is enhanced if there is a male to help to protect the young and to supply food. Hence a female is aided by a prolonged and (almost) exclusive relationship with a good provider, that is, by monogamy. Moreover, a woman's reproductive errors are more expensive than a man's. Women have therefore 'developed strategies (such as coyness and . . . slow erotic arousal) to resist seduction and to capture and retain a man'. The statement about 'strategies' presumably implies a fixed propensity ('instinctive', or 'genetically coded'). It therefore disregards the universal human ability to adapt behavior to circumstances. Some feminine conduct during courtship represents intelligent adaptation to particular social demands, and hence can be changed according to need.[24]

A woman is limited to only about twelve children at most, regardless of the number of her partners; and, as we know, in the gatherer–hunter or simple agricultural stages the number is usually less, as a result of prolonged breast feeding. In contrast, a man can substantially increase his biological fitness by having even one or two children by women other than his wife. (Granted, in most societies, the hazards attached to sleeping around require an additional term in the equation; their net effect on a man's reproductive success is, however, difficult to calculate.) Ideas such as love, affection, even duty, prominent in everyday life, do not appear in this account. If this portrait of ourselves were a true one, the best we could hope for would presumably be a precarious truce between the sexes.

The writers on this theme also blind themselves to the drastic social changes now in progress. Even if human beings were genetically uniform, we could not (as we know) hope to predict how they would behave unless we knew what environment they would be living in. Moreover, environments are sometimes an outcome of intelligent planning; and the knowledge needed for such planning is steadily increasing. Whether it is used will depend, in part, on whether people reject statements that they are bound to the wheel by their biology.

It is not, in any case, entirely clear why women should stop at monogamy and moderate chastity. Here is a version of the ideal neo-Darwinian woman proposed by Michael van Ghiselin:

> She should mate with every male in the neighborhood, thereby assuring herself of genetically diverse offspring. Immediately after mating she should eat her mates, providing herself and her offspring with food and keeping other females from reproducing.[25]

This intentionally outrageous portrait still omits an important trait of *Homo egoisticus*: deceitfulness. Alexander tells us that he

> can conceive of only one possible argument against the notion that society is based on lies . . . Consider . . . 'Thou shalt love thy

neighbor as thyself' . . . this admirable goal is clearly contrary to a tendency to behave in a reproductively selfish manner. 'Thou shalt give the impression that thou lovest thy neighbor as thyself' might be closer to the truth.[26]

Among those who emphasize the role of deceit in human affairs, R.L. Trivers has a leading place. Trivers supports his argument by reference to actual observations of animal camouflage and other *deceptive* effects in nature.[27] Hence color patterns that make an animal difficult to see, or especially attractive to a mate, are equated with lying. This seems to be an extreme case of being deceived by analogy. There appears to have been little discussion on whether we could have been 'selected' for truthfulness; yet it is possible to imagine a case for it. After all, we mostly tell the truth, and scholarly studies are a search for truth. (We also tend to blush when we lie.[28]) Any such hypothesis would, of course, be just as untestable as the others.

In an extension of his argument about deceit, Alexander states that self-deception, too, is a necessary consequence of our evolution by natural selection. Pierre van den Berghe similarly writes: 'we have evolved as a species so uniquely equipped for deceit that we have a seemingly unlimited capacity to deceive ourselves.'[29] There is, however, no serious discussion of what these statements imply. On the face of it, the idea of self-deception is, as Sissela Bok points out, contradictory, or paradoxical. 'How can one simultaneously know and ignore the same thing?'[30] In modern times the most influential answer has been that of psychoanalysis. We are said to have an unconscious mind that contains thoughts, desires and emotions hidden from the conscious self. The mind is divided into compartments. Concealment is achieved by repression. Unwanted feelings are kept from awareness by a censor. Psychoanalytic writings are dripping with such metaphors. They provide an effective, and sometimes illuminating, *description* of some of the strange things that people do. Bok cites, as examples of self-deception, 'the anorexic girl . . . who thinks she looks fat in the mirror, and the alcoholic who denies having a drinking problem'.[31] But psychoanalytic descriptions do not answer Bok's question: they do not explain what they describe, though they sometimes purport to do so. These anomalies have indeed no satisfying or generally accepted explanation, psychological or physiological.

An important feature of the psychoanalytic interpretation is that it treats such phenomena as pathological – as something to be cured, not as part of an unalterable human nature. The aim of psychoanalysts is to bring unacknowledged feelings and desires into consciousness. In doing so, they accept a principle which has been urged for at least two and a half millennia: know thyself. Human beings no doubt often go in for various kinds of deceit in a big way; and we may behave as though, while talking to ourselves, we were uttering falsehoods. More generally (to introduce another metaphor), in our outlook we often seem blinkered. Writers preoccupied with features such as deceit concentrate their attention narrowly on our (and their own) failings. Yet, in doing so, they illustrate also the human capacity for self-knowledge. Such knowledge can benefit both our health and our morals. Its existence contradicts the belief that our conduct is ruled by genetically fixed impulses.

Sociobiological pessimism does not stop at deceit. Trivers also provides a neo-Darwinian treatment of conflict between parents and children. To appreciate his position fully, we have to look back to the time when group selection was accepted by evolutionists. It was supposed that behavior which conferred an advantage on the community could be selected, even at the expense of the actor. Given that presumption, according to Trivers, a child 'could be viewed simply as the agent of its parents' self-interest, created and molded to do what was good for the parent.' (Presumably he means the parents' genes.) But rejection of the possibility of group selection 'had the beneficial effect of drawing attention to the self-interest of *all* actors in a social scene . . . ' A child has 'an independent self-interest, different from the parents but equally legitimate.'[32]

The use of the term legitimate suggests a reference to a child's rights – a moral or legal concept. But for Trivers legitimacy depends on genetical advantage; and he assumes that a child is favored by a long period of parental care, while a parent's chances of contributing to later generations are enhanced by a shorter period. The result is something like a class struggle within each family.

These theories are not founded on any systematic analysis of human motivation, or on anthropological or historical findings; and of course they have no genetical basis at all. On the contrary, familiar facts are often disregarded; or, if they are inconvenient, they are explained away. 'Confucianism', writes van den Berghe, 'emphasizes filial piety as the ultimate virtue' but also 'stresses parental responsibility'. But he also states that Confucianism is 'essentially one-sided kin selection', and so tries to force a social phenomenon into an inappropriate biological frame.[33] 'One-sided' refers to the emphasis on inheritance in the male line, whereas kin selection is concerned with all genes, regardless of their source.

In many societies, particularly where most people are peasants, the livelihood of whole families depends on the work of every member, and land and other assets are passed on from generation to generation. But, as we know, the forms of family (and property) relationships vary greatly (Chapter 4). No doubt, in any society, some parents exploit their children and some children rebel against their parents. It is also possible to find opposite behavior. To say that these phenomena are coded in our genes adds no information (what is *not* so coded?): one might just as well say that the behavior is due to an innate parental or filial impulse; to do so would be to revert explicitly to a kind of 'explanation' popular in the middle ages (page 53) but now usually rejected.

Among the many sets of customs difficult to reconcile with neo-Darwinism are some of our sexual practices. Extreme examples are voluntary celibacy and socially sanctioned homosexuality. In ancient Greece, as in a number of other societies, much sexual activity was wasted (biologically speaking) in friendships between men; and some homosexual conduct seems to occur in most communities. The simple calculus of genetical cost-benefit 'predicts' that this cannot happen. We might therefore conclude that the calculus should not be applied to human action in all its variety; but this does not satisfy sociobiologists.[34] Instead, they fall back again on the concept of inclusive fitness (page 94). Conduct that lowers a person's chances of reproducing can be favored, or 'selected', if it sufficiently improves the chances of near relatives. The supposition is that, if not today then in the past, homosexuals have been so helpful to their brothers, sisters, cousins, nephews or nieces that their genes have been more successful than those of people with no homosexuals in the family. There is no evidence for this: our knowledge of human genetics tells us nothing relevant about the survival value of homosexuality or friendship. And, as usual, an alternative theory is available: Richard Dawkins proposes a deliberately fantastic 'sneaky male' hypothesis, according to which

> homosexuality represents an 'alternative male tactic' for obtaining matings with females. In a society with harem defence by dominant males, a male who is known to be homosexual is more likely to be tolerated by a dominant male than a known heterosexual male, and . . . may be able, by virtue of this, to obtain clandestine copulations with females.[35]

The scope for ingenuity in this field is infinite. Among universal human activities which appear most obviously 'unDarwinian' are those related to enjoyment of the arts and to creative or playful performances undertaken for their own sake. But Alexander, like other sociobiologists, declines to accept such phenomena as 'evidence against a theory based on inclusive-fitness-maximizing'. He attempts no serious analysis of why people create works of art or why others enjoy them. Instead he writes:

> Consider the relationship between status and the appreciation of fashion, art, literature or music. What is important to the would-be critic or status-seeker is not alliance with a particular form but with whatever form will ultimately be regarded as most prestigious. . . . *Alliance with a powerful or influential person's opinion is not arbitary, even if the opinion itself is.*[36]   [Original emphasis.]

So, if Alexander's hypothesis is correct (and if I understand it), I like the music of Vivaldi because doing so has enhanced my social status. (Would I have done still better to go in for Stravinsky?)

The outbreak of writing on these themes began in the 1970s, and might be regarded only as a passing fashion, like the crinoline or the foxtrot, and an outcome of the views of a few eccentrics. But it is more than this, for two general reasons. First, the ideas of sociobiology have, like the myths about aggression, spread far outside the learned journals in which specialists argue: they are regularly taught, in universities and schools, as accepted doctrine; and they appear even in the popular press. Second, these ideas raise fundamental questions which will continue to be debated long after current controversies have been forgotten.

The popular imperative

Just how far the influence of sociobiology has spread is a question for future social historians. They might begin with a long, well written article, published in *Time* magazine in 1977, on what the writer calls 'a new and highly controversial scientific discipline'.[37] A summary of sociobiological teaching includes the 'biologically inevitable' conflict of children with parents, genes for 'conformism, homosexuality and spite', and self-deception as a product of evolution 'simply because a cheater can give a more convincing display of honesty if he lies to himself as well as to his neighbor'. A leading theorist is quoted as making 'a bold prediction: Sooner or later political science, law, economics, psychology, psychiatry and anthropology will all be branches of sociobiology.' A prominent psychologist, we are told, believes that 'religious teachings have evolutionary importance – an idea that a few theologians have picked up from sociobiology.' A Unitarian theologian is quoted as saying that 'the truths in religion have been selected because they are necessary and essential to man.' (It would be interesting to know how this assertion is reconciled with the ability to choose between good and evil.) Criticism of sociobiology is also recorded. 'Opponents', we are correctly told,

> have been slow to mount a scientifically based counterattack . . . : few critics feel competent to cut across all the disciplines involved, from ethology and mathematics to anthropology and game theory.

Nonetheless, the impression left on readers is likely to have been misleading. To quote a philosopher,

> The cover of *Time* magazine's sociobiology number, showing a couple of lolling puppets making love while invisible genes twitch the strings above them, gives the level at which the message can be expected to reach the general public.[38]

A similar comment could be made more strongly of other stories in the press. Readers of the London *Daily Mail* in 1978 were told of the 'explosive theory that rips apart economic and political thought'.

> The father who saves his child from a burning house is motivated not for love of the child but a determination to protect the genes which his child inherited from him. . . .
>
> *The Conservative case has long recognised that man is essentially a competitive animal and that this combativeness if properly harnessed can and will lead to a healthy society.*   [Original emphasis.]

Man, we are also told, is 'an oppressive animal': we are 'greedy, rapacious, self-serving individuals, out to get what we can for ourselves and Devil take the hindmost.' I give more examples of the spread of such teaching in the next chapter. Meanwhile, for those who prefer a less blinkered view of our species, here is Francis Bacon's version of parental attitudes.

> The joys of parents are secret, and so are their griefs and fears. . . . Children sweeten labours, but they make misfortunes more bitter: they increase the cares of life, but they mitigate the remembrance of death. The perpetuity by generation is common to beasts; but memory, merit, and noble works are proper to men.[40]

Adaptability and tradition

The most obvious of the fundamental questions raised by sociobiological stories is the role of reason in human affairs. Pierre van den Berghe puts it in modern terms. 'Sociobiology', he writes,

> predicts that we shall continue to reproduce, consume resources, and destroy each other with abandon because we are programmed to care only about ourselves and our relatives. So far there is little evidence to show that sociobiology is wrong.

The last sentence seems to brush off modern knowledge of contraception and the world-wide movement for natural conservation and disarmament. But the author adds:

> The ultimate challenge of humanity is to prove sociobiology wrong . . . through self-conscious change in our behavior.[41]

Here is an example of a repeated theme in traditional interpretations of humanity: the clash of reason and morality, on the one hand, with unreasoned impulse and egoism, on the other – in the plainest words, between good and evil.

When we use our intelligence in social affairs, we are influenced by the traditions, some ancient, some recent, of our community. Traditions are 'inherited' by non-genetical means. To quote the biological mathematician, A.J. Lotka:

> In place of slow adaptation . . . by selective survival, increased adaptation has been achieved by the incomparably more rapid . . . *exosomatic* evolution.[42]   [Original emphasis.]

P.B. Medawar, distinguished both as an experimentalist and as a philosopher of biology, regards exosomatic change as a 'liberating conception':

> It means that we can jettison all reasoning based upon the idea that changes in society happen . . . under the pressures of ordinary genetic evolution; [we can] abandon any idea that the direction of social change is governed by laws other than laws which have at some time been the subject of human decisions. . . . That competition between one man and another is a necessary part of the texture of society; . . . that division of labour within a society is akin to what we can see in colonies of insects; that the laws of genetics have an overriding authority; that social evolution has a direction forcibly imposed upon it by agencies beyond man's control – all these are biological judgements; but . . . bad judgements based upon a bad biology.[43]

The concealed presumption is the usual one of genetical fixity: that sex roles and reproductive practices will be unaltered in any environment. In contrast, we have the fact of modern control of contraception, which represents a refusal to accept the physiology 'preprogrammed' by our evolution. Women today in some countries are living in an environment without precedent: for the first time, a woman can confidently decide for herself when (or whether) she shall have her first child, or subsequent children. Correspondingly, the decline in fertility in many countries is difficult to reconcile with the presumption, derived from evolutionary theory, that we all tend to maximize our Darwinian fitness.

Whenever we are told of some supposedly fixed propensity, patriarchy or anything else, we are free to respond, with Karl Marx, that the point is to change it. We create novel environments for ourselves and can rapidly alter them. Among the favorable upheavals of our time is that due to modern medicine, especially public health. In rich countries the expectation of life at birth is over 70 years – a figure never before attained. A transformation of at least equal importance is occurring in the relationships of the sexes: the change in the condition of women may seem to our descendants the most notable social achievement of our century.

These upheavals are due partly to conscious intention. They represent the ways in which traditions can change. They are sometimes likened to an evolutionary process; but it is more important to *contrast* genetical and non-genetical transmission. A person is endowed with a set of genes at conception, and that set cannot (at present) be altered in later life; but cultural and other environmental influences arise and disappear, gradually or suddenly, over a lifetime. The present book gives many examples. Hence the survival of a tradition requires social interactions different from those needed for the survival of genes. An example is the relationship of teacher and pupil in the learning of a complex skill: in an advanced society, the pupil is rarely a member of the teacher's family; information is commonly transferred to people who are not even remote cousins.[45] Such a process is wholly different from the passage of genetical material at conception: its study is the material not of biologists but of social scientists and historians; and there is no good reason to think they will be materially helped by genetics or the theory of natural selection.[46] As before, when we make a serious attempt to analyze human affairs, we have to use kinds of knowledge which fall outside the natural sciences.

Questions of morals

The attempt to interpret the human condition in neo-Darwinian terms leads to neglect, not only of history, but also of morality. Words that usually have moral meanings are prominent in such writings but are used to refer to phenomena that belong to population genetics. In casual conversation these usages would pass as word play, but in texts intended to be serious they have wider implications. A critic of *Homo egoisticus* has written:

> If human beings could be shown on scientific grounds to be innately selfish, instinctively self-centered and uniformly self-regarding, the appeal of much moral philosophy would be greatly diminished.[47]

Above I cite passages by influential writers that convey just that message. The message is also contained in the misuse of terms such as altruism, egoism and selfishness, kin and nepotism.[48]

Of these, altruism is the most important. In the debate on how natural selection could produce animals that help others, this word has come to be regularly used for conduct that decreases the fitness of the helper but increases that of another animal of the same species. Such behavior I call bioaltruistic. Fitness, of course, refers to some measure of an individual's contribution to later generations (page 94). When we talk of bioaltruism in this sense, we are concerned with the *effects* of an act on the survival or reproduction of others. The concept is an objective one, and appropriate for study of the populations and social interactions of other species. It does not entail any reference to feelings or intentions.

In contrast, the complex moral concept of altruism concerns a person's intentions. An action is called altruistic when it is *designed by the actor* to help another, and is performed with no expectations of reward (other than pleasure, perhaps, that results from helping). The word altruism itself usually refers to a rule of conduct: 'concern for others as a principle of action' (*OED*). This lexical definition implies that altruistic conduct is accompanied by empathy for the person helped: the helper may feel vicarious anguish on seeing somebody in pain or difficulty; or joy when another's end is achieved. Altruistic actions are not, however, only a matter of feeling: they may also be an outcome of reasoning.[49] A person may ask of another (or of him or herself): 'how would you like it if someone did that to you?' Or 'how would you feel if somebody failed to help you in such a predicament?'

Similarly, the moral concept of the opposite principle, egoism or selfishness, concerns motives rather than the effects of action on others: an egoist's reasons for action are personal desires. In some evolutionary writing self-sacrificial caring for one's children is called selfish, when in ordinary speech it would be called altruistic. In this case the slapdash use of words and imagery has led to a reversal of meaning.

Another key term in neo-Darwinian theory, kin, has – like altruism – been stripped of its moral implications. Kin, as used for instance in the expression kin-selection, refers only to genetical relationship (Chapter 6). There is no objection to this usage in the context of evolutionary genetics; but error arises when kin in the biological sense is equated with the concept of kin that we meet in human societies.[50]

It is hardly news that we tend to favor our near relatives, especially our children; but our helpful actions, both altruistic and reciprocal, go far beyond this. R.M. Keesing, an anthropologist, in a survey entitled 'Simple Models of Complexity', comments on the importance of common membership of a settlement or neighborhood – a phenomenon we have already met among Australian aborigines (pages 42–4). Such membership is an everyday experience, and is often independent of biological relationship: our neighbors or companions at work are usually not parents, children or even cousins. As Keesing shows, even anthropologists have difficulty with the diversity of human relationships. He writes:

> We do not, I think, understand friendship anthropologically. What friends do together slips through our analytical net . . . Yet friendship is so basic that . . . most of us take it for granted,[51]

In friendship, people assume obligations of kinds that are taken for granted within families. But what we accept as a family obligation varies among communities and is often quite unbiological. We receive an equivalent set of genes from each parent and so, on biological grounds, parents should provide equally for their daughters and sons; but as we have seen, they rarely do so. In patrilineal societies fathers provide primarily for their sons; but other communities have a matrilineal succession in which the mother is all-important. Both systems are kin-centred, but neither corresponds to expectations based on genetics. Sometimes, the significant relationships are economic. E.R. Leach gives an example from a Sri Lankan village in which ownership of property determined the obligations others expect within families. Hence, as he writes, in anthropology the term kinship is most conveniently used to cover much more than 'blood' ties: 'it refers rather to a widely ramifying pattern of *named* relationships [which] link the individual members of a social system in a network.'[52]

Among these relationships are those of adoption. Taking children of other parents, and treating them as one's own, is widespread, perhaps universal, in human communities. I give examples from Australian aborigines in Chapter 4. A survey of kinship among the diverse peoples of Oceania provides a good idea of the range of practices concerning adoption and the flexibility of the idea of kin. 'Kinsmen may be added or eliminated through transactions in behavior, residence, and land rights. Adoption is one of the processes involved in these transactions.'[53] On the opposite side of the world, and in a distant time, were the people who gave rise to the civilization of ancient Greece. In Homer's *Odyssey* (about 800 BC), Odysseus is formally accepted by the king and queen of the Phaeacians, that is, by the rulers of a group of strangers. In this way the poet describes a feature of the tribal life of his time. Today, adoption is often regulated by law rather than unwritten custom, but is still more widespread. In every community of which we have knowledge, the unit of the family is modified, distorted, expanded or even rejected in ways which cannot be predicted on the basis of any simple model. As usual in human affairs, reduction to biology omits essentials.

The extension beyond biology of the concept of the family is recognized in many everyday expressions. Male trade unionists call each other 'brother'. (It has been frivolously remarked that Englishmen do not marry their sisters; yet many English doctors do just that.) Pierre van den Berghe supports the idea of 'a genetic underpinning . . . for human sociality'; but he also writes:

> Clearly, for fifty million Frenchmen or a hundred million Japanese, any common kinship that they . . . share is highly diluted. Similarly, when twenty-five million Afro-Americans call each other 'brothers' and 'sisters', they are greatly extending the meanings of these terms.[54]

All major ethical systems include the idea of the 'Family of Man'. However dreadfully we fail to live up to this ideal, it remains powerful. In this half-century it has been given expression in the organizations of the United Nations which, rarely headlined and often obstructed, concern themselves with food, health and education for the whole of humankind.

The failure to take account of friendship, and of the many manifestations of altruism (in its primary sense), makes *Homo egoisticus* sound like a strange species. Yet the writers are themselves members of the species they describe. A reader may wonder whether these writers believe that they are describing themselves, and that the moral principles urged and acted on (however imperfectly) in most societies are totally spurious. Of their predecessors in the late nineteenth century T.H. Huxley had apparently few doubts:

> ... men who are accustomed to contemplate the active or passive extirpation of the weak, the unfortunate, and the superfluous; who justify that conduct on the ground that it has the sanction of the cosmic process, and is the only way of ensuring the progress of the race; who, if they are consistent, must rank medicine among the black arts and count the physician as a mischievous preserver of the unfit; on whose matrimonial undertakings the principle of the stud has the chief influence; whose whole lives, therefore, are an education in the noble art of suppressing natural affection and sympathy, are not likely to have any large stock of these commodities left.[55]

Occasional passages by widely read modern authors seem to call for comment by another Huxley. In a lecture on human values E.O. Wilson writes:

> To us the savageries of Tamerlane and Simon de Montfort are merely diverting curiosities, or at best data for abstract historiography. In the vastly altered world of a thousand years hence, the histories of Hitler and Stalin will mean nothing more. All that will matter to our distant descendants is whether we made it possible for them to enjoy health, security, freedom and pleasure. Indeed, if it were perceived that our benighted existence was the essential evolutionary step leading to their own liberation, they will be glad we suffered.[56]

But Wilson gives no evidence on how many people look on the cruelties of the past (or the present) as diverting. It is possible to regard them with horror.

More important, he and others display a curious dissociation in their statements about their own species, that is, about themselves. Not one, as far as I can find, has said anything like this: 'I am aggressive, competitive, selfish, spiteful, deceitful and ruthlessly bent on contributing largely to the next generation: I therefore maintain a harem.' Instead, both their misanthropy and their preoccupation with genetics are qualified.

When Wilson discusses human evolution he states that ethical systems can 'create complex, intractable moral dilemmas', and that 'these, of course, are the current condition of mankind.'[57] Hence, on the one hand, we are presented as puppets manipulated by natural selection; but, on the other, if we face moral dilemmas (as we do), we have both the ability and the obligation to make moral choices.

Richard Dawkins, who begins his widely read book by telling us that we are born selfish, ends it by declaring: 'We, alone on earth, can rebel against the tyranny of the selfish replicators' (that is, genes).[58] Pierre van den Berghe similarly regards proving sociobiology wrong as being 'the ultimate challenge of humanity'; and in a slightly later work he denies that 'everything or even most things are . . . reducible to biology', or that 'biology is destiny in a rigid, immutable way'.[59] (Presumably biology is destiny only in a flexible, changeable way.) R.D. Alexander urges (as we know) a grand analysis of human conduct 'in evolutionary terms'. Yet he acknowledges that recent accelerated social changes cannot be a result of genetical change and agrees that satisfactory biological analyses of cultural variations do not exist.[60] But his most notable example of ambivalence is the couplet from Borges with which he ends his book:

> Beyond my anxiety, beyond this writing,
>   the universe waits, inexhaustible, inviting.

The realm of discourse to which this belongs is not that of population genetics.

A paradox

The neo-Darwinians just quoted have evidently not fully worked out where their arguments lead. Human action is reduced to a consequence of natural selection; but the limitations of evolutionary theory are overlooked, and the possibility that the theory cannot account for important aspects of human existence is disregarded. Then, faced with disconcerting implications, the writers recoil.

An uncompromising advocate of the application of neo-Darwinism to humanity might regard the recoil as merely a display of weakness: it reflects, he might say, no more than a distaste for an unpopular, but correct, conclusion. M.T. Ghiselin has written what seems to be a forthright statement of such a position. He heads it 'Some Unpopular Conclusions'. 'The evolution of society', he tells us, 'fits the Darwinian paradigm in its most individualistic form. . . . The economy of nature is competitive from beginning to end' and fully explains human social life. 'No hint of genuine charity ameliorates our vision of society, once sentimentalism has been laid aside. What passes for co-operation turns out to be a mixture of opportunism and exploitation.' And he concludes: 'Scratch an "altruist", and watch a "hypocrite" bleed.'[61]

In fact, however, the neo-Darwinian account of the human species leads not only to rejection of morality (and lack of charity) but also to a logical impasse, of a kind that arises also in other narrow interpretations of humanity. A movement in political science, sometimes called historical materialism, has been influential in the modern period. In its extreme form, this doctrine

states that there are fixed laws of historical development, which are sometimes thought of as analogous to the laws of organic evolution; although,
as we know, those laws are still a matter of debate. In Chapter 6 I describe
how some biologists have deified natural selection. Similarly, to quote E.H.
Carr, historical realists hold that 'whatever succeeds is right, and has only
to be understood to be approved'. But such a belief, he adds, 'must, if
consistently held, empty thought of purpose, and thereby sterilise and ultimately destroy it'.[62]

In biological accounts of our species, the paradox arises, once again,
from reduction. If we are *nothing but* vessels for self-replicating genes (or
gene-complexes) then the sole determinant of all our actions is their effect
on our inclusive fitness, or on the survival of bits of chromosome. If talking
rubbish increases our chances of contributing to the next generation, then
– as Darwin might have said – talking rubbish will be favored. The same
applies to talking sense. Indeed, the means of deciding what is rubbish and
what is sense disappears, for the statements of neo-Darwinists or of their
critics are not founded, in any significant way, on evidence or on logic –
or, if they are, that is an accident of the need to survive. 'So there is no
argument, only the universal war of gene against gene'.[63] Arthur Meltzer
asks how a sociobiologist explains the existence of anti-sociobiologists. He
believes that most people regard themselves as 'basically generous, caring
and unselfish'; and he suggests that, even if this view is incorrect, it may
be biologically advantageous to hold it; perhaps, therefore, sociobiology
will be eliminated in the struggle for existence.[64] But, of course, on strict
neo-Darwinian principles, there is no criterion by which the correctness of
the view can be judged.

It may be objected that the preceding paragraph merely sets up a straw
man: that neo-Darwinists, whatever they write about humanity, do not,
in the end, regard themselves and their friends as nothing but competing
bags of genes. Perhaps, after all, they accept the existence of logical laws
and moral principles which are valid *regardless of their contribution to
fitness*. (This would allow them to acknowledge, for instance, that people
who martyr themselves for a cause do not necessarily do so solely for the
benefit of their near relatives.) And, indeed, I do not doubt that the writers
in this field all frequently behave altruistically (in its primary sense). But
if so the neo-Darwinian movement, applied to humanity, no longer appears
as a great new principle: its contribution dwindles to less than epoch-making
statements on comparative anatomy and physiology, and entertaining, but
untestable, speculations on our evolution.

While writing this chapter I came upon the following passage in a recently
published work of fiction:

> By introducing . . . variations into their genetic structure, he removed all traces of emotions such as love, pity, fear and compassion. . . . [He] had created a race of creatures motivated by hatred and totally devoid of conscience.

Could this be a parody of the humanoid creatures depicted by neo-Darwinians? It occurs in a description of a popular British television program for children. The villain is a depraved but brilliant scientist, whose plans are foiled (but only incompletely) by the protagonist, also a scientist and polymath.[65] The story could therefore be interpreted as a rejoinder to sociobiology.

Parody or not, we are reminded once again that scientists work in a community, and influence the ideas of a large public. Descriptions of the human species by biologists are judged in the first place by whether they qualify as scientific: by whether the observations are correct, and the conclusions logical. When the conclusions are related only tenuously to authentic findings, and inadequately supported by logic, we are entitled to look at them primarily as part of a historical process outside the natural sciences. That is the topic of the next chapter.

9 Darwinism, genetics and politics

The ethical progress of society depends, not on imitating the cosmic process, still less in running away from it, but in combating it.

T.H. Huxley[1]

This chapter could have been entitled 'Social Darwinists and their Successors'. The expression, social Darwinism, is used for attempts to explain human society by evolutionary principles and, in particular, to support a conservative political outlook by reference to biological findings. The theories of evolution and of genetics have both influenced and been influenced by the social changes going on in the world outside. Here is an example.

R.A. Fisher and eugenics

One of the principal founders of modern biological mathematics was R.A. Fisher. Though he was an innovator, even his most original contributions depended on the work of many forerunners. Statistical analysis originated, in the seventeenth century, partly in response to problems connected with public health and with death rates. Among the questions examined by the pioneers was what premiums should be paid by persons who wished to insure their lives. Fisher's immediate predecessors included Francis Galton (1822–1911) and Karl Pearson (1857–1936). Both these brilliant men were energetically involved in applying mathematics and scientific findings to social issues.[2]

Fisher's early work also included important contributions to the new science of genetics. The first successes in this field depended on the discovery of variation of genetically simple origin: for example, whether a fruit fly has red or white eyes. Fisher showed how features such as stature or body weight (or intelligence, if that can be measured), which vary continuously, could be analyzed in the same 'Mendelian' framework. Hence his researches fell within the internal development of mathematics and biology: they arose logically from previous findings.

But Fisher was also, like Galton and Pearson, involved in public affairs: he was active in the eugenics movement in England soon after it was founded.[3] He held that differences in ability are mainly determined genetically, and that the fate of societies depended on whether the breeding of superior persons could be encouraged. B.J. Norton describes the relationship between early work on genetics and the social process of which the eugenics movement was a part. (By eugenics I mean the improvement or preservation of human populations by controlled breeding.) Norton gives evidence of Fisher's intention to expound the overriding importance of heredity in human variation. Fisher himself wrote: 'An examination of the best available figures for human measurements shows . . . little or no indication of non-genetic causes', and he commented on 'the practical importance of the predominant influence of natural inheritance'.[4] Norton is concerned to 'place' Fisher's work in its historical context: 'We should see it as a stunning contribution to eugenics'.[5] (In view of the obvious environmental influences on variation in features such as stature, 'stunning' seems an appropriate adjective.)

I must now restate some principles which, though obvious, are sometimes overlooked. To classify Fisher's work in this way says nothing about the validity of its scientific content: that depends on its empirical findings and logic. As a contribution to science it is not affected by its social implications or by its author's motives, or by the social influences acting on him. The eugenists in Britain and elsewhere, as we see below, were a political pressure group, and misused genetics to support their cause; but this does not invalidate genetics.

The intrinsic worth of a scientific statement does not depend on the social or psychological processes associated with it. But our interest in such processes is enhanced when we encounter supposedly scientific assertions that lack foundation, such as those of Fisher on human variation. The preceding chapters give many other examples. We ask why these assertions have been put forward and – more important – why they have been accepted; and to answer we have to examine the social circumstances in which they were made. In the present chapter we again face statements on human heredity and evolution; and we find that some of these statements do not belong primarily to biology, but instead reflect a historical process independent of the natural sciences.

The central notion is of a fixed heredity and hence of predictable behavior. Abilities and other features of personality are described as if transmitted intact, like property, from parents to offspring. Moreover, we are all supposed to have in common an unalterable and disagreeable nature. In 1983 the doctrine was crudely expressed by a leading British conservative politician: 'man', he said, or 'man's nature' does not change; original sin is something with which children are endowed at birth; and he added, rather unnecessarily, that he was not a great believer in progress.[6] As we know, since Darwin biological ideas have been used to justify pessimism of this sort. Biology has been applied especially to three issues of fateful consequences for us in the twentieth century: the differences between social classes and their significance for education; the differences between races, and their significance for civil liberties, colonialism, immigration policy and, again, education; and, most important, the differences between the sexes.

Improving the breed

The Darwinian account of evolution was first put before a large public in 1859, in *On the Origin of Species*. It was founded on the notion of improvement by selective breeding, and soon afterwards its implications for human beings began to be debated. Darwin himself questioned the effects of medical treatment and of state welfare (Poor Laws) in allowing 'the weak members of civilized societies [to] propagate their kind'.[7] But the eugenics movement, which later became influential especially in Britain, the USA and Germany, was founded by Darwin's cousin, Francis Galton.

Galton was 37 when the *Origin* was published, and it came to him 'almost as a revelation': he concluded that 'the furtherance of human evolution in desirable directions' was an overriding necessity, and that such a program was possible because both physical and mental qualities were inheritable.[8] Galton was already a notable mathematician and an advocate of the use of statistical methods in biology; he later invented the method of identifying people by their fingerprints, and helped to initiate the use of weather maps; he also interested himself in an extraordinary variety of researches, including unconventional ones such as the statistical investigation of the efficacy of petitionary prayer. (He found no evidence that prayer is effective.)

Galton's proposals were based on a feature called civic worth (or, sometimes, natural ability). Although this characteristic was never measured, it was treated as though it could be recorded in the same way as stature or body weight. In Galton's system, the largest group in the population was that of the respectable working class, and their civic worth was near the average for the whole population. Above them were the less numerous skilled workers, small tradesmen and similar persons; and still further up were the much fewer large employers and members of the professions. Below the middle group were the poor and low paid, and below them (again, fortunately, fewer) criminals and paupers. So far we have the numbers of different kinds of men, classified by occupation; but Galton assumed that variation in civic worth was genetically determined. In his time able people could still seriously hold that both criminality and poverty are inherited characters, and that income is an accurate measure of an individual's value to the community.[9]

Galton therefore urged strict measures to prevent the breeding of lunatics, the feeble-minded, persistent criminals and paupers. Measures designed to benefit the poor, such as state education, old-age pensions and free meals in schools, were held to encourage the improvident lower classes to breed more and so to produce an excess of persons of genetically inferior worth.[10] When the Eugenics Education Society (later the Eugenics Society) was founded in London in 1908, Galton became its first president. The policy of the society was 'to promote the fertility of the better types . . . , whilst diminishing the birth-rate amongst those which are inferior'.[11] The words are those of Leonard Darwin (1850–1943), fourth son of Charles Darwin and a later president of the society. Shortly afterwards, a Professorship in Eugenics and a Eugenics Laboratory were founded in London University.

Most members of the Eugenics Society were academics, physicians or lawyers. (Many physicians, especially medical officers of health, on the other hand, were hostile to eugenic ideas.) The eugenists feared that their own class (considered as a hereditary caste) was dying out owing to the higher birth rate of the working class. The eugenics movement has been described as 'a passionate protest against the hard fact that the meek do inherit the earth'.[12] The eugenists, however, were divided about the landed aristocracy: some held that 'blue blood' was accompanied by an instinct for being public spirited and by inherited skill in administration. Others treated the lords with scorn. At one period even socialists, such as G.B. Shaw, Sidney and Beatrice Webb and H.G. Wells, toyed with eugenics, but they did not join in the praise of the professional classes.

Among those who scorned the aristocracy was Galton's most distinguished follower, Karl Pearson – another polymath, and London's first professor of eugenics. Pearson not only produced a massive output of research in mathematics and biology: he also wrote poetry and a play, and published work on history, politics and philosophy.[13] In his time, in England, an inherited title still conferred status and even power on the holder, but Pearson considered aristocracy a sham. Inherited wealth led to idleness, selfish luxury and a weakening of healthy racial instincts; inheritance by eldest sons was disgenic; and the then common custom of lords marrying heiresses led to degradation of the stock. Pearson combined his contempt for aristocracy with a powerful defense of his own class against his 'inferiors'. For him, if an able-bodied man were unemployed, he must be unemployable, and his children would probably be unemployable too.

Another eugenist of academic distinction, George Adami, vice-chancellor of Liverpool University, explained, in 1921, what he meant by the 'true aristocracy'. At the age of 18 every man and woman should be allowed to submit themselves to a test, approved by the state, of physical fitness and intelligence. The best in both measures would be A1, and their achievement published. 'Think', he writes, of the start in the world it would give to a man or a woman to be able to refer to his or her record as belonging to the A1 class . . . Consider the preference the A1 man or woman would have in marriage, how parents before giving their consent would require that he who sought their daughter's hand should produce his Eugenic Society certificate . . .

And he goes on to foreshadow that the results of such tests, accumulated over the years, 'would become the human stud book'.[14] Adami's plan was not adopted. Nor was the proposal of R.A. Fisher, urged until as late as 1943, that family allowances (paid to parents to help them to rear their children) should be at higher rates for the rich than for the poor.[15]

The opinions and practical proposals of the eugenists justify classifying their movement as a political one, in which a group of the upper middle class tried to defend their privileged position, and supported their arguments by an appeal to biological science. In the early part of the twentieth century the poor were breeding much faster than the rich. Hence their Darwinian fitness seemed to be superior to that of the middle class. The eugenists wished to change this.

More important, during much of the period since the birth of Galton, the influence of the richer classes had been threatened by an increasingly powerful and vigorous labor movement[16]; and the wealth of even the modestly rich began to be eroded by taxes used in part on behalf of the health, education and pensions of the poor. The last quarter of the nineteenth century was a period of vigorous political activity. In the words of a historian, 'After 1890, class-war made the basis of politics in Europe.'[17]

The squalor and degradation of the poor in the growing cities and industries of the nineteenth century are familiar features of the history books: the employment of young children for long hours in brutal conditions, and the employment of women – even during pregnancy – in the mines, were among the extreme examples of ruthless exploitation. It may be doubted whether the slaves on the plantations of Alabama were worse off. One response in mid-century was the formation of trade unions – at first illegal and violently suppressed – and, later, of workers' parties. In trade unionism, Britain, the USA and Australia led the world.[18] But political parties representing the interests of industrial workers were founded, in the last decades of the century, in Germany, France, Italy, Belgium, Spain and Britain, in that order.[19] Some of these parties were socialist: that is, they aimed to dispossess the owners of banks, industry and other means of production, and to administer the 'commanding heights of the economy' through the agency of an elected government.

The opponents of labor and of the socialists were divided. In Britain people spoke of the 'hedgers and ditchers'. The latter, the more conservative, favored defending privilege to the last ditch, and regarded an income tax of one shilling and sevenpence in the pound (8 percent) as the end of the world as they knew it. The others, the liberals, in Britain and elsewhere in Europe, made concessions: they legalized trade unions; and in Britain they founded the welfare state, with old-age pensions, health and unemployment insurance and other measures. State schools had been introduced earlier.[20]

Education and intelligence

Hence the British eugenists and their allies failed to prevent the beginnings of the welfare state; but later they did influence policy on education. In the present century Galton's concept of civic worth was replaced by a measure designed to be much more objective and 'scientific', the intelligence quotient, or IQ. When the French psychologist, Alfred Binet (1857–1911), designed the first modern intelligence tests, he wished to identify children who were lagging badly behind the majority, and therefore might benefit from special help. Binet's interest was in environmental action. His work was quickly taken up by eugenists in Britain and the USA, but they presented the tests as measuring an innate or inherited quality, general intelligence. According to them, *whatever the environment*, intellectual development, measured by the IQ, develops inexorably, in each individual, within a narrowly defined range. As we see later, some still take this stand today.

The concept of the intelligence quotient has had an extraordinary influence on ideas about learning and teaching. Its limitations have also been well analyzed, but are still sometimes disregarded.[22] The tests were designed primarily to show how well a child could cope with reading and writing his or her own language, and with arithmetic. The standard was (and still is) the average performance in the child's own population: is the child's score above or below the average for the child's age and, if so, by how much? Assuming that a child continues with conventional study after the test, the IQ gives some guidance on the academic ability the child is likely to show, at least during the next year or two.

Three common misunderstandings about the IQ are important. First is the long-standing presumption that it remains the same from year to year. Such constancy is not only *a priori* unlikely: it is in fact not found. IQ can change substantially (say, by 20 points) during childhood. Changes are greatest among the youngest children, but occur at all ages. We have nothing like a complete account of how the environment influences intellectual development; but, as the studies of fostered children show, all aspects of school performance are affected by upbringing (Chapter 7). Moreover, as we see further in Chapter 10, the development of brain function depends in intricate ways on stimulation from the environment.

Intellect, therefore, is no more fixed at conception than is health. The tremendous advances in medicine, especially public health, of the nineteenth and twentieth centuries, were due to those who rejected the belief (formerly widespread) that the urban poor were stunted and disease-ridden by nature or heredity. Ignorance and lack of skill should similarly not be thought of as beyond remedy.

The second misunderstanding concerns the heritability of intelligence quotient. Much has been made of the fact that the heritability of IQ is high. But, as we know, heritability varies with the environment: a feature may have a high heritability, yet change greatly when the environment alters (page 106). As an example, the heritability of stature in most populations is very high; but a new environment can produce an astonishing change in height in one or two generations.[23] The children of Japanese emigrants to the USA are much taller than children of families that remain in Japan: for boys of 15 years, the difference may average about 10 centimeters. Such changes are not confined to emigrants. In European countries during the early decades of the century adult stature rose steadily; and, more recently, maximum height has been reached at an earlier age. The causes may include better diet and less infectious illness.

Still more enigmatic changes have occurred in intelligence quotient.[24] They make an odd story. In the 1920s and later, British eugenists raised the alarm about the intellectual state of the country: intelligence, they said, was inherited; the poor were more fertile but less intelligent than the rich; hence the nation was losing intelligence genes and the national intelligence was declining. Their hypothesis was extensively tested. In 1932 and 1947 nearly all Scottish school children aged 11 answered tests of IQ; and, contrary to expectation, the later cohort scored *higher*. Similar rises have been recorded in England and other countries. The largest was in the USA: during the period 1932 to 1978, performance by children on standard tests rose by about 14 points.

Fears about intellectual decline have not, however, been calmed. The recent decline in adult literacy in the USA (page 112) has caused more, and more justified, concern than the supposed loss of intelligence genes. Should hereditarians therefore return to the notion of declining national intelligence? Scepticism about the genetical interpretation of the IQ, and about the IQ itself, would be more appropriate. As the reader may have gathered, we know nothing of practical use about the genetics of intelligence. Figures of heritability are in any case irrelevant for social action: we live in changing environments in which ability is enhanced by good health and schooling – and impaired by television. Sources of variation in the environment are all-important.

The third misunderstanding concerns the scope of a series of tests. If a universal human quality, general intelligence, is being measured, the tests should apply evenhandedly to all human types, from Aleuts to Zulus. The

Wechsler test, one of the most used, has 11 sections of which, writes Jerome Kagan, all 'use words, pictures, and materials that are more familiar to middle-class, white Americans than to any other . . . group.'[25] Questions include: Who wrote *The Iliad*? What is the Koran? What does *plagiarize* mean? Kagan also gives examples from the Dove test, 'selected to be familiar to urban, poor blacks': What is a gashead? What does 'handkerchief head' mean? Whom did 'Stagger Lee' kill in the famous blues legend? (I should be rated as severely retarded by this test.) Recall also the work of A.R. Luria, described in Chapter 7, who found great differences in style of reasoning between unlettered peasants and those who had had some years of schooling. There is no ground for thinking that the differences observed by Luria were genetically determined; but they would certainly have influenced the ability to respond to intelligence tests.

The tests are, in fact, not 'culture free'; nor do they assess a mysterious, internal quality which exists and grows regardless of experience. A person's score depends largely on the kind of information to which he or she has been exposed.

Binet designed his system as an objective and reliable way of assessing how well individual children were doing at school, in comparison with others of the same age. He did not consider that a single quotient could adequately sum up so complex a quality as intelligence. Nor did he regard IQ as a measure suitable, by itself, for ranking schoolchildren. His method can still help teachers to decide what kind of teaching will benefit particular children; but later versions of his tests have been used to make misleading comparisons of kinds of people, especially poor and rich and (as we see below) groups of different color, nationality or custom. Differences between such groups are presumed to be genetical, and it is inferred that attempts to improve the intellectual development of the poor, or otherwise disadvantaged, cannot succeed. The evidence against this belief, some of which is given in Chapter 7, is now overwhelming. Yet it is still influential. As late as 1978, a British government report rejected much of the evidence put before its authors, and emphasized the 'genetic factor' as an important cause of poverty.[26]

For a long time it was assumed, not only that the IQ is genetically fixed, but also that it is closely related to earned income. Eventually, in the early 1970s, the difficult and laborious task of studying the relationship was attempted.[27] The general question asked was what factors in a child's upbringing could be identified as contributing to 'success' – that is, to earnings. In the USA, a significant factor proved to be the kind of family in which a child is reared. Parents tend, not always intentionally, to bias their children toward life in their own social class. The family 'background' (which, to a child, is more than a background) has a substantial influence on a child's progress at school; and this influence is independent of IQ or any other measure of intellect; it also has a measurable effect on income. A second significant factor is duration of schooling: the longer it is – again regardless of 'intelligence' – the greater a person's income is likely to be. The third factor is intellect, as measured by the IQ; but, to quote Christopher Jencks, 'the genes that influence IQ scores appear to have relatively little influence on income.'[28]

The dominating influence of the intelligence quotient is therefore doubly paradoxical: the IQ does not measure what it has been supposed to measure; and it is not highly correlated with the most usual index of achievement. Some factors, important for worldly success, are evidently quite independent of what is measured by intelligence tests. Among them may be traits, difficult to assess numerically but often easily recognized, such as drive, assertiveness and ambition. Noam Chomsky speculates bitterly that 'wealth and power tend to accrue to those who are ruthless, cunning, avaricious, self-seeking, lacking in sympathy and compassion, subservient to authority and willing to abandon principle for material gain'; and that the heritability of these traits might be as great as that of the intelligence quotient.[29]

Even as an index of skills, the IQ has severe limitations. Tests of the kind designed by Binet and his successors do not measure originality (or creativity), or how easily a new skill or idea is learned. These are serious deficiencies. Correspondingly, intelligence testing gives us no insight into how we or our children think, or how we can *improve* performance in intellectual tasks. In fact, no system of measuring abilities is both soundly based and generally accepted. Some psychologists, with good reason, recommend that tests of academic promise should put more emphasis on logical skill, less on vocabulary, and that reasoning, memory and other qualities should be estimated separately. One person may have a superb memory but be weak on reasoning, another may be the opposite, and so on. But such proposals are still far from matching the range of human accomplishments. Among qualities found to some degree in nearly all human beings are musicality, muscular skills (including those of dancing) and, above all, sensitivity to the feelings of others.[30] But these, as we see later, are features of *Homo sapiens*, not of *H. egoisticus*.

The American scene

The idea of intelligence as genetically fixed has had even more influence in the USA than in Britain; for, when the eugenics movement spread, it took hold especially in North America.[31]

American eugenic pronouncements include accounts of two families, the Jukes and the Kallikaks, of which the second has an invented name. The histories of both have appeared repeatedly in supposedly serious publications. In both, an initial pairing is described as giving rise, solely because of bad heredity, to a line of degenerate, defective or criminal descendents.

The story of the Kallikaks states not only that the union of a soldier with a defective girl had this outcome, but that the marriage of the same man to a respectable young woman founded a family of upright, hard-working citizens. In each lineage, it was supposed, the genes of the soldier's consort determined the qualities of the issue unto the third and fourth generations (or later). The methods by which worth was assessed did not, however, bear examination.[32] The story of the Jukes also illustrates the tactics of the eugenists. R.L. Dugdale, who first studied this family, put much emphasis on their adverse environment. Only later, when eugenists were energetically searching for such pedigrees, was heredity held out as all-important.[33]

The scope of bad heredity, in the estimation of American eugenists, is indicated in a measure passed by the Iowan legislature in 1913: it lists 'criminals, rapists, idiots, feeble-minded, imbeciles, lunatics, drunkards, drug fiends, epileptics, syphilitics, moral and sexual perverts, and diseased and degenerate persons'.[34] At this time complex character traits were said authoritatively to be 'inherited' in a simple, 'Mendelian' manner. C.B. Davenport, a geneticist and the leading figure in American eugenics, not only attributed membership of the poorer classes to inferior genes: he also stated that violent temper was a dominant condition; that alcoholism, laziness and truancy were similarly due to single genes; and that drapetomania, or the tendency to run away (with its high incidence among slaves), was a simple inherited character. Desirable traits, also said to be genetically simple, included musical ability, inventiveness and thalassomania (the tendency to run away to sea – appropriate for sailors).[35]

In 1905 American universities began to teach eugenics to undergraduates: in 1914 there were 44 such courses; it is estimated that by 1928 seventy-five percent of tertiary institutions were presenting eugenics in some form. The influence of this teaching was reflected in legislation: during the first third of the century, 24 states passed laws under which people held to be social misfits were sterilized. In the same period, about 20 000 underwent the operation. By the end of 1958 the number had risen to about 61 000.[36]

Initially, a number of leading American geneticists supported eugenics but, after 1920, they began to withdraw. It was becoming evident that few significant human traits are genetically simple. More important, the interaction of heredity and environment began to be analyzed, and the scope of environmentally caused variation was becoming clearer. It came to be realized, at least by the clearest-minded geneticists, that to talk of a gene 'for' a character is always likely to be misleading: the effect of a gene, or set of genes, may vary a great deal according to the environment. Hence a measure such as the intelligence quotient should never be treated as an estimate of *innate* worth.

There were also serious moral and legal problems. As M.H. Haller writes, 'The enthusiasm of some eugenists for sterilization was matched by the fear of others that neither existing knowledge, common decency or public opinion justified such laws.'[37] Among the most formidable attacks on eugenic programs was that of J.B.S. Haldane, published in 1938. He presented the relevant genetics (summarized in Chapter 7), and also commented on the way in which sterilization laws were used.[38] Even where sterilization is voluntary in law, it may be difficult to guarantee that it is so in fact. Haldane cites the case of an agricultural worker, John Hill, in the State of Washington, who stole some hams to feed his wife and five children, all of whom were described by the judge who tried the case as mentally subnormal. Sterilization was suggested and explained to Hill, and he accepted it. Haldane writes:

> We are not told whether the consent would have been obtained so readily had the suggestion been made by a man who had not the power to send Hill to prison for fifteen years [for grand larceny]. Nor is it clear what tests were employed to detect the mental subnormality of the Hill family. Some quite intelligent people do not appear at their best in a criminal court.

Haldane also quotes another judgment, on a man named McCauley, sentenced to compulsory sterilization by the same judge.

> 'This man . . . is subnormal mentally and has every appearance and indication of immorality. He has a strain of negro blood in his veins, and has a disgusting and lustful appearance.'

Haldane comments that, clearly,

> Hill would not have been sterilized had he possessed an independent income. And it is unlikely that McCauley would have been, had . . . his appearance [been] more in conformity with Judge Holden's aesthetic standards. In my own judgement at least one well-known cinema star 'has a disgusting and lustful appearance', but I claim no scientific basis for this opinion, nor do I suggest that it be used as a basis for eugenical sterilization.[39]

As the scientific and moral defects of eugenical ideas became clear, support for them continued to decline. In Haller's historical survey, published in 1963, the eugenics movement is described as almost without influence.[40] But, six years later, an academic incident, minor in itself, revived the old arguments. An American university journal published a long article by a psychologist, A.R. Jensen. At that time special efforts were under way, in the tradition of Binet, to help children who seemed to have been handicapped by the poverty of their environment. Jensen's argument implies that such efforts must fail: intelligence, he holds, depends on 'intelligence genes', and is therefore largely determined by the genotype acquired at conception.[41]

Later, in a lengthy treatise, Jensen raises his concept of intelligence almost to the status of a transcendental principle – just as others have treated natural selection, the gene and social dominance. He gives a picture of a scale of being, in which the animal kingdom appears as an array of types, from *Ameba* to extra-terrestrials, each given its place according to its intelligence quotient.[42] The idea of an intelligence quotient for *Ameba*, a jellyfish or even a flatworm, let alone a visitor from outer space (Dr Who?), sounds like a satirist's attempt to reduce the notion of the IQ to absurdity. In any case, the various abilities we find in the animal kingdom cannot be described in terms of a single quality, called intelligence, adaptability or anything else: each species, or group of species, has its typical range of abilities; these are related, like other traits, to the species' particular mode of life.[43] A honey bee can learn to fly to one color rather than another; but a blowfly – also a marvel of insect organization – has no such ability. An octopus has learning abilities that resemble those of mammals but, unlike mammals, cannot learn to distinguish objects by their weights. A cat, that hunts at night, can easily learn to respond 'intelligently' to sounds, but a monkey, which is active in daylight (and hears well), cannot. Much more important, human beings have a number of kinds of ability (or intelligence), of which only one is (imperfectly) measured by intelligence tests.

Jensen's conclusions were originally founded on the evidence, from studies of twins, of the high heritability of IQ. Much of this work was published by Cyril Burt, a professor of psychology in London University. For many years Burt was influential both as a teacher and as a public figure.[44] In 1938 he advised a British government committee that the tests measured all-round intellectual ability, and that this quality is innate. Partly as a result, children in British schools were segregated, at the age of 11, into separate schools according to their test scores.[45] As late as 1971, Burt contributed a chapter to a volume on conservative educational policy, in which he retains references to innate ability.[46]

Since Burt's death, his work on twins has been shown to have been faked. This notorious scandal has, however, merely complicated the debate that Jensen began: the 'hereditarian' position of Galton, Pearson, Burt and their successors rests not on fraud but on errors. The errors arise partly from neglect of the principles of genetics, especially the fact that variation has environmental causes (Chapter 7), and partly from the limitations of the intelligence quotient. Yet, despite these limitations and much protest, traditional 'intelligence tests' retain much of their hold. In 1985 an American critic of the tests writes that the USA 'remains in the unrelenting grip of the IQ-testers'; and he remarks that testing is part of a multi-million-dollar industry.[47] The IQ is therefore still important in relation to educational policy. Attitudes to the IQ may determine presumptions about the educability of the poor and, above all, of members of different races.

Preserving the breed: Britain and the USA

For the early eugenists were not only conscious of belonging to a privileged class: they were also confident that they were members of a superior race. Here is Karl Pearson again:

> It is a false view of human solidarity, a weak humanitarianism, not a true humanism, which regrets that a capable and stalwart race of white men should replace a dark-skinned tribe which can neither utilise its land for the full benefit of mankind, nor contribute its quota to the common stock of human knowledge. . . . There is cause for human satisfaction in the replacement of the aborigines throughout America and Australia by white races of far higher civilisation.[48]

But Pearson also remarks, in a footnote, that his views 'must not be taken to justify a brutalizing destruction of human life'. The 'lesser breeds' might be incapable of benefiting from higher education, but white men should treat them kindly.

A genetical account of psychological differences between races had already been forcefully put forward by Galton.[49] The special temperaments of Hindus, Arabs, Mongols, Teutons and others, he held, are passed on, from generation to generation, in the same way as structural features. The traits typical of North American Whites, too, were a product of natural selection: these people, bred from 'restless and combative' Europeans, were 'enterprising, defiant, and touchy; impatient of authority; . . . very tolerant of fraud and violence; possessing much high and generous spirit, . . . but strongly addicted to cant.'[50] Members of savage races, such as Negroes, retain an innate, untameable quality even when reared by Whites, and often revert to their natural, wild ways.

Attitudes of this kind were accepted among the most influential and even humane persons at the time, and remained prominent for decades. They appear in the learned journals, where we meet, once again, the idea of the recapitulation of ancestral characters. In 1866, the following appeared in the *Anthropological Review*:

> As the type of the Negro is foetal, so that of the Mongol is infantile. And in strict accordance with this, we find that their government, literature and art are infantile also. They are beardless children whose life is a task and whose chief virtue consists in unquestioning obedience.[51]

Ethnology (which later became social anthropology) consisted principally of accounts of 'primitive' or 'backward' groups set in an evolutionary framework; gatherer–hunters and people living by simple agriculture were held to represent recent stages in the evolution of the most advanced races; the latter were represented by Pearson and his stalwart friends.

The imperial outlook was reflected in school history books, with their maps in which the British Empire – a 'quarter of the world' – was shown in red. It was also present in anthropological teaching. Undergraduates of the University of London, in the first decade of this century, were given examination questions such as this: "What are the principles that should guide the investigation of mental process in races of low culture? Consider how far these correspond with those appropriate to the investigation of child psychology.'[52] (Compare page 59.)

There was a marked discrepancy between the principles put forward by the most influential workers in this field and their practise. Pearson, like Galton before him and his successors afterwards, emphasized the need for accurate measurement and exact, scientific analysis.[53] Yet, as we know, 'civic worth' was never measured. Similarly, statements on racial differences were no more scientific than the later ravings of Hitler and his Nazis. The eugenics movement adopted opposition to 'miscegenation', or the marriage of members of different races, as part of its policy, without any scientific analysis of its effects; and a 'eugenic' campaign against Jews was launched in advance of study of Jewish characteristics. The opposition to Jewish immigrants in Britain had one success, in the Aliens Act, passed by Parliament in 1905. Twenty years later a report on the health and abilities of Jewish children from Poland and Russia failed to show the expected lower intelligence. (This finding was itself worthless by modern standards, for the tests used were not based on the careful methods of Binet.) Pearson, however, was able to state that Jews were dirtier and physically feebler than Gentiles.[54]

In the present chapter, and elsewhere in this book, we see how supporters of certain political and social doctrines have grasped at biological findings to hold up their ideas. The persecution of Jews, a shameful tradition of Christendom, was one of the longstanding customs given support (by false argument) from biological science. Attitudes to Jews also, however, illustrate confusion in eugenic thinking. G.R. Searle writes: 'As even Pearson admitted, the Jews had much to offer other races in the matters of sex hygiene and race culture. The eugenists especially admired the pride that Jews took in their family and their ancestry . . . In many respects the Jews were a model of what eugenists were seeking.' Some eugenists regarded the Jewish community as evidence of the correctness of their 'science'.[55]

As usual, a blind eye was turned to cultural or, more generally, environmental influences. In the first centuries of the Roman Empire, Jews spread throughout the world. Since then, Jewish communities have kept to their religious and other traditions, but Jews have not retained a uniform appearance. In each country, they tend to resemble the non-Jewish inhabitants: Chinese Jews are Chinese in appearance; Abyssinian Jews resemble other Abyssinians; Indian Jews are much more like their southern Indian neighbors than (say) the descendants of exiles from Polish or Russian ghettos. In the last case genetical differences in the structural traits of a small population evidently resulted from intermarriage, but cultural features persisted.[56] (The latter can, of course, change quickly. For nearly two millennia, Jews have lacked a military tradition; but, now that they once again have a country, they also have a formidable army.)

Anti-semitism was also a component in American hostility to immigrants.[57] The complaints concerning Jews were of their low intelligence combined with membership of radical labor unions. A demand for a selective immigration policy was met by the Immigration Act of 1924. This Act was founded partly on the evidence of academic psychologists that there were marked differences in the character of people of different nationality, and that these differences were 'inherited'. It restricted not only the total number of immigrants, but also regulated their national origins. The objective was to maintain 'racial homogeneity' and a basic stock derived from the 'Nordic race'.[58] The characteristics of this race differed from those attributed to Americans by Galton: Nordics were 'rulers, organizers, and aristocrats . . . individualistic, self-reliant, and jealous of their personal freedom'.[59]

H.H. Goddard, the author of the story of the Kallikaks, had in 1912 reported an incidence of feeblemindedness of 80 percent or more among Russian, Hungarian and Jewish immigrants; Italians scored 79 percent. (Many of these people knew little or no English; all had just finished a long and exhausting voyage, and were tested in strange and confusing surroundings.) As a result, many aliens were deported on the grounds of mental deficiency.

The Eugenics Research Association welcomed the curb on immigration. The members were alarmed at what they saw as an invasion by a degenerate horde of unfit, high-grade imbeciles and persons from the Slav and Latin races with a marked propensity for pauperism and sex offences; 'they think with the spinal cord rather than with the brain'.[60] Such expressions were used during the hearings, in 1923, of an American government committee on immigration. The journal of the Eugenics Research Association was edited by H.H. Laughlin, a biologist. 'Humorless and dogmatic', writes K.M. Ludmerer, 'he pursued eugenics as if it were a . . . crusade.'[61] Laughlin was a leading worker for sterilization laws (of which the successes are described in the previous section). His own list of the virtues especially prized by 'American stock' was presented to the government committee in 1924; it included 'truth-loving, inventiveness, industry, common sense, love of beauty, responsibility'. These, he stated, are all items 'of a biological order', and could be studied by the methods of biological science. No scientific findings on them were, however, presented.[62]

Such findings came from different sources. While American geneticists were withdrawing their support without public fuss, social scientists forcefully criticized the dogmas of the eugenists. In particular, anthropologists, led by Franz Boas (1858–1942), argued against preoccupation with evolution and genetics.[63] More important, evidence was produced. In 1935 Otto Klineberg summed up authentic knowledge on race differences.[64]

The first important findings were from the intelligence tests given to army recruits during the First World War. Within each state the Negro average was always lower than the White. But there were differences between states; and these were so great that Negroes from some northern states scored *higher* than Whites from the south. Since the southern states were backward in educational matters, there was evidently an environmental effect arising from variation in schooling. The lower average of Negroes in each state might, it follows, be due to the fact that nearly everywhere in the USA Negroes are economically and socially inferior to Whites, and have correspondingly inferior facilities for their children. In some places, Negroes reached White average levels, or even surpassed them. In 1923, 500 Negro children tested in Los Angeles had a mean IQ of 104.7, a little above that of the Whites. Klineberg gives much additional information, critically analyzed, and sums up: 'as the environment of the Negro approximates more and more closely that of the White, his inferiority tends to disappear.'[64]

As we know, the intelligence quotient has many limitations even as a measure of intellectual abilities. Few other important traits have been measured on a similar scale. Crime has been prominent in eugenic propaganda, but statistics of crime are difficult to interpret.[65] For instance, attempts have been made to compare the criminality of Jews with that of non-Jews. In Germany and other European countries, during 1883 to 1916, Jews were less likely than others to commit crimes of violence, sexual crimes, theft or embezzlement; murder and theft, in particular, were about three times as common among non-Jews. But Jews had a worse record for fraud and forgery. Does this finding signify that Jews are 'naturally' dishonest? In fact it gives us no valid evidence on this question: an exceptionally high proportion of Jews were shopkeepers or employed in business, and correspondingly few in manual work. Manual workers, unlike people in business, have little opportunity for fraud or forgery.

There is, in fact, no good evidence for 'racial' or genetical differences in criminal tendency between large groups. On the other hand, nationality, and hence upbringing, has a marked effect. In Chapter 5 I give examples of its effect on violence. To take an extreme case, what in most communities would be condemned as murder is in a few regarded as virtuous and necessary; in Sicily and parts of India vendetta has been a normal element in social life.

> The wildest dreams of Kew are the facts of Katmandu,
> And the crimes of Clapham chaste in Martaban.

But families which have moved to a new environment may rapidly change their ways. During the late nineteenth century, crime rates in the USA declined steeply. As a result, early in the present century Italian and Irish immigrants had a much higher incidence of convictions for homicide than had Whites born in America; but the homicide rates among their children resembled the American norm. In the same period, Irish immigrants were seldom convicted of rape or gambling, but again their children resembled the rest of the white population.[66] By the 1960s, however, national patterns had changed. The incidence of murder in the USA was twice that in India, five times that in Australia, and ten times that in France and Spain. Such differences between nations, and changes in a few decades, cannot be genetically caused. The cultural causes may range from unemployment to inadequate schooling. The formidable task of identifying them is only obstructed by preoccupation with genetics.

Preserving the breed: Germany

The cult of race had its supreme expression in Germany under the rule of Hitler. The pseudo-Darwinian origins of this movement have been traced in detail. The story begins with E.H. Haeckel (1834–1919).[67] Ernst Haeckel made massive contributions to embryology and to the systematics of invertebrate animals. He was also a dedicated Darwinist, an atheist and a romantic. He thought of evolution as driven by a cosmic force which he discussed with religious passion. He thus managed to combine an emphasis on scientific methods with a revolt against reason. Despite this confusion, Haeckel had a vast influence, especially through his evolutionary philosophy, called monism, which rejected the conventional separation of mind (or spirit) from matter. His major work on this theme, *The Riddle of the Universe*, first published in 1899, had a sale of 100 000 in its first year. It went through many editions; by 1933 half a million had been sold in Germany alone.

Like more recent pseudoscientific works, *The Riddle of the Universe* succeeded despite severe criticism by scholars. Daniel Gasman writes that even

> members of the academic and intellectual community . . . became deeply attached to Haeckel and to his ideas. Indeed, it is not really possible to understand Haeckel and [his] role in German intellectual life without being aware of the magnetic hold which he exerted on his adherents. He was regarded by them as a singular religious and national prophet.[68]

In 1905 a Society for Race Hygiene was set up – a German precursor of the Eugenics Society in England. And a year later the Monist League was founded. As Gasman remarks, at first, on the surface, the Monist League favored science, optimism and progress; but there was also a component of 'moody pessimism': Germany was represented as heading toward cultural disaster.[69] The members of the League therefore designed an all-embracing social and political program for their country, founded on their interpretation of Darwinism. Here is Haeckel's summary of their position.

> The theory of selection teaches that in human life, and in animal and plant life, . . . only a small and chosen minority can exist and flourish . . . The cruel and merciless struggle for existence which rages throughout living nature, and in the course of nature *must* rage, . . . is an incontestable fact. . . . This principle of selection is as far as possible from democratic; on the contrary it is aristocratic in the strictest sense of the word.[70]

A central presumption was the superiority of the White, Germanic races, or 'Aryans': these were held to display the greatest difference from our ape-like ancestors. Negroes were considered incapable of higher intellectual development or of becoming civilized. Other inferiors included the Japanese. (Later, the Nazis were obliged to make their Japanese allies honorary Aryans.) Haeckel and many members of the Monist League were also strongly anti-semitic, and urged the disappearance of the Jewish nation. (Jesus, however, was stated to be only half Jewish: his father, they said, was a Roman who had seduced Mary.[71])

Like eugenists elsewhere, the Monist League held crime and much illness to be genetically determined. Alcoholics, beggars and similar undesirables were to be eliminated, but superior types were to breed more so that Germany should be great. Hence women were to be liberated, but only so that they could be free to have children and be good mothers.[72]

The program of the Haeckelians was reinforced by the philosophical movement represented by Oswald Spengler, the unmitigated pessimist who nonetheless had a large following, especially in the literary world (Chapter 2). Spengler's gloomy *Weltangst*, or world anguish, is supported by elaborate imagery, rather than systematic argument.

> A civilization is born at the moment when, out of the primitive psychic conditions of a perpetually infantile humanity, a mighty soul awakes and extricates itself . . . This soul comes to flower on the soil of a country with precise boundaries, to which it remains attached like a plant. Conversely a civilization dies if once this soul has realized the complete sum of its possibilities . . . and thereupon goes back into the primitive psyche from which it originally emerged.[73]

Spengler was obsessed with the desire for power – and with some justification, for he was active during the rise of Italian Fascism and German

Nazism. Some of his views fitted well with the Nazi philosophy: he held, for instance, that the superior whites were threatened by intermarriage with black and yellow races. And in his last book he wrote this.

> Man is beast of prey. I shall say it again and again. All the would-be moralists and social-ethics people who claim or hope to be 'beyond all that' are only beasts of prey with their teeth broken . . . If I call man a beast of prey, which do I insult: man or beast? For remember, the larger beasts of prey are *noble* creatures, perfect of their kind, and without the hypocrisy of human moral due to weakness.[74]

Both the animal metaphor and the incoherence are characteristic.

In the 1920s Spengler tried to become directly influential in politics, but he did not succeed in taking part in public affairs. His importance is that he represents, and helped to create, the climate of opinion in which the Nazis flourished. The doctrines of racial superiority, of genetical purity and of incessant struggle culminated in Hitler's accession to power in Germany in 1933. Here is Hitler, in 1928, echoing – even plagiarising – the Monists and Spengler:

> The idea of struggle is as old as life itself. . . . In this struggle, the stronger, the more able, win, while the less able, the weak, lose. . . . It is not by the principles of humanity that man lives or is able to preserve himself above the criminal world, but solely by means of the most brutal struggle.[75]

Hitler was consistent. A macabre example is given by Albert Speer (1905–1981), Hitler's armaments minister. During the last days of the Second World War (and of Hitler) Speer tried to save something of his country from the wreck. Here is Hitler's response:

> If the war is lost, the people will be lost also. It is not necessary to worry about what the German people will need for survival. On the contrary, it is best for us to destroy even these things. For the nation has proved to be the weaker, and the future belongs solely to the stronger eastern nation.[76]

During the 12 years that ended with the melodrama in the *Führerbunker* in Berlin, the Nazis had put their version of evolutionary biology into practice. In 1933 the Monist League was replaced by the *Ernst Haeckel Gesellschaft*; and this organization was sponsored by the Nazi *Gauleiter* of Thuringia, Fritz Sauckel, who after the war was condemned to death for his crimes.[77] In the same year a sterilization law was enacted, and within 12 months 56 000 persons, supposed to be genetically defective, had been sterilized; the number rose to more than a quarter of a million before the end. In a further step toward ridding Germany of 'undesirables' a decree of 1939 legalized euthanasia. In two years 50 000 people were killed under this ruling. K.M. Ludmerer remarks that this experience was a useful preparation for the murder of millions of others.[78]

At least five million were victims, during the 1940s, of the Nazis' 'Final Solution' – the decision to kill all Jews and to seize their property, including the gold from their teeth and the hair from their heads.[79] It is impossible to measure the contributions of racist and eugenical propaganda to this unimaginable crime; but we know that its scale could have been much reduced but for systematic obstruction of attempts at rescue. Among the accessories, through inaction, to the Nazis' program of mass murder were a Pope, a Mufti of Jerusalem, a British Foreign Secretary, an American President and Secretary of State, and many senior public servants in several countries. A.D. Morse ends a book on this subject by asking whether, after the holocaust, genocide is now unthinkable; or 'are potential victims somewhere in the world going about their business, devoted to their children, aspiring to a better life, unaware of a gathering threat?'[80]

After Hitler
The question asked by Morse is still to the point. Hostility to persons of different color or national origin remains widespread; so does unfounded belief in the inferiority of certain types, or in the need to keep people of different types separate. Those who express such attitudes continue to claim support from biology and psychology. The second half of the present century has indeed seen a revival of racial and other forms of intolerance: a few prominent behavioral scientists have supported a narrow genetical interpretation of differences in ability and social worth; a new group of writers on political and social science has resurrected biological images of their own species; and extremist political movements have made use of such notions, and have even formed or influenced governments.

Among the behavioral scientists, the prolific popularizer, H.J. Eysenck, writes:

> All the evidence to date suggests the . . . overwhelming importance of genetic factors in producing the great variety of intellectual differences which we observe in our culture, and much of the difference observed between certain racial groups.[81]

The book from which this passage is taken has been called 'generally inflammatory' and insulting to 'almost everyone except WASPs and Jews'.[82] In another popular work, Eysenck states that

> the whole course of development of a child's intellectual capabilities is largely laid down genetically, and even extreme environmental changes . . . have little power to alter this development.[83]

As we know, this statement is in conflict with the principles of genetics, especially the interaction of heredity and environment. Nor did it correspond to the facts even when it was written. For practical purposes it implies that remedial education, and all future attempts to improve teaching methods, or other aspects of rearing, are doomed to fail. This is the message that reached a large public. Yet elsewhere in the book many qualifying statements leave a reader in a state of bewilderment. Among them is reiteration of the need for more research. As we know, by 1979 some of the research had been done, and had confirmed the favorable effects of remedial teaching (pages 110–113).

The public impact of the hereditarians, in the 1960s and later, is represented by an article in the popular magazine, *Fortune*. The title is 'The Social Engineers Retreat Under Fire'; and the theme is the criticism by certain writers of 'reformists' among social scientists.[84] This well written account refers to 'ancient propensities . . . inextricably rooted in our genes'; its main message, when shorn of such metaphor, is the usual one of heredity as destiny. The work of A.R. Jensen (described above) is invoked, especially the notion that 80 percent of human intelligence is inherited, and the emphasis on the gap of 15 points of IQ between Negroes and Whites. There is also a reference to disappointment with 'social engineering' – that is, with attempts to help disadvantaged people (especially black Americans) to do better at school and college.

Jensen and his views have had extensive treatment in other widely read journals, including *Life* magazine, and *The New York Times Magazine*. *Newsweek* has printed an article entitled, 'Is Intelligence Raised?' and *Time* one on 'What the Schools Cannot Do'.[85] These are more than ephemeral journalism: they represent a significant trend in the politics of the late twentieth century. Jensen's original article was read into the record of the United States Congress, and was distributed to members of the government. At the beginning of 1970 President Nixon reported on compensatory education programs and, in the words of a journalist, gave them 'almost straight F's for failure, with only a desultory and lower case e for effort'.[86]

There are, however, no longer societies of leading academics dedicated to asserting the genetical superiority of their own group:[87] in the late twentieth century, prejudice and discrimination have rarely been promoted by anthropologists or mathematicians. But they remain an important element in the outlook of neo-Fascist parties and of the more significant movement sometimes called the New Right. Even European countries with the strongest democratic traditions have small Fascist parties.[88] They are systematically anti-intellectual in most of their activities, and they promote the narrowest forms of nationalism, persecution of minorities and obedience to the state. They assume the superiority of the 'white race' and are often anti-semitic. In the USA the most prominent of such movements is the John Birch Society. Some publish distortions of neo-Darwinian theories about humanity. Sociobiology, their members are told, shows human society to be founded on genetical instincts of dominance, aggression, territoriality and, above all, the racial impulse of in-group solidarity.[89]

Parties that propagate such views are often regarded as beneath contempt. One should, however, remember that, until 1933, Hitler was commonly dismissed as a mere mountebank. To quote a work on anti-semitism, 'it is a great mistake to suppose that the only writers who matter are those whom the educated in their saner moments can take seriously.'[90] This warning, published in 1967, had become still more appropriate two decades later, when 'neo-conservatives' had published longer and more elegantly worded works on similar themes. In their pronouncements on race we find two major elements: assertion of the inferiority of certain types; and a belief in the need to keep different types separate, sometimes coupled with a mystical attitude to 'blood'.[91]

On inferiority, the most prominent recent example (described above) is the campaign in the USA to present persons with dark skins as less intelligent than those with pale ones. The idea of genetically fixed inferiority has, however, been so belied that few prominent people now state in public that they accept it. More emphasis is put on the supposed need of different groups, racial or national, to be separated. Such notions are not confined to South African Whites. In a collection of conservative essays, a British Member of Parliament warns of impending 'social tension and antagonisms' which, he holds, inevitably result from the immigration of people with dark skins. In the same collection, a prominent journalist states that 'massed immigration' is 'potentially destructive' and is being used by people 'openly committed to overturning our political and social arrangements' but does not say who they are; and another complains of 'modish hypocrisies' which ignore 'the instincts of the nation'.[92] Similarly, a conservative political scientist writes of 'natural prejudice' in an apparent attempt to condone hostility to persons of different appearance or custom.[93]

The doctrines of racial separation and of 'instinctive' intolerance are put foward without regard to the findings of history or of social science. The implication of inevitable antagonism between groups that differ in physical type or in custom does not fit the facts, for co-operation, too, occurs on a large scale. Today, in countries such as Brazil, despite a history of European conquest, diverse types intermarry and live and work together without regard to 'blood' ties. A traveller sums up the Brazilian scene when he describes 'a blond Negro talking Spanish to a red-headed Chinese'.

Justified concern with intolerant behavior should lead to questions on its causes. Some are obvious. Wars of conquest may end with the defeated under the rule of a victorious minority; and the losers may become slaves, serfs or 'wage slaves'. Today, slavery in its traditional forms hardly exists and revolts by slaves no longer occur: a source of discord has been removed. Others could be.

References to 'instinctive' xenophobia imply that such attitudes are fixed; but in fact, like the violent and pacific forms of conduct described in Chapter 5, they are much influenced by what people are taught, especially in early life. As a result, even in communities in which intolerance is common, some people are free from it. They accept the doctrine of humanity as a single family which has been prescribed for at least two millennia. In the first century AD, Paul of Tarsus urged on the Collossians forbearance and charity: 'there is neither Greek nor Jew, circumcision nor uncircumcision, Barbarian, Scythian, bond nor free'.[94] When widely read writers announce an opposite message, they are helping to create the social divisions of which they complain.

Social Darwinism and the sexes

Of all existing inequalities, beyond even those of race and class, the status of the sexes leads in importance; and it is a subject on which prominent Darwinists, neo-Darwinists and neo-conservatives have much to say.

In its application to ideas about the sexes, Darwinism was caught up in a complex historical process in which religions have played a leading part. Darwin himself grew up in a society in which Christian doctrine was still influential. The celebrated outburst of the Scottish theologian, John Knox (about 1514–1572), on the monstrous regiment of women, represents a persistent Christian theme. It contains the following passage:

> To promote a woman to beare rule, superioritie, dominion or empire aboue any realme, nation, or citie, is repugnat to nature, cotumelie to God, a thing most contrarious to His reveled will and approved ordinace, and finallie it is the subversion of good order, of all equitie and justice.[95]

This is perhaps an extreme case; but Christianity and other major religions, strict or lax, monotheistic or not, have required a rather uniform set of virtues in married women: these include chastity, gentleness, orderliness and skill in cooking and other household tasks, but not intelligence or learning.[96] A woman was sufficiently learned if she could distinguish her husband's bed from another's.[97] Subservience to the husband is always implied. Barbara Tuchman, in a vivid account of fourteenth-century Europe, remarks on the common male presumption that a woman is either a scold or a shrew. She quotes Thomas Aquinas (1225–1274), the most influential Christian theologian after Augustine, on woman as by nature subject to man, for in man reason predominates. The upbringing of a child must be guided principally by the father, not the mother. 'That women reacted shrewishly in the age of Aquinas was hardly surprising.' And 'perhaps scolding was her only recourse against subjection'.[98]

Religious exhortation has been reinforced by the enactments and pronouncements of lawyers. William Blackstone's celebrated *Commentaries on the Laws of England* (1765) has influenced the legal status of British, American and other women to our own day.[99] According to Blackstone, marriage suspends the existence of a woman as a legally recognized person: in law, husband and wife are one. The husband is obliged to provide for his wife, and to pay her debts. The law is held out as protecting and benefiting the wife; but she is always a subject of her husband. It is doubtful whether a woman under Blackstone's system was better off than she would have been in Greece of the fourth century BC, when Aristotle likened the position of married women to that of slaves.[100]

Hence Darwin's theories of evolution were worked out in a society in which wives were, both in custom and in law, subject to their husband's authority. In Britain, on marriage, a woman's property became the property of her husband. This law was changed only in the year of Darwin's death. Conventional attitudes were supported by bizarre beliefs. Educated people, especially medical men, believed the brain of a woman to be not only smaller than that of a man, but also structurally and functionally inferior. In the middle of the nineteenth century the frontal lobes were described as the centers of intelligence, and so the female frontal lobes were said to be smaller in proportion than those of the male. Later, it was the turn of the parietal lobes to house intelligence, and the female parietal lobes were stated to be underdeveloped.[101] (In fact, one cannot identify sex from brain structure.) Accordingly, in the late nineteenth century, illness among the few women university students was regularly attributed to excessive brain work; and women were said to be too weak to stand the strain of examinations.[102]

This was part of the social environment in which social Darwinism was founded. The founder was Herbert Spencer, the 'mediocre thinker whose influence was . . . greater than any other anywhere in the world'.[103] Spencer had no conventional schooling, but he was a competent mathematician; and he had a practical training as a civil engineer for a railway company. He first became interested in evolution when he saw fossils taken from railway cuttings. Later, his attitude toward evolution became one of almost religious fervor: for him, our systems of morals are a product of evolution, and they owe their existence to their survival value.[104] Spencer's influence was, however, above all in social science: in the USA, in the last decades of the nineteenth century, his works, especially his *Principles of Sociology*, were sold in hundreds of thousands.[105]

Spencer, not Darwin, was the originator of that notorious tautology, the survival of the fittest.[106] Here he is in 1876:

> Not simply do we see that in the competition among individuals
> of the same kind, survival of the fittest has . . . furthered production
> of a higher type; but . . . to the unceasing warfare between species
> is mainly due both growth and organization. Without universal
> conflict there would have been no development of the active pow-
> ers.[107]

His confidence in the beneficial effect of the cosmic process, or of evolution
resulting from conflict, was reflected in his recommendations on social
policy. He opposed state schools for all on the ground that the poor were
innately ineducable. Health services were similarly pernicious: they preserved the unfit, and weakened the race. Spencer's outlook had much in
common with that of Galton and his followers; but Spencer did not advocate
eugenic breeding, for he believed in the 'Lamarckian' transmission of 'acquired' characters.

Spencer's writings, justly regarded as forerunning sociobiology, are
explicit on the social roles and abilities of women.[108] Compared with men,
women were more emotional, more childish, more liable to depression and,
of course, less intellectual and imaginative. They were, however, superior
to men in their intuitive powers, in compassion and in gentility. For about
a quarter of a month, when not pregnant, they were ill and fit only for
light duties. Like many others, Spencer expressed concern at the excessive
use of the brain by intellectual women, especially those of the upper class.

The argument against education for women was held out as scientific
and founded on biology and physiology. Yet most of those who used it
avoided an obvious implication: women of all classes have monthly periods,
and so should be released from all arduous work for about one week in
every four. Such a policy, applied to a whole population, would have caused
a social upheaval, in which disruption of the domestic arrangements of the
rich would have been prominent. Some anti-feminists were open about
their concern for the upper classes: education for superior women would,
they feared, have a sterilizing effect and result in an excessive proportion
of children from the lower strata.

Spencer himself was compassionate and, if not genteel, at least gentle.
He pointed out that women even in civilized countries had commonly been
brutally treated, but he held that this custom should be honored in the
breach.[109] A few years after his death British women who had demonstrated
in favor of votes for women were put in prison, where some were violently
assaulted, with permanent damage to their health.[110] Probably, Spencer
would have been horrified at this conduct. He would, however, also be
startled, if not dismayed, at the changes that have taken place in this
century. Reay Tannahill quotes an American senator speaking in 1866:

> When the women of this country come to be sailors and soldiers,
> when they come to navigate the ocean and to follow the plow;

> when they love to be jostled and crowded by all sorts of men in
> the thoroughfares of trade and business; when they love the
> treachery and the turmoil of politics; . . . then it will be time to
> talk about making the women voters.[111]

All these have happened, and women have the vote. But there is still energetic
opposition; and in some countries the liberation of women has not begun.

This is the social context in which neo-Darwinian concepts have been
used, in the late twentieth century, to analyze sex roles. A universal human
nature is described or, rather, two natures, male and female: a man is fated
to be energetically competitive (or aggressive) in the attempt to achieve
high status, and keen on his job; he is also inherently promiscuous, and
dominates women. A woman is compulsively nurturant, home-loving and
(moderately) chaste, and subservient to men. And there is an unending
conflict of interest between the sexes.

The implications of such writings are most clearly shown in the popular
press, as in this passage from an article by E.O. Wilson in the *New York
Times Magazine*:

> In hunter–gatherer societies, men hunt and women stay at home.
> This strong bias persists in most agricultural and industrial societies
> and on that ground alone, appears to have a genetic origin.[112]

(As we know, women gatherer–hunters are in fact not stay-at-homes; and,
even if they were, it does not follow that modern women must be. And
the second sentence, if it has a meaning, is also a *non sequitur*: nearly all
people wear clothes; has clothing oneself therefore a 'genetic origin' and,
if so, in what sense?) Wilson further believes that even if men and women
had the same education, 'men are likely to play a disproportionate role in
political life, business and science'. Social change, in this regard, is presumed
to be impossible.

A psychologist, writing in *She*, goes further: 'most modern biologists'
assume that 'our bodies and brains are transient machines geared to the
preservation and proliferation of our genes'; hence

> some of the traditional differences between men and women, *such
> as their novelty drive*, active versus receptive nature of their libidos
> and the speed of their sexual response cycles, are fundamentally
> built into their biology.[113]   [Emphasis added.]

The article in *She* appears to be a sequel to an earlier one in the fashion
magazine, *Vogue*, in which Richard Dawkins tells the reader that a woman
is 'a throwaway survival machine for [her] immortal genes'.

> All details of evolutionary history are inscrutable, but in any case
> I don't care much about details. The *principle* that we are survival
> machines for our selfish genes . . . is a fundamental truth whose
> implications we must face.   [Original emphasis.]

Among these implications is 'ruthless selfishness even within close families', especially 'the selfish exploitation game which I have called the battle of the sexes'.

> The theory says that females should care for their children more than males do, and males should have a natural tendency to be more promiscuous than females.

Unfortunately, as Dawkins also writes, according to theory the male ought to be the gaudy sex. Yet

> our own culture breaks the . . . rules. It is the *female* who dresses up, . . . paints her face, studies fashion.[114]

Perhaps therefore, before the days of the welfare state, 'women competed for men as economic providers'. Here Dawkins seems to imply that the difference between men and women of 'our own culture' and those of other cultures in which men do dress up is genetically determined. There is no evidence for anything of the sort; and all experience of the effects of upbringing points in the opposite direction. Dawkins therefore adds: 'To be honest, I suspect that the truth is rather more complex since culture is involved.' So the reader is left in doubt: should she wear a skirt, or trousers?

Other such articles have appeared in American journals ranging from *House and Garden* to *The National Observer*.[115] Occasionally, persons with political power unguardedly reveal similar attitudes. In 1979 Patrick Jenkin, a senior member of the conservative British government, combined theology and politics in the following statement:

> If the good Lord had intended us all having equal rights to go out to work and to behave equally, you know He really wouldn't have created man and woman.

And in 1985, Patrick Buchanan, a United States presidential aide, stated unequivocally that nature intended women to stay at home.[116]

As we know, sociobiological sex struggles are expressed in the terms of cost benefit rather than theology. Correspondingly, these ideas have seeped into the world of commerce. An article in *Business Week* describes how a 'highly controversial theory of human behavior has been swirling through campuses' and 'paints a disturbing, even repulsive, picture of human behavior'. Bioeconomics, an offspring of sociobiology, is – we learn – held by some economists to provide 'the genetic basis for human desires' and so to tell us 'which economic policies will work and which will not'.[117] The reference to the campuses recalls how, earlier in the century, American universities took up the teaching of eugenics. I have found no comprehensive survey of sociobiological teaching in universities, but the presumption that ethology and evolutionary theory are sources of enlightenment on our social lives seems to be widespread. As an instance, the Calendar of a Canadian university states that 'One modern aspect of Zoology is animal behaviour, the study of which helps us to understand human behaviour in warfare, politics and other activities.'[118]

The influence of sociobiological doctrine may, however, have been greatest in school teaching, especially in North America. In 1975 an article in a journal for biology teachers urged that sociobiology should be taught in schools, for the study of animals could help to solve the problems of human societies.[119] This advice was widely accepted. A notorious item is a film, *Doing What Comes Naturally*, of which the awfulness has to be seen to be believed. Dogma on the relationships of the sexes is presented in an extreme form: in war

> you loot and pillage, but you also grab up the women and you either inseminate them on the spot or you take them back as concubines. You kill off the adult males; you sometimes castrate young boys and bring them back as servants. So . . . warfare has traditionally had a strong sexual counterpart to it, *which is certainly biological*, and you don't have to look far to see that *there's that tendency running today*.   [Emphasis added.]

And in another passage the audience is told that men are much more interested in science and politics than are women, and that this is an outcome of the biological 'background' we share with all other vertebrates. All this is interspersed with frequent (and misleading) references to the conduct of baboons. Similar messages are conveyed in a widely used program for teaching, *Exploring Human Nature*, distributed in the USA by the Educational Development Corporation.[121]

These teachings are echoed in serious journalism. In 1978 the *New York Times* published five prominent articles on women in society. The first four were on the recent progress of women toward equal opportunity and status. The last was on the limits said to be imposed on this progress by the biological differences between the sexes; and sociobiological arguments are quoted as showing that these differences 'will forever leave men the dominant sex'.[162]

Such doctrines exemplify, in an extreme form, the defects of a narrow biological interpretation of humanity. They combine misuse of animal metaphor, misinterpretation of animal behavior and disregard of history. The diversity of the social lives of animal species permits few generalizations even within a single class, such as that of the mammals. If, for example, a feminist wished to invoke another species to support her freedom to go out to work, she could call upon the queen of beasts. Lionesses not only look after the children but also bring home the bacon: they combine to hunt down prey for their cubs and themselves. But she should be warned that, once the meal is ready, the males come in and take the lion's share. Perhaps there is, or has been, a human society analogous to that of lions;

but, if so, it illustrates the unpredictable variety of our customs. The biological arguments against freedom for women use the same logic as those against freeing slaves: they equate one changeable set of customs with what is natural, proper and inevitable for our whole species.

Whatever the private views of individual scholars, the stories about the fixity of human sex roles are, in effect, attempts to put back the clock and to delay the freeing of women from subjection. To many, the opportunities already open to women are highly disconcerting. Advances in physiology and in manufacture now allow women in rich countries to control conception: they can confidently plan to have their babies when they wish. Correspondingly, attitudes toward contraception have drastically changed. As well, advances in public health and in the standard of living have greatly reduced death rates in infancy and childhood: a woman need no longer have many children in order to ensure the survival of a few. The most usual number of children for a modern fecund woman may become three. Dorothy Dinnerstein suggests a period of six months of absence from 'normal activity' for each child; or, as she puts it, three percent of the period from 15 to 65 years.[123] Others might prefer different figures. Working hours for both sexes are becoming shorter and more flexible; hence, once breast-feeding is over, parents can share equally in child care.

These changes, in progress as I write, have gone with the appearance of women in occupations from which they were, not long ago, forcefully excluded: women are leading figures in activities as diverse as the sciences and politics; they join the police, pilot aircraft, design houses, make furniture and edit newspapers. To say, in the late twentieth century, that women should be barred from such callings is to turn one's back on everyday experience, and to reject an enlargement of freedom which surpasses even the abolition of slavery. The movement also widens opportunities for men, as we see from the rising numbers of men as primary school teachers and hospital nurses.

Hence freedom for women means much more than husbands doing the washing up. Its many social consequences have still to be fully revealed, but some can be foreseen. Nuclear families, consisting of parents and children on their own, will probably become much less common, with benefit for people of all ages and both sexes. Economic changes will also follow. J.K. Galbraith has described the 'crypto-servant role' assigned to women in modern rich countries and some of its economic consequences. A married woman is responsible for the care of her children, the purchase of clothes and household goods, the management of the house, and cooking. She therefore represents both a large and important class of consumers and is also a major source of services. As her role changes, many services must be increasingly provided from outside the home, often by small concerns, that is, by authentic private enterprise.[124]

Such changes are not adverse to the human desires for sex or friendship, for parental care of children or for home-making. On the contrary, they make them easier to achieve. Moreover, women's independence leads also to liberation for men. Men are borne, suckled, mothered and taught by women; later they live and work with women. In 1869 J.S. Mill wrote that 'what is now called the nature of women is an eminently artificial thing . . . no other class of dependents have had their character so entirely distorted from its natural proportions.' Men are slowly finding that they too benefit from having mothers, wives, daughters and colleagues with characters not 'distorted from their natural proportions' – who are not treated as domestic animals or slaves, but are properly respected and free.

Equality

Proposals for the enlargement of our liberties are still strenuously resisted. In the late twentieth century, some neo-conservatives (the 'new right') have restated their great-grandparents' support for privilege and have revived their objections to equality for women. As a prominent English neo-conservative historian, Maurice Cowling, has written: 'it is not freedom that conservatives want; what they want is the sort of freedom that will maintain existing inequalities or restore lost ones.'[126] Their proposals are linked to belief in the genetical determination of socially important qualities. Such a stand can be taken only by turning a blind eye to modern genetics: it makes the fundamental error of ignoring the vast influence of differing environments. Not long ago, literacy was confined to a small minority of clerics or clerks in each civilized population. The difference between the clerks and the rest cannot have been genetically determined for, when universal schooling was introduced, general literacy followed. Arithmetical skill has a similar history. As the resources given to education are increased, and teaching techniques are improved, the levels of literacy, numeracy and other skills rise, in both sexes.

Such transformations can take place in a few decades. Like the recent increases in stature and growth rates observed in many countries, they are due not to genetical variation but to successful efforts to improve environments, especially those of children. During the past century, we have learned how to feed our children and ourselves well, and how to prevent most of the major infectious diseases. The knowledge is far from fully applied, but it has already benefited millions. International bodies have now accepted the obligation to make these benefits universal. In doing so they acknowledge the scope of environmental action in promoting human welfare. The scientific foundation of teaching is less secure, but much is known about the effects of schooling: children can respond to improved education as they can to improved diet.

Entailed in this advance is the principle of equality of opportunity: that all children should, as nearly as practicable, have similar opportunities to be healthy and educated. To apply the principle is not easy. Our attempts must be influenced by the variation in skill and motivation (whatever its origin) found in any population. Consider any socially valuable activities from, say, computing to stockbreeding: special training in such arts must be confined to a few, selected, regardless of parentage, from those who show promise and who wish to specialize in them.

The contribution of genetical differences to the variation is not known, and is irrelevant. Political genetics, with its opposite message, systematically underrates not only particular kinds of people but also the whole of the human species. The ignorance and incompetence, even criminal conduct, found in any population are held to be fixed by heredity and therefore preserved, generation after generation, indefinitely. That at least is implied, though usually not clearly stated, by the exponents of *Homo egoisticus*. The capacity of human beings to learn from experience, and to use example and teaching to transmit what has been learned, is brushed aside. In the next chapters, therefore, we turn to models of humanity founded on attempts to analyze our intelligence.

Part 4

Homo operans:
the greedy species

What is a man,
If his chief good and market of his time
Be but to sleep and feed? A beast, no more.

Shakespeare: Hamlet

The self portraits derived from Darwinism reduce us to automata. An alternative movement, associated with the name of Pavlov, reduces our intelligence to mechanical 'conditioning'. An influential successor to 'Pavlovism' explains all human action by the effects of rewards and punishments. These images pay attention only to behavior: feelings and even thoughts are excluded. They owe their impact partly to being based on experimental findings. Experiments have shown how we can regulate some of the activities of laboratory mammals in cages. Similar methods have been recommended for controlling human beings.

Problems that behaviorists have tried to solve include mental illness, the treatment of criminals, the education of children, the management of employed persons, and the design of an ideal society. But all attempts at such behavioral engineering have eventually had to face the complexities of human beings as persons: we have not only 'senses, affections, passions' with which behavioral engineering cannot cope, but also the ability (however imperfectly developed) to reach conclusions by independent investigation and reasoned argument.

To be useful, the results of experiments have to be combined with our commonsense, everyday knowledge of ourselves and of our history. When this is done, experimental studies can enlarge our understanding of ourselves in our social lives and in our work and play.

175

10 Conditioning and improvising

> *The confusion and barrenness of psychology is not to be explained by calling it a 'young science'; its state is not comparable with that of physics in its beginnings. . . . For in psychology there are experimental methods and conceptual confusion. . . . The existence of the experimental method makes us think we have the means of solving the problems which trouble us; though problem and method pass one another by.*
>
> *Wittgenstein[1]*

In the epigraph of Chapter 7 I quote the Sokratic question: is virtue innate in a person, or must it be acquired by experience? In the dialogue that follows, Plato turns to asking whether knowledge is innate, and concludes that, in a sense, it is. But for more than two centuries it has been more usual to reject the idea of a store of wisdom (or of original sin) available at birth. Sociobiological doctrine is in this respect exceptional.

The two images outlined in the present chapter try to explain how knowledge arises from experience. The first was designed to give a physiological account of behavior and even of the mind, and so was a development of the notion that the body is a machine. Physiologists have been successful because they study what can be directly observed: they measure heart rate, blood pressure, electrical changes in nerves, the rates of secretion of glands and much else. When learning was to be similarly studied, it was clearly learned *behavior* that could be recorded and measured. Hence arose the movement called behaviorism. Nineteenth century psychologists had been much occupied with feelings and emotions reported by their subjects. Such introspection, writes E.G. Boring,

> had produced no agreement about feeling. Is feeling a sensation, an attribute of sensation or a new element? Does it really exist? . . . Psychology claimed to be science but it sounded like philosophy and a somewhat quarrelsome philosophy at that.[2]

Thoughts, expectations, intentions and beliefs are directly accessible only to the person who thinks, expects, intends or feels: they are not public; according to the behaviorists they should therefore be extruded from their science.[3] Scientific psychology could and should be reduced to the study of overt behavior (including verbal behavior).[4] Private events, writes a modern behaviorist, should be regarded not as sources of action but as the outcome of agencies which act on the person from outside.[5]

Hence, according to this doctrine, if people are allowed to be conscious at all, it is only in the restricted sense of *responding to stimuli*. A behaviorist's procedure may be used in examining a patient with brain injury: the surgeon looks for contraction of the pupils in response to light, or for a reaction to pressure on the skin or to the patient's name. For clinical purposes this is useful. But to confine all psychological enquiry in this way is to omit the ideas of the autonomous human being and of self-consciousness.

Such a philosophy encourages objective study; and it can reasonably be held that a science of *animal* behavior requires some form of behaviorism. But in human affairs behaviorist metaphysics withdraws from everyday experience. Most of us – even, perhaps, behaviorists themselves when off duty – presume that statements of feelings, intentions and reasons for action are more than by-products of external stimulation. We do not find ourselves in the dilemma of a strict behaviorist who feels obliged to greet a colleague by saying, 'You are very well! How am I?'

A Russian revolution

The first major behavioristic systems were Russian. In 1863, a physiologist, I.M. Sechenov (1824–1905), presented a book with the title, *An Attempt to Introduce a Physiological Basis for Psychological Processes*. The tsar's censor objected; and the book, slightly modified, eventually appeared as *Reflexes of the Brain*.[6] Later, the book was read by the young Pavlov. The censor's response was not surprising: the ideas it contained were revolutionary. Sechenov, though principally an experimenter, was impressed by Darwinism, and especially by the notion that the brain and behavior of human beings are products of evolution. This encouraged him to draw conclusions about his own species from observations on animals. He held that the problems of psychology could be solved only by physiologists, and that their subject matter should be reflexes. Learning, he said, was the result of the association of external stimuli with the movements of muscles. Had he lived later, we might have had a reflex model of humanity based, not on Pavlov's dogs, but on the frogs which provided Sechenov's most notable findings. Sechenov rejected will and self-consciousness as the sources of human action. The apparent ability to make moral choices, he held, is a delusion. As G.A. Kimble remarks, Sechenov was 'a mechanist whose match has never been met in Western psychology'.[7] This, perhaps, makes him sound unattractive. It is therefore instructive that, in Boring's words, he was 'sympathetic, responsible, generous [and] public-spirited'.[8]

In the West Sechenov's work was long neglected. Another experimentalist who anticipated Pavlov, though only by a few years, was celebrated in his own time. Jacques Loeb (1859–1924) was born in Europe but married an

American wife and in 1891 emigrated to the USA.[9] In 1912 he wrote an (uncensored) work, *The Mechanistic Conception of Life*, in which he held that comparative psychology would advance through the work of biologists trained in the physical sciences. Donald Fleming comments that his 'mechanistic animus' made him 'shameless even in his prime'. Loeb had already, in 1899, written of 'associative memory' as a criterion of consciousness.[10] Loeb's most important researches were, however, on the orientations of insects. The direction of these movements is often precisely regulated by light, gravity or other physical features: they have an automatic, mindless quality that encourages a mechanistic interpretation.[11]

Apart from Pavlov and Loeb, the most notable mechanist of the late nineteenth century was the Russian neurologist and psychiatrist, V.M. Bekhterev (1857–1927) who, though younger than Pavlov, took his medical degree two years earlier.[12] B.P. Babkin describes him as a man of tremendous energy, 'an outstanding figure in the scientific world, at least in Russia', who 'looked more like a coachman than a professor'.[13] His work and ideas have had little recognition outside his own country; yet his behavioral researches began about 20 years before those of Pavlov's school; and, like Pavlov, and at about the same time, he observed what he called 'associative reflexes'. Moreover, his presumptions, like Pavlov's, were physiological and objective, not psychological and mentalistic. He used a method, later adopted by many American experimenters, of applying a slight shock to an animal's foot; the shock causes withdrawal of the limb. If shock is paired with a harmless stimulus, such as a sound, the new stimulus by itself soon comes to provoke withdrawal. He did not fasten his animals in a harness, or implant tubes to collect glandular secretions.

The simplicity of his method allowed Bekhterev to experiment on human beings: shocks were given to hand or foot, and were accompanied by 'neutral stimuli', including sounds and visible changes. These had effects similar to those observed in experiments on animals. Bekhterev therefore proposed a 'new science', which he called reflexology. This strictly objective study was to replace psychology: consciousness and the knowledge a person has of him or herself could not be studied experimentally, and so were rejected. Bekhterev, in fact, proposed an authentic image of humanity, *Homo reflexus*; and this image is absent from our list only because of the historical dominance of Pavlov.

Pavlov, like Sechenov, was a physiologist, and remained staunchly so even during the last three decades of his life. During that time he made his massive contribution to behavioral science. He was both an exceptionally ingenious experimenter and also a source of original ideas. Many of the experiments described in his writings were carried out by pupils or colleagues, but they remain shadowy figures. Pavlov had a dominating as well as an inspiring personality.

Homo pavlovi

Our account of the image we owe to Pavlov begins, appropriately, with a picture. On the left a dog stands on a table. The animal, immobilized in a harness, is connected, by several cables, to recording equipment; a tube leads from its mouth to a glass cylinder. On the right, separated from the dog by screens, a neatly dressed young man, the experimenter, sits at another table; the experimenter can see the dog without being seen, and has at his fingertips controls which allow him to test the dog's response to various stimuli; in particular, a plate with food on it is ready to be swung into place in front of the dog. The response most studied is the amount of saliva secreted by one of the salivary glands. The drawing is from Pavlov's *Lectures on Conditioned Reflexes*[14] and, though crude, diagrammatic and unattractive, in the half-century since 1928 it has been reproduced again and again in histories and textbooks of psychology and physiology.

The image of humanity we now face is represented, not by the experimenter, but by the dog. The experimenter's actions are very complicated indeed, but the dog's reactions are, superficially at least, quite simple. The apparent simplicity arises from Pavlov's method. The dog has fasted for some hours (or, in plain words, is hungry). In a standard experiment, a bell rings. The dog pricks its ears and turns its head. This, the orienting reflex, is often left out of brief accounts. Seconds later, a plate with food appears, and the dog eats. While it does so, its mouth waters, and by an ingenious arrangement the amount of salivation is accurately recorded. The scene is repeated on a number of occasions, and the dog's behavior gradually alters. The orienting reflex disappears: instead, the dog watches the place where the plate appears; it may also paw the ground and lick its lips. It seems to be expecting the food. But expectation, being a mental event, is known to human beings by introspection; it is therefore not usually included in the conventional objective descriptions of these experiments. Yet the anticipatory character of the behavior is a central feature of what happens: when trained, the dog salivates *before* the food appears; indeed, it salivates copiously even if no food is presented. And that is the conditional reflex.[15]

Pavlov and his associates studied the conditional reflex in immense detail. In doing so, they provided concepts and a vocabulary for subsequent studies of habit formation and learning. One concept is that of generalization. One says that a dog (or a person) responds to 'the sound' of a bell, but in fact no two sounds (and no two responses) are exactly the same: the responses are to a population or class of sounds. If an animal is trained to respond to, say, sounds of a given frequency (such as that of high C on the piano), and then another, similar note is sounded, this elicits a similar response, though usually a smaller one. Such generalization is clearly essential in everyday learning. The obverse is discrimination; and discrimination can be made progressively narrower by training: a subject of exceptional acuity might learn to respond to high C, but to ignore C sharp.

Another important feature of conditional reflexes may be called distraction. Pavlov named it external inhibition. During his early experiments a pupil might, after weeks or months of labor, establish a conditional reflex, and wish to demonstrate it to the professor; but as soon as Pavlov arrives in the laboratory, the conditional reflex fails. Later, experimental subjects were prevented from seeing or hearing the experimenters, and external inhibition was studied formally. A dog is trained to salivate to a tone. Then, at the time the tone is sounded, a novel stimulus is also presented. The salivation fails, and the dog instead displays the orienting or 'What-is-it?' reflex: it pricks its ears and turns its head. With repetition of the new stimulus, the orienting reflex, as usual, fades, and the salivary response returns. These phenomena are unsurprising. Their importance lies in the precise and repeatable form in which they were made.

Pavlov also founded the study of what we now call extinction. A conditional reflex (again, salivation in response to a sound) is first established. Next, the tone is sounded, but no food is given; nonetheless, salivation occurs. Unrewarded presentations of the tone are repeated, and the response progressively lessens; eventually, no saliva is secreted: the response has undergone extinction. Once again, such a result is to be expected: it is no use salivating if there is no food. But the next day, the tone is again presented without the food, and salivation occurs. Evidently, extinction is not an undoing of previous learning: it is learning not to respond. And *this* learning is lost overnight, leaving the original conditional reflex (almost) intact. Only if such extinctions are repeated, day after day, is the conditional response finally wiped out.

These examples give a glimpse of some leading findings that emerged from the early use of Pavlov's method. As so often in biological research, it was partly the accumulation of minutiae that impressed other experimenters and theorizers. Pavlov's central behavioral principle was the influence of association. To show how association acts, observations were made in strictly regulated conditions and the phenomena recorded were measurable.

The social impact

Pavlov's method has been applied to human beings. As experimental subjects we are superior to dogs: we do not have to be maintained, on a research grant, in expensive kennels; and if, say, undergraduates are politely asked by a professor to take part in research, they are likely to agree without prolonged training, and without having to be restrained during the experiment. If a student hears a musical tone, and then has lemon juice squirted into his mouth, he salivates. After four days of training, some of the subjects salivate at the sound of the tone by itself. Extinction soon occurs; but this at least seems to be a case of a classical or Pavlovian conditional reflex. There are other examples.[16]

Homo pavlovi has therefore seemed to simplify human learning, and so to give hope of controlling it, perhaps by relating it to what goes on in the brain. As we see below, the experimental findings were in fact far from simple. Yet crude images of humanity were derived from them. In a popular essay on mind and matter, one of the century's leading philosophers, Bertrand Russell, writes:

> . . . the most essential characteristic of mind is memory, using this word in its broadest sense to include every influence of past experience on present reactions. . . . The influence of past experience is embodied in the principle of the conditioned reflex . . . For example: if you wish to teach bears to dance, you place them on a platform so hot that they cannot bear to leave a foot on it. . ., and . . . you play 'Rule, Britannia!' on the orchestra. After a time, 'Rule, Britannia!' alone will make them dance. Our intellectual life, even in its highest flights, is based upon this principle.

Russell also wrote of conditional reflexes as 'above all . . . characteristic of human intelligence.' And in a widely read introduction to philosophy he states, quite wrongly, that a child learns to speak by acquiring conditioned reflexes.[17]

In a similar vein, a distinguished biochemist and humanist, Joseph Needham, states that Pavlov's findings lead to 'many important conclusions for the future of humanity'; these include rejection of the idea that 'human nature' is beyond improvement.[18] Fortunately, however, the future progress of humanity does not depend on our resemblance either to drooling dogs or to dancing bears.

An index of Pavlov's influence is also found in textbooks of physiology. For several decades these works, written for medical students, regularly followed the account of the brain with a chapter on conditional reflexes. The impression given was that reflexes and conditional reflexes comprise all that a physician needs to know about human behavior – or at least that they fully represent, in Pavlov's phrase, the 'higher nervous activity'.[19]

The spread of the idea – better called a myth – of conditional reflexes outside the academic world led George Bernard Shaw to ridicule Pavlov in his brilliantly witty fantasy, *The Adventures of the Black Girl in her Search for God*, published in 1932. Thirteen years later, exasperated by a broadcast eulogy of Pavlov, he described Pavlov as 'for the moment the Pontifex Maximus of biological science'; and he dismissed researches on conditional reflexes as demonstrating the obvious and as a 'reduction of Science to absurdity'.[20] Shaw's criticism is not coherent; but it reflects both the social standing of the 'conditioned reflex' and the distaste aroused by the mechanistic animalitarianism of behaviorists.

Sechenov, Bechterev, Loeb, Pavlov and his followers all tried to reduce complex and enigmatic phenomena to simple components. Pavlov himself wrote:

It is obvious that the different kinds of habits based on training, education and discipline of any sort are nothing but a long chain of conditioned reflexes.[21]

And P.K. Anokhin describes how at first Pavlov used mentalistic expressions, such as 'psychic secretion', when he induced salivation in response to a ringing bell, but gradually surrendered all conventional, psychological terms.[22] The traditional psychological concepts of intention, will and consciousness were discarded. Mind-body dualism was rejected. The powerful methods of the physical sciences were being applied with immense success in physiology. Now, it seemed, the same methods could solve problems hitherto debated under the headings of mind and intelligence. Pavlov's grand design was to create a science of behavior in which all actions could be explained by the workings of the nervous system. To understand learning, he held, one must study the making and breaking of new connexions in the brain.

The complexity of 'conditioning'

The social impact of the conditional reflex has then arisen from its apparent simplicity. Yet the principal interest of conditional reflexes arises from the complexities they reveal. One common omission from short accounts is a description of the dog's state, and of what the dog does (apart from salivating). In the early stages of training, when the bell sounds, heart and respiratory rates rise; the animal may paw the ground and lick its lips. It seems tense and worked up. As training proceeds, it calms down.

The most notable complexity in Pavlov's findings, the experimental neurosis, seems to be an outcome of the internal agitation that accompanies the training. A dog can be trained to salivate copiously on hearing a tone but not to salivate when it hears a tone of different frequency. Now suppose the animal has to make increasingly difficult discriminations: the tones, initially perhaps an octave apart, are made progressively closer. When, eventually, the animal's discrimination begins to fail, its total behavior alters. Instead of running to the laboratory and jumping on to the table, it may resist the experimenter, struggle and howl. It may go off its food; it may be incessantly restless; or it may remain motionless in a corner for long periods. The form of the 'neurosis' depends on the dog's personality.[23]

In everyday language, the dog is confused and frustrated; and, being strapped in a harness, it cannot escape. Perhaps we ought to hesitate to speak of confusion and frustration, because these are words for states we know in ourselves by introspection; but, whatever words we use, something important is going on internally, not only the making and breaking of connexions between stimulus and response: we are faced with a phenomenon that obliges us to go beyond stimulus and response.

There is much other evidence of internal complications. I take an example from experiments in which a slight electric shock is used to cause an animal to 'freeze' (or to display fear).[24] Suppose an animal, on several occasions, hears a tone and at the same time sees a light switched on. A human being might remark that tone and light are associated or linked; but there is at this stage no evidence, from the conduct of the animal, that it has learned anything. The next step is to sound the tone several times, and to follow it, each time, with a shock; after such training, the tone by itself induces freezing. This represents the classical conditional reflex. Lastly, the *light* is switched on alone: and that induces freezing too. Clearly, an association has been silently established during the first stage when light and sound were presented together – silently, in the sense that it developed without any behavior to show what was going on and, moreover, without any reward. This enables the animal to put two and two together when, later, the occasion arises for it to do so. If a human being behaved in an analogous way, we should say that he thought that light 'meant' shock. And so psychologists write of the cognitive element in such training.

These are American experiments. A modern Russian example also illustrates the importance of what an animal *expects*. A dog is as usual trained to salivate to a sound by giving it meat on a dish. It is separately trained to salivate to a light by giving it bread on a different dish. When fully trained, it is given bread again, but on the meat dish. Like a *bon vivant* offered claret with fish, the dog now shows signs of disturbance: salivation fails, and the animal looks from one dish to the other.[25] Whether we talk of 'expectancy' or use some other expression that sounds less as though the dog were human, the behavior is not that of a machine programmed to establish or cancel simple connexions.

In experiments on human beings, expectancy becomes still more obviously important, and partly depends – as common sense would predict – on what the subject has been told. People communicate by language. (Russian Pavlovians call this the second signalling system.) The influence of this social component has been shown in many formal experiments. One much used method is to record electrical changes in the skin (the galvanic skin response) that occur when one is suddenly hurt or alarmed.[26] Subjects may respond in this way, on the turning of a switch, as a result of being told that they would receive a shock, even when the current is not in fact switched on.

There is, however, much individual variation: some people, as W.W. Grings points out, are suspicious. Other sources of difficulty are inconsistent results. In one series of experiments, subjects were told that light-on would be followed by shock; and so it was, on 20 successive occasions. They duly developed a skin response to the light alone. They were then told that light would *not* be followed by shock, and the electrodes were removed. But, although the subjects believed, correctly, that they would not now be shocked, their skin response still at first occurred. (Loss of the response, or extinction, was, however, hastened by the reassurance they received.) In other experiments, subjects were told to expect a powerful shock after a signal; and they responded as though shocked, though none was given. They were now told (like the others) that there would be no further shock, and again the electrodes were removed; and this time the response (the fear) disappeared at once.

Evidently, verbal information has a profound influence on how a person responds. Indeed, a subject given misleading or incorrect information about the rewards to be offered is likely to behave accordingly, and to ignore the objective situation if it contradicts what the experimenter has said.[27] Such findings are incompatible with any simple concept of conditioning.

Conditional therapy?

The practice of behavior therapy points the same way. Some psychotherapists, perhaps influenced by their textbooks of physiology, have sought help from Pavlov in solving the inscrutable problems of mental illness. C.M. Franks formally defines the resulting 'behavior therapy' as 'the beneficial modification of behavior in accordance with experimentally validated principles based upon stimulus-response concepts of learning';[28] but, as we see below, the existence of any such principles is debatable.

Pavlov himself held the experimental neuroses of his dogs to be related to human 'nervous disturbance'; but he said that they 'should not be taken as ... explaining the incalculably complex symptoms observed in man'.[29] There is in fact no good reason to think that experimental neurosis tells us anything useful about human afflictions.[30] Nonetheless, behavior therapy is much used. The underlying presumption is that 'conditioning', or the making and breaking of stimulus connexions, is a fundamental element in the illnesses usually called 'mental'. Early theories in this field have been called 'breathtakingly simple and straightforward'.[31]

In psychological medicine, disturbed behavior is usually regarded as a sign of emotional conflict which may be concealed from the sufferer as well as from others; but in behavior therapy the unwanted conduct is treated directly: no inner conflict or emotional state is allowed for. In treating anorexia nervosa, the irrational refusal to eat, the question why the patient (usually a young woman) refuses food is not asked.[32] Instead she may be put in a bare room and rewarded, if she eats, with praise or with music or other entertainment. This method is described as being sometimes effective, at least for a short time. Similarly, a child's phobia – for instance, an irrational fear of furry animals – may be treated by giving the child enjoyable food in the presence of an animal; the frightening object is at first at a distance.[33] This may have a lasting effect.

Irrational fears are, however, more usually treated by desensitization without reward. This, the most widely used method of all those called behavior therapy, is also recommended for stuttering, impotence and frigidity. In the treatment of a phobia, such as a horror of snakes, the patient is exposed to progressively more alarming representations of snakes until inured to these animals. Thus psychoneurotic traits, and other unwanted conduct, are treated as if they had been learned much as an experimental animal develops a new habit; and so must be unlearned. But densitization is quite unlike anything done in laboratory conditioning.[34] Moreover, some therapists also give training in relaxation at the same time, and so help the patient, through his or her own efforts, to be calmer in the presence of the terrifying object.

These are attempts to form positive associations, but in therapy the breaking of connexions is more usually emphasized. In aversion therapy an attempt may be made to put an alcoholic off drink by, in effect, punishing the consumption of liquor. A similar method has been used to treat behavior such as sexual perversions. Conventional behavior therapy has both advocates and opponents. According to some reports, after treatment 60 per cent of alcoholics abstain from drinking. But 'practical disadvantages' include the dangers of the drugs used and 'the debilitating nature of the treatment that can lead to both patient and staff resistance'.[35]

A patient to be treated in a strictly Pavlovian manner should be directed to a laboratory with no people or even dogs, only machines. But in practice doctors, nurses or other therapists are always present: they speak to the patient and betray, intentionally or not, their attitudes to the patient and to the treatment. Experiments have shown the value of reassurance: patients under treatment, for instance for phobias, do better if they are encouraged to expect success.[36] A healing attitude and manner are always helpful, whether the therapist's allegiance is to Pavlov, to Freud or other authority, or to none.

This rejection of simple behaviorism, which may sound like common sense to the reader, is now widely accepted even by psychiatrists who use some form of behavior therapy: that is, by those who concentrate on their patients' unwanted actions, and on getting rid of them by retraining, instead of looking for a buried cause.[37]

In yet a further move away from behaviorism, much emphasis is now sometimes put on what the patient can do for him or herself. In 'covert sensitization' for alcoholism, patients are first trained in relaxation, and are told that by self-relaxation they can reduce their need for alcohol. Then they are persuaded to imagine, in minute detail, their actions as they are about to drink their favorite beverage, and also to imagine nausea as accompanying the first sip.[38] There are great complexities, both in treatment and in analysis of the results; but a success rate of nearly 70 per cent may be achieved when the procedure is carefully followed.

The connexion of covert sensitization with Pavlov's experiments is hardly visible. The procedure is primarily verbal; it also makes use of the patient's powers of imagination and, especially in relaxation, of self-control. Therapists who use this method have completely discarded the mechanistic behaviorism represented by *Homo pavlovi*: they have in effect restored the idea that a person has responsibility for his or her own conduct. The autonomous human being reappears.[39]

Although behavior therapy, in its many forms, sometimes succeeds, it is uncertain why it does so. There is no longer any simple theory – indeed, there is no theory at all – to account for all the facts. Edward Erwin suggests that behavior therapists now agree on only one thing, the need for exact experimental tests of the methods they use.[40] But it must be doubted whether such tests are possible.

The crude, popular image that I name, *Homo pavlovi*, like other images represents the desire for simple explanations of complex and sometimes bewildering phenomena – those concerned with how we learn skills, including social skills, and habits, including bad habits. Looking back on nearly a century of effort, we can see notable achievements. But a trained and restrained dog, salivating at the sound of a bell, is of primary interest only to those who study a special branch of physiological psychology. The general importance of the Pavlovian image is not in its scientific content, but in the strange attraction it has had for people in our historical period.

Trial and error

The experiments on conditional reflexes were carried out in an age of machines. They both reflected the outlook of the time and also helped to create that outlook, especially in the USA and the USSR. Although the Russian findings, fragments excepted, were not available until 1927, there was by that time a substantial American literature on 'conditioning'. J.B. Watson (1878–1958), one of the leading founders of the behaviorist movement in America, unlike Pavlov did not emphasise physiology. Here are the opening words of a review published in 1913:

> Psychology as the behaviorist views it is a purely objective experimental branch of natural science. Its theoretical goal is the prediction and control of behavior. Introspection forms no essential part of its methods, nor is the scientific value of its data dependent upon the readiness with which they lend themselves to interpretation in terms of consciousness. The behaviorist, in his efforts to get a unitary scheme of animal response, recognizes no dividing line between man and brute. The behavior of man, with all of its refinement and complexity, forms only a part of the behaviorist's total scheme of investigation.[41]

And in a later work, *Behaviorism*, Watson stated that 'belief in the existence of consciousness goes back to the ancient days of superstition and magic.'[41]

The enthusiasm evoked by Watson's works makes an interesting study. His *Behaviorism* was praised in the learned journals, and more than praised in the press. In London a reviewer in the *Nation* wrote of a system that 'would revolutionise ethics, religion, psychoanalysis – in fact all the mental and moral sciences'. The New York *Times* said that it marked 'an epoch in the intellectual history of man'. And the *Tribune* went further: 'Perhaps this is the most important book ever written. One stands for an instant blinded with a great hope.'[42] But, as we see below, though the collective blindness lasted for more than an instant, Watson's behaviorism has hardly marked an epoch.

Nonetheless, despite its firmly adopted limitations, American experimental psychology took a step further than Pavlov. When the Russian work became widely known, an obvious question concerned the scope of the findings on conditional reflexes: to what extent do the responses of a dog immobilized on a table reflect what is happening when a dog (let alone a human being) is moving freely in a natural environment? Our next and last image is again based on laboratory experiments; but in them the animals are not strapped down: they are allowed to use their muscles to perform positive actions.

Again we begin with a picture. The animal is now a laboratory rat. It is in a cage or box furnished with a lever, a small bowl and a source of light; the floor is a grid that can be electrified to give the animal a slight shock. The animal (often called 'S' for 'subject') presses the lever; there is a click, but nothing else happens. The animal presses the lever again, and again. On, say, the tenth occasion there is a slightly different sound, and a small sugary pellet drops into the bowl. The animal eats the pellet: it has been rewarded for its exertions; pressing the lever has been reinforced.

The animal continues. Each pressing of the lever and each delivery of food is recorded automatically. The pressing is called an operant. There is no experimenter (or 'E') in this picture. If E's research grant is large enough, the output from the equipment goes straight to a computer programmed for statistical analysis. An important advantage of this method over Pavlov's is the ease with which laboratory animals, such as rats, mice, monkeys and pigeons can be induced to press a bar or to peck a target. Vast numbers, especially of rats and pigeons, have been observed in these conditions, and hundreds of PhDs have been earned by writing theses on the results.

The method is highly pragmatic. Those who use it do not, as a rule, study behavior itself, but only the outcome of behavior. An animal may press a bar with one paw or two, or with its snout or even by sitting on it; the pattern of movements may be constant or varied; and the amount of energy used may vary too. In some conditions, the animal may lick or gnaw at the lever.[43] None of this signifies: what counts is the operant, especially the rate at which it is performed.

The first major survey of operant conditioning was Skinner's *The Behavior of Organisms*.[44] The title reflects the author's intention to provide a system applicable to all animals, including human beings. In a later work he states that pigeons, rats, dogs, monkeys and children 'show amazingly similar properties of the learning process'.[45] The main findings concern the effects of reward and punishment on the pattern of bar-pressing. The system is therefore a form of behaviorism with special limitations. What a mammal or bird does is determined by its previous experience of the results of action: by selecting a suitable pattern of rewards, behavior can be controlled and molded at the will of the experimenter. Primary rewards or reinforcers include food and warmth; but if any other event, such as the ringing of a bell, regularly accompanies experience of a primary reinforcer, the new stimulus too becomes reinforcing. Such secondary reinforcers can in turn be associated with further reinforcers. In this way an animal may be trained to perform a series of acts, such as running along a passage, raising a door, ringing a bell and finally pressing a bar. Only the last act is followed by the arrival of a primary award, such as a fragment of food.

Most of the resulting observations are, like Pavlov's, of detail. For example, if a hungry rat receives food once in every five bar-pressings, it regularly presses the bar five times in rapid succession, eats the food, presses five times again, and so on. But if food is given only once in 100 pressings, long periods without bar pressing are likely. Food may also be delivered,

A model of humanity? A laboratory rat in a 'Skinner box' eats a sweet pellet – a reward for pressing the lever on the right.

say, *on average* once in 100 pressings, but the number of responses before each reinforcement may be varied at random: a human being would then say that one cannot tell when the pay-off will occur. On such a schedule, an animal is less likely to have long pauses after eating.

Much ingenuity has gone into elaborating observations such as these; but the restricted character of the operant has usually prevented the findings from throwing much light on the learning process.

The method has, however, been put to good use in answering special questions. One is: just what sort of things will an animal work for? An obvious answer is water, food, warmth and other necessities for survival, and for the opportunity to reproduce and to rear young. But in fact a mammal will work for a much wider range of changes. If, for instance, an animal is in a dark box, but can switch on a light for a few seconds, it will do so repeatedly. Perhaps, then, it is working for light. But, if it is put in a lighted box, it will work to turn the light off. It will also work to dim or to brighten a light that is on all the time. And it will behave in a similar way when the only pay-off is a slight noise.[46]

Curiosity and exploration

Working for change and variety is widespread. Darwin referred to it in 1874: 'Animals', he wrote, 'manifestly enjoy excitement, and suffer from *ennui*, and many exhibit *Curiosity*.'[47] It took nearly a century for behavioral scientists to catch up with Darwin. Yet the relevant activities are obvious enough. Suppose a mammal such as mouse, cat or monkey finds itself in strange surroundings. It explores: a mouse moves around sniffing; a monkey handles strange objects. (These movements are the initial source of operant behavior such as bar-pressing.) We may ask what provokes such behavior, and also what use it is to the animal.[48] When we do so, we enter a region of behavior which does not fit the neo-behaviorist framework.

The short answer to the first question is: novelty. We are now concerned with the movements of animals whose immediate, obvious needs have been satisfied: they are not thirsty or hungry, too warm or too cold, deprived of companions, or diseased. A mammal in this potential nirvana may go to sleep; but it is quite likely first to patrol a region around its nest or lair. In an artificial environment, in which movements are automatically recorded, patrolling can be observed after a meal and a drink. Typically, places least recently visited are first approached; and if a new corner of the environment is opened, it provokes extra attention. Mammals will indeed *work* for access to a complex, structured environment. Rats will learn the way to a furnished room, in preference to one similar but unfurnished. Caged monkeys will learn how to exert themselves merely to watch moving objects or to hear sounds outside; they will also spend a lot of time solving puzzles for no reward except the achievement itself.

The tree of knowledge

As for human beings, we are capable of being restless, inquisitive and innovatory from an early age. Attitudes to these qualities, however, have varied. For Greeks of the classical age, inquiry was good in itself. Euripides, in the fifth century BC, praised the man 'who has knowledge that comes from inquiry': he is said, optimistically, to bring no trouble to his fellow citizens, while he 'surveys the unageing order of deathless nature, of what it is made, and whence, and how.'[49]

A little earlier, and slightly to the east, we seem to find an opposite outlook. In the most famous of creation myths, the founders of humanity are forcefully ordered away from the tree of knowledge. But, of course, Eve would have none of this.

> When the woman saw that the tree was good for food, and that it was pleasant to the eyes, and a tree to be desired to make one wise, she took of the fruit thereof, and did eat, and gave also unto her husband with her; and he did eat. And the eyes of them both were opened.[50]

Eve's independence has conventionally been held to be deplorable, but revision of this adverse judgment is long overdue. Today, we often not only encourage curiosity, we even study it. One of the achievements of modern psychology has been the analysis of the human desire for novelty, change and stimulation.

The liking for variety is evident almost at birth. Jerome Kagan has described what attracts the attention of infants, aged two days, by showing them lights that moved or went on and off; or he presented them with patterns that contained sharp black and white contrasts. Infants at this age have had little experience of things seen; and they respond especially to change, as of a moving or flickering light, and to contrast, such as the edge of a black patch on a white background. Hence even at this age infants already seek changing stimulation.[51]

At two months infants have seen many objects. They are now attracted by slight differences in something familiar. When we see something (or somebody) we know, we recognize it at any angle and at a great range of distances. We take this remarkable ability for granted. But it is remarkable. We know that a coin is round, but at most angles what we see is an oval object. We develop this ability in infancy. Of course, most of the important things we see are much more complicated than a coin – for example, a parent's face. Infants evidently help themselves by giving special attention to appearances a little different from what they expected. (Perhaps parents can help by making faces at their infants.)

In Kagan's experiments, the infants did not respond to variety only in things seen. Some heard an ascending scale of eight notes played a number of times. Then the same notes were played, but in a new order. This aroused extra attention: the infants' heart rates went down, and the infants smiled. Many parents give their children something moving to watch, such as a mobile. J.S. Watson and others rigged a mobile that could be made to move by pressure on a pillow. Given the opportunity, infants soon discovered this, and wriggled about to produce the movements; and, when they saw the movements, they laughed, gurgled and kicked.[52]

Later, a child's search for variety becomes more various. H.S. Ross gave children aged one year a number of choices: familiar or new toys; one toy or several; a familiar room to play in or a strange one. Mother was always present, but the experimenter was out of sight. All the kinds of novelty induced more play or exploration.[53]

The desire for variety is still present when one is supposed to have 'put away childish things'; but the most notorious findings on adults concern the obverse, that is, the dislike of monotony – of being *denied* stimulation.[54] The subjects of the first researches on 'sensory deprivation' were in a sense victims of the wickedness of our times: the experiments were provoked by reports of what happened to political prisoners kept in isolation.

Young, active men were paid to lie on a couch, in a warm, quiet room, with translucent material over their eyes and with their hands covered. A few hours of these conditions proved intolerable. Some of those who endured longest temporarily lost the capacity for coherent thought and

Fruit of the tree of knowledge: exploration at one year.

developed hallucinations, such as the feeling of being divided in two. Prolonged, monotonous stimulation has similar effects. If, however, the subjects are free to take exercise, the ill effects are much less severe. In addition, as usual, what the subjects have been told, or have learned, influences what they feel: if they expect distress they are more likely to experience it.

If we turn to the positive desire for novelty among adults, the problem is where to begin and what to choose. The most obvious fields are in sports and play, and in the fine arts and music. One may therefore ask whether the exact methods of laboratory science can be applied to the human liking for stimulation through the arts.

The answer is yes, up to a point. As an instance, R.G. Heyduk has made a systematic test of the liking for musical complexity. His stimuli were four specially composed variations on a theme, from simple to elaborate: the elaboration was in the chord structure and in the amount of syncopation. One hundred and twenty people rated the pieces on a 13-point scale, from extreme dislike to extreme liking. As expected, the complex variations were preferred. Some subjects also heard one piece several times; again as expected, they usually took longer to get bored with the more elaborate compositions.[55]

So the initial hypotheses were neatly confirmed. But these findings, from a well designed and analyzed project, also expose the limitations of behaviorism: they convey nothing about the experience of music. To say this, however, is to enter the forbidden field of introspection, for experience of enjoyment, boredom or dislike is part of our knowledge of ourselves. It is usual to regard such feelings as being among the causes of what we do; but, for a neo-behaviorist, private events are not to be regarded as 'prime movers of behaviour':[56] such events are recognized, but only as by-products of external (rewarding or punishing) agencies, and perhaps also of our genes.[57]

Before I deal with this quandary, I must answer the second question: what is the use of being curious, or of working for stimulation? The answer is in two kinds of concealed effects of exploration.[58] As an animal moves around, it learns about its environment. At the time, the learning cannot be detected, hence it is said to be latent. But, later, the animal may need to move from one place to another by the shortest route. Many experiments have shown how an animal that has previously wandered around an area can quickly learn the way from a lair to a new source of food. It can also rapidly find cover on the approach of a predator. In ourselves, we take such spatial skills for granted. We can often, at need, improvise a new route which brings us to a goal without error. Such spatial abilities are sometimes included among intelligence tests.

Exploration has also a less obvious function: especially in early life, it promotes the general ability to learn. Young mammals allowed to wander around complex environments later solve spatial problems more quickly than others reared in boring cages. Such effects are not confined to path-finding abilities. Rats reared in featureless cages have been compared with others that, from birth, had their cages decorated with triangles and circles. As adults, all were given a problem in which one shape indicated the presence of food, and another, no food. The rats reared with modern decor learned more quickly.

The scope of early experience, in the development of skills, has only recently been realized.[59] For a long time it was assumed that an animal such as a cat was born with the ability to identify and distinguish simple shapes; and that the same applied to human beings. (In old-fashioned terms, such abilities might have been called instinctive.) Experiments have shown otherwise. Kittens have provided strange and instructive findings. Some, in early life, were allowed to see only vertical lines, in the form of black and white stripes; others saw only horizontal lines. Later they were tested by being shown a thin stick, in conditions in which any ordinary kitten would respond playfully. The kittens that had experienced vertical lines played with the stick when it was held vertically, but ignored it when it was held horizontally, and vice versa.

There are now many such findings; and occasionally accidental 'experiments' have revealed similar processes in human beings. One form of congenital blindness is due to cataract, which makes the cornea opaque. When it became possible to remedy this condition surgically, some people, blind from birth, had sight conferred on them as adults.[60] The result was dismaying: at first, they could understand nothing they saw; objects, perfectly familiar by touch, were unrecognized by sight. Sighted people go through a process in infancy, of learning to see, that they later forget. These adults had to learn to see, like infants, when they had already adapted to blindness.

Hence for mammals, including ourselves, the development of normal function depends on variety of experience in early life; and variety in turn often depends on exploration. Here is yet another aspect of the importance of studying individual development, or ontogeny, already emphasized in the discussions of instinct and of genetics (Chapters 4 and 7).

Understanding space

Our knowledge of exploration, and the learning that goes with it, could be used to illustrate how much human beings have in common with other species; but if we look outside the psychological laboratories we find, as usual, much that is unique to humankind. The ability revealed by study of exploration is one aspect of spatial intelligence.[61] This term covers a range of skills, from perception of the form of objects to various aspects of navigation.

Recognition of form is an everyday experience. We instantly identify a familiar face (or even posture) in a crowd. With experience, we distinguish the topographical features and the species of plants and animals of the moorland, river, jungle or park near which we live; for gatherer–hunters such proficiency is vital. People who live by breeding cattle know each animal just as they know members of their family. We recognize objects, or kinds of objects, seen at a great range of distances and from a great variety of angles. Specially trained people identify important groupings of features, as in the diagnosis of disease by clinical methods.

Howard Gardner gives examples of tests of comprehension of spatial relationships. Here are two:

> Take a square piece of paper, fold it in one half, then fold it twice again in half. How many squares exist after this final fold? Or . . . A man and a girl, walking together, step out with their left feet first. The girl walks three paces [while] the man walks two. At what point will both lift their right feet from the ground simultaneously?[62]

Such tasks may be regarded as testing spatial imagery, an ability possessed in high degree by architects and sculptors but not confined to specialists. We all have some capacity for carrying a map in our heads. In Chapter 3 I give examples of the role of metaphor in creative thinking. Such metaphors often entail spatial imagery.[63] The picture of the atom as a miniature solar system, though fanciful, has helped both physicists and their pupils.

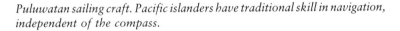

Puluwatan sailing craft. Pacific islanders have traditional skill in navigation, independent of the compass.

The understanding of spatial relationships is also crucial in technically simple cultures. The Eskimo, in their barren, seemingly almost structureless environment, identify features, undetected by strangers, such as the shapes and orientations of small snow drifts and cracks in the ice. (They also make high scores in formal tests of spatial ability.) Kalahari gatherer–hunters know

> every bush and stone, every convolution of the ground, and have usually named every place in it where a kind of veldt food may grow, even if that place is only a few yards in diameter, or where there is only a patch of tall grass or a bee tree.[64]

In the entirely different environment of a vast ocean with many scattered islands, the Puluwatans of the Caroline Islands are skilled navigators. Their special knowledge enables them, without even a compass, to make safe journeys, out of sight of land, of more than 80 kilometers. The information used includes the patterns of stars, waves and weather.[65] Such abilities depend, of course, not merely on exploration and individual learning but also on systematic teaching; in this they are quite unlike anything achieved by animals.

An American devolution

By reducing human action to what can sometimes be observed in restricted conditions, neo-behaviorists have tried to present a comprehensive portrait of humanity. In the words of Derek Blackman in 1981, they 'seek to understand human behaviour in terms of the . . . pay-offs for what we do.'[66] Perhaps it has been this simplification that has given neo-behaviorism its popularity. The metaphysical foundation of neo-behaviorism has been clearly put by Skinner.

> A scientific analysis of behavior must, I believe, assume that a person's behavior is controlled by his genetic and environmental histories rather than by the person himself as an initiating, creative agent. [67]

The principle stated here is explicitly a presumption: it is a starting point for experimental investigation, not a finding from research. But Skinner, in another place, goes further, with a striking *non sequitur*.

> Those who reject the scientific [that is, Skinner's] conception of man must, to be logical, oppose the methods of science as well.[68]

In many such writings Skinner, with unjustifiable arrogance, holds out not only behaviorism but his own version of it as the only truly scientific psychology. In the sentence last quoted he also dismisses his critics as either unscientific or lacking in logic.

It is, however, not 'scientific' to ignore all the complexities outlined in the previous pages. Moreover, the attempt to explain all we do by the attainment of reward and avoidance of pain – that is, by pay-offs – runs into logical as well as psychological difficulties. Consider enjoyment. It might be supposed that behaviorism has nothing to say about our enjoyment of the arts or, indeed, of anything else. Skinner, however, has much to say. He asks, for instance, what can be meant by expressions such as, 'I like Brahms' or 'I enjoy Brahms'. He answers that they 'can be regarded as statements that the music of Brahms is reinforcing.'[69] To an irreverent reader, this might suggest an experiment in which a human subject is settled in a comfortably furnished Skinner box in which pressing the bar switches on the music of Brahms. As far as I know, this experiment has not been performed. Hence we do not know, for any class of persons, how often Brahms would be switched on – or how soon boredom would set in: in the long run, the effect might be like that of the experiments on sensory deprivation.

More seriously, saying that something is reinforcing adds nothing to the statement that it is enjoyable. By avoiding reference to enjoyment, one merely omits reference to the inner experience of what is enjoyed. As Noam Chomsky remarks, 'The real content of Skinner's system can be appreciated only by examining such cases, point by point.' He adds:

> We can get a taste of the explanatory force of Skinner's theory from such (typical) examples as these: a pianist learns to play a scale smoothly because 'smoothly played scales are reinforcing'; 'A person can know what it is to fight for a cause only after a long history during which he has learned to perceive and to know that state of affairs called fighting for a cause'; and so on.
>
> Similarly, we can perceive the power of Skinner's behavioral technology by considering the useful observations and advice he offers: 'Punishable behavior can be minimized by creating circumstances in which it is not likely to occur'. If a person 'is strongly reinforced when he sees other people enjoying themselves . . . he will design an environment in which children are happy'. If overpopulation, nuclear war, pollution, and depletion of resources are a problem, 'we may then change practices to induce people to have fewer children, spend less on nuclear weapons, stop polluting the environment, and consume resources at a lower rate, respectively'.[70]

The quotations are from Skinner's *Beyond Freedom and Dignity*. This strange work was received with widespread interest, and often respect. A reviewer in *The Times Literary Supplement* described it as 'a serious contribution to modern thought . . . well written, provocative and thoughtful'; and another, in the *Economist*, found it 'persuasive, coherent, fluent, gripping stuff'. As Chomsky's examples show, much of the argument is vacuous: we can encourage conduct that we like by arranging that it occurs, discourage what we dislike by preventing it from happening. The emptiness is especially clear in the treatment of the central concept of reinforcement: in effect, an action is said to be positively reinforcing (or rewarding) if it is repeatedly performed; and it is repeatedly performed because it is positively reinforcing.

In Chapter 5 I describe how the theory of natural selection is, in one aspect, a tautology, or system of tautologies. As sometimes stated, that theory, too, is empty. But, the mathematics of evolutionary theory can guide investigation of the real world. In contrast, neo-behaviorist tautologies lead nowhere.

The blinkered preoccupation with reinforcement is also reflected in the concept of behavioral engineering. Skinner, during most of his life, has looked forward to a future when his notion of scientific psychology would vastly improve the human lot. A radio discussion with a philosopher, Geoffrey Warnock, included the following exchange:

> SKINNER: I would attribute my optimism to some rather substantial demonstrations that things can be done that we used to think impossible.
>
> WARNOCK: Yes. That's still going to leave us with the question whether we should do them or not.
>
> SKINNER: Yes: I think that is a question. And if I am convinced that we should, then it is up to me to apply my behavioral engineering to convince you and others and get things done.
>
> WARNOCK: I would rather you tried to persuade me.

But persuasion implies attention to inner feelings, and it may require logical argument. Neither is recognized as significant by Skinner: opinions are not to be altered, only behavior, by reinforcement.

Since, for strict neo-behaviorists, human beings are only objects, for them a statement of intention lacks its usual significance. Suppose a person, asked why she is going into the garden, replies that she wants to pick some flowers. An objective observer might have inferred this from previous experience of the person's behavior. But what of the actor herself? According to neo-behaviorism (assuming that it is consistent), she too is inferring what she is about to do from her own previous experience.[71]

But in fact a statement of intention is of a kind quite different from one based on observation. 'If you see', writes Norman Malcolm,

> someone rummaging about . . . on his desk, and remember that . . . on previous occasions the rummaging had come to an end when he grabbed hold of his spectacles, you might reasonably conclude . . . that he is now looking for his spectacles. But it would be weird if *he* were to reason as follows:

'Here I am rummaging about on my desk. When I have done this
in the past my activity has terminated when I have caught hold of
my spectacles. Therefore, I am probably looking for my spectacles.'[72]

If a person made such a remark, one would suppose him to be either joking
or deranged. A statement of intention is not a prediction based on a series
of earlier experiences. John Shotter, in the same vein, asks us to imagine
a situation in which he says to a woman, 'I love you'. He writes that, in
saying this,

> I usually do not mean that I have carried out an extensive observa-
> tional analysis . . . of my inner states and outer behaviour – my
> heart rate, breathing, sexual arousal, time spent in looking at her,
> and attempting to be near her, . . . and have made up a report
> which sums up the results.[73]

As he says, there *are* occasions when one resorts to such self-analysis – to
observing oneself as if from outside; but they are the exceptions.

It is, of course, possible to fit at least some personal statements into a
behaviorist's framework. If I say I want my supper, and still more if I say
I am hungry, that can be related, at least in principle, to the state of my
blood, stomach and other organs; and the statement that I am hungry is
likely to be followed by easily predictable behavior: I eat. Even so, what I
say about my hunger remains the best source of information for others
about my state. 'People', writes Malcolm,

> tell us things about themselves which take us by surprise, things
> which we should not have guessed from our knowledge of their
> circumstances and behavior. A behaviorist philosopher will say that
> if we had known more about their history, environment, and
> behavior, we should have been able to infer this same information.[74]

But as Malcolm adds, this is a presumption, not a conclusion founded on
evidence.[75]

Because neo-behaviorism does not allow the existence of persons, it has
no place for personal responsibility. Hence it rejects morality. In *Beyond
Freedom and Dignity* Skinner refers, for example, to the rule, 'Thou shalt
not steal'. It is usual to presume that a person may choose whether to
accept the rule or to reject it. Skinner does not agree that a person is free
to decide such questions: what he calls 'autonomous man' is incompatible
with his philosophy. Whether a person obeys the rule, he says, 'depends
upon supporting contingencies'.[76] He is referring, of course, to the person's
previous experience of reinforcement. The implication, as usual, is that all
our apparently independent decisions are determined by previous rewards
and punishments. We never decide for ourselves, on principle, to turn down

an attractive offer; nor do we stick to our guns in the face of disagreeable consequences. In this passage, as elsewhere, human conduct is reduced to that of a confined animal controlled by the immediate effects of reinforcement by food or drink, or by the avoidance of pain.

A manipulator's utopia

Nonetheless, neo-behaviorists have, paradoxically, prescriptions for social action. The 'possibility of free choice' is explicitly questioned; yet we are urged to 'improve our understanding of human behavior' so that we can solve the problems that face us.[77] As Max Black remarks, 'The spectacle of a convinced determinist *urging* his readers to save the human race is bound to be somewhat comic.'[78] If all we do is determined by the external forces of reward and punishment, regardless of anything we believe we want, recommendations that imply choice are pointless. Indeed, the recommendations themselves, being only the consequences of reinforcement on the person making them, have no rational foundation. (A similar paradox arises from the neo-Darwinian image of humanity, described in Chapter 6.)

When we judge the neo-behaviorist prescriptions for social action, a crucial question is who controls whom. A collection of neo-behaviorist articles, entitled *Control of Human Behavior*, is introduced with the statement: 'all those who seek to modify human behavior can benefit by a knowledge of the principles . . . derived from . . . experimental analysis.'[79] But the writer does not tell us which of the many kinds of control he advocates. Control of children by parents (and even of parents by children) is both necessary and (usually) benign. Another often welcome kind of control is exerted by an expert, such as a physician or a lawyer; in this case the client is not, as a rule, forced to consult the specialist, or to accept the advice given. Other usually acceptable kinds of control include those exerted by the police and judiciary of a free society. Sinister implications of 'control' arise from the actions of a dictator, a tyrannical boss or jailer, a slave owner or even a certain kind of teacher: control is then arbitrary and determined by such factors as the desires of a ruler, or the presumptions of an authoritarian.

These are all examples of external control: they say nothing about self control. Skinner, as we know, recognizes only external controlling forces. 'A scientific analysis of behavior', he writes, 'dispossesses autonomous man and turns the control he has been said to exert over to the environment.'[80] But Skinner's writings often suggest that, in effect, he is not concerned with control in any usual sense, but only with action and reaction as in – for example – a chemical system: when two substances combine to make a third, we do not say that one kind of molecule controls the other. Skinner writes:

> In noticing how the master controls the slave or the employer
> the worker, we commonly overlook reciprocal effects and . . . are
> led to regard control as exploitation, or at least the gaining of a
> one-sided advantage; but the control is actually mutual. The slave
> controls the master as completely as the master the slave, in the
> sense that the techniques of punishment employed by the master
> have been selected by the slave's behavior in submitting to them.[81]

Many readers (not only slaves and exploited workers) will, I suppose, find
this passage ludicrous; but it is consistent with the author's presumptions.
A slave, faced with the choice between submission and risking death or
flogging by running away, would not accept Skinner's analysis; nor would
a master, who could choose between keeping and freeing the slave. But
these statements imply that people have desires and intentions that influence
what they do, and that they have some freedom to choose. Skinner and
his followers reject such commonsense presumptions.

Yet the methods of neo-behaviorism continue to be used and recommended. S.B. Stolz, in a work published in 1978, shows how people are
manipulated by 'behavioral technology'; her method uses rewards (sometimes only tokens) or punishments, but not explanations: there is no appeal
to reason or to principle, and she looks forward to its use on a larger scale
as a 'prospect for the future'.[82]

This is a frightening portent. 'Because neo-behaviorism has no theory of
man', writes Erich Fromm, 'it can see only behavior, and not the behaving
person.' A man, he goes on, may smile because he wants to hide a feeling
of hostility; a salesgirl may smile because she has been instructed to do so;
or one may smile with pleasure on seeing a friend; but for neo-behaviorism
'a smile is a smile' regardless of its context. And Fromm asks whether a
neo-behaviorist can really be unaffected by such differences: is he 'so alien-
ated that the reality of persons no longer matters to him?'[83]

The original neo-behaviorist proposals for a new society are in Skinner's
novel, *Walden Two*.[84] In an autobiographical sketch Skinner writes of
Francis Bacon's utopia, *New Atlantis*. Bacon (1561–1626) is justly regarded
as the social philosopher who announced the coming of an age dominated
by the natural sciences. Science was to be the guarantee of human progress
and enlightenment. *Walden Two*, we are told, is Skinner's *New Atlantis*.[85]
Laboratory methods of controlling behavior are to be used in a comprehensive design for human betterment. The book describes a community in
which behavioral engineering, in the hands of experts, is to make possible
a social order founded on objective knowledge. Socially acceptable conduct
is to be systematically induced, and even ensured, by reward (not punishment). The system is, however, not represented as complete: more information is needed, and is to be acquired by experiment.

In this imaginary society,

> all that happens is contained in an original plan, yet at every stage
> the individual seems to be making choices and determining the
> outcome. . . . Our members are practically always doing what they
> want to do – what they 'choose' to do – but we see to it that they
> will want to do precisely the things which are best for themselves
> and the community. Their behavior is determined, yet they're free.[86]

The speaker in this passage is the man who has planned and is managing
the community; and another character appropriately accuses him of playing
god. But, despite this element of self-deprecation, and the inconsistencies
already mentioned, the implications of Skinner's plans for a 'scientific'
community are clear: he looks forward to a tidy society in which everyone
is happy and managed; the inhabitants are to be manipulated by a benign
– and remarkably well informed – dictator or perhaps an all-powerful
committee. Here is another comment by Fromm:

> Skinner's implied optimism alone would not have made his ideas
> so attractive were it not for his combining of traditional liberal
> ideas with their very negation.
>
>   In the [present] cybernetic age, the individual becomes increas-
> ingly subject to manipulation. His work, his consumption, and his
> leisure are manipulated by advertising, by what Skinner calls 'posi-
> tive reinforcements'. The individual loses his active responsible role
> in the social process . . . What a relief [to him] to learn that condi-
> tioning is the best, the most effective solution. Skinner recommends
> the hell of the isolated, manipulated man of the cybernetic age as
> the heaven of progress.[87]

The neo-behaviorists' Utopia looks forward to steady improvement in
the human condition: yet its authoritarian presumptions ensure that in fact
progress will be difficult or impossible. To quote K.R. Popper,

> The authoritarian will in general select those who obey, who believe,
> who respond to his influence. But in doing so, he is bound to select
> mediocrities. For he excludes those who revolt, who doubt, who
> dare to resist his influence.[88]

It is just the 'divine discontent' of these excluded persons that prevents
stagnation and holds out hope of progress.

Real communities

Like most Utopias, that of Skinner is an exercise on paper; but
some of his followers have had the courage to found and to live in colonies
ostensibly run on the principles of behavioral engineering. Their exertions
enable us to test *Walden Two* against the real world. R.E. Ulrich, an

American experimental psychologist, describes how he became interested in 'behavior modification', and turned from research on rats and monkeys to organizing an experimental community, Lake Village, in western Michigan. 'I really believed', he writes,

> that behavioral engineering could make profound changes in human life and that, with a little caution, these changes would surely be for the better. . . . For the most part, Skinner was our leader. . . . Utopia seemed at times just within reach.

And he looks forward to 'expanding the experimental laboratory into the less controlled environment of our everyday society.'[89] Ulrich also refers to the increasing use of 'behavioral strategies', by industry to raise productivity and by the army and the prison system to regulate the conduct of soldiers and convicts.

> Virtually every institution in the United States has been touched by behavior modification, if only in feeling a need to erect defenses against it.[90]

When Ulrich and his associates set up their colony, they had already had experience of applying behavioral engineering in a day-care center for 60 children. The care of young, energetic individuals presents problems of personal relationships which are easily imagined. As well, Ulrich refers to 'problems with fire marshalls, zoning laws, and public health inspectors'; and, he adds,

> since our program is run by persons who some would call freaks, we are being watched in other ways.[91]

When the community of Lake Village began, there were also formidable difficulties in financing it. A reader may say impatiently that such obstacles are irrelevant: what we are concerned with is the role of behavioral science. But the design of a community, not on paper but in actuality, requires attention to medical, legal, economic and other administrative matters, as well as those supposed to be covered by schedules of reinforcement.

In their behavioral engineering, the colonists met difficulties which, in retrospect, seem predictable. In *Walden Two*, the community is managed by its founder, Frazier. In Lake Village, Ulrich tells us, there is no Frazier, that is, nobody who has full control and complete understanding of what is happening. And on regulating the activities of members through calculated rewards, he writes:

> If I sat down at one of our Sunday morning meetings at Lake Village and proposed a system for monitoring and differentially reinforcing everyone's social interactions, I'd be laughed out of the house.

Ulrich also remarks on the need for trust and affection among colleagues, and on the questions that arise because we are not at all sure what makes people happy. He concludes: 'There is no Walden II nor will there ever be.'[92] Thus, in his report on Lake Village, he and his friends appear as persons, not as preparations responding to schedules of reward.

Despite some degree of disillusion, Ulrich retains a belief in neo-behavioristic methods – which makes his frank account especially valuable. For here we see once again how a system that reduces human action to single, narrowly restricted principles fails when faced with reality. Above I describe attempts to use the methods of 'Pavlovian conditioning' to explain learning and to treat mental illness. They fail because simple explanations do not match the facts even of animal behavior, let alone ours. The same applies to the neo-behaviorism of Skinner.

It does not, however, follow that all attempts to design human communities are bound to fail. Not all have had such a narrow foundation.[93] I take two modern instances.

In the 1930s M.K. Ghandhi (1869–1948), the 'maker of modern India',

Farmer in a prosperous rice-growing region of Karnataka, southern India. What will the world's vast peasant populations become in the twenty-first century?

founded an All-India Village Industries Association, and himself went to live in a village in the Deccan. There he set up an *ashram*, or settlement, in which he experimented on village agriculture and other industry, and on education. Gandhi had already travelled widely in India, and seen the village life of many regions in intimate detail. His political views included both a distant vision of a possible Indian society and severely practical proposals for immediate action. The moral foundation of his ideas was a form of pacifism or non-violence, combined with a rejection of the notion that wealth and technical progress are necessary for happiness. He had, to some extent, an answer to the question that baffled Roger Ulrich.

Gandhi's proposals include village industry, such as the use of hand looms for making cloth from materials grown locally. Every adult is to contribute to the production of necessities. Children are first to be taught crafts, only later the conventional letters and numbers. Much emphasis, with good reason, is put on teaching hygiene. Education is to continue in adult life.

To some agronomists accustomed to modern 'agribusiness' these proposals have seemed doctrinaire and reactionary. But technically advanced farming uses little labor – a plentiful commodity in India; and it is greedy for fuel and electricity. Such a technology, applied to most of the world's peasant populations, would destroy the social structure and the environment; it would also soon run out of fuel.[94] Gandhi's proposals were therefore in this respect realistic. The extent to which they are applicable is, however, still debated. India has more than half a million villages, and Gandhi's principles have been applied in only a fraction of one percent. But, valid or not in their original form, they cover a great range of aspects of village life, from the aesthetics of traditional Hinduism to modern social medicine. Their many-sidedness is one reason for their value.

Our second case comes from Israel. The Israeli agricultural settlements (some of which are now also industrial) began before the modern state of Israel existed. Today the land of a kibbutz is owned by the nation, and is worked by a group of people whose property (apart from such items as clothing) is held in common.[95] There are now more than 200 kibbutzim. Like the Gandhian settlements, they are founded on strongly held beliefs in the moral value of labor and in a person's responsibility for other members of the community.

Some of the Israeli settlements have continued for half a century, changing with the members' experience and with innovations in agriculture. Most have only a few hundred members; few have more than a thousand. They continue to be egalitarian and self-managed; and they resemble an *ashram* in being quite unlike any 'commune', real or imagined, controlled by a single leader or by an elite. They also differ sharply from groups set up so that the members can escape the threats and responsibilities of the normal world, for they are complete, viable communities. Today their members include many who have lived all their lives in a kibbutz, and have been educated, even to tertiary level, in them. Such people accept communal living and a socialist economy as natural and normal.

The survival of the kibbutzim depends, however, on what happens in the state of Israel, and hence on economic and political events outside it. Nonetheless, they have succeeded for long enough to contradict some common presumptions. They show how people can work together in a community without the incentives of riches or self-aggrandizement. (Compare the account of the San people on pages 66–7.)

The cynic's conventional response to such optimism is the cliché that one cannot change human nature: it is, he holds, our nature (and his) to exert ourselves only for the rewards of wealth, power and prestige. This libel on our species is obviously false, but it does raise difficult questions: why do we work, or fail to do so? For that matter, why do we play? In the next chapter I suggest some answers.

11 Work and play

'What is a cynic?'
'A man who knows the price of everything and the value of nothing.'

Oscar Wilde

The biological images of humanity represent us as consistently serious: we relentlessly and aggressively further the survival of our genes, and adapt our responses with computer-like regularity to our experience of punishment and reward. Yet we are in fact rather a frivolous species. We have called ourselves not only rational and tool-making (*Homo sapiens* and *faber*), but also *H. ludens*, or gamesman.[1]

We are, however, accustomed to distinguish sharply between work and play; and, according to proverbial lore, all work and no play makes Jill a dull girl. But in practise, as we see below, it is often difficult to decide whether an activity is one or the other. The result has been difficulty not only for biologists but also for economists. Human beings work for food and other necessities, but not only for them. Sometimes they fail to work when, on any simple, rational calculation, they should do so; and, more interesting, sometimes they work without need. One result is that economists' models or images of humanity have limitations strangely similar to the biological images. Economists try to reduce the complexities of human motivation to simple principles; and their stories also resemble the biological images in having political implications. The present chapter continues the account of the discrepancies between simplified accounts of human action and what we actually do; and we begin with those of economists.

Economic persons

According to Latin grammarians, '*artifex* and *opifex* common are to either sex '. The same applies to Economic Man. This creature is competitive, yet rational, and is so well informed (and energetic) as to be able always to buy at the lowest price and to invest in the most profitable way.[2] Economic Man is a cousin of yet another prominent image: 'The Common Law of England', writes a lawyer, 'has been laboriously built about a mythical figure – the figure of "The Reasonable Man".[3] Economic Man, like *H. egoisticus* bent on maximizing fitness, is often a starting point for calculations. He has been called the totem of economists, but not all economists regard him as a useful account of themselves and their fellows. According to Amartya Sen, the image entails the presumption that,

> when asked a question, the individual gives that answer which will maximize his personal gain. How good is this assumption? I doubt that in general it is very good. ('Where is the railway station?' he asks me. 'There', I say, pointing at the post office, 'and would you please post this letter on the way?' 'Yes', he says, determined to open the envelope and check whether it contains something valuable.)[4]

Another economist, D.M. Bensusan-Butt, describes several possible types of society. In one there is a bias toward non-economic activites, such as religious practises, and a strong emphasis on social duties beyond as well as within the family. (This description could apply to the European Middle Ages.) For Economic Man, he says, such a community is 'ultra-stupid'. Bensusan-Butt himself regards Economic Man as a 'mischievous and misleading myth'.[5] If he is right, here is a further marked resemblance to *H. egoisticus* and *operans*.

Yet simplified economic portraits of humanity are still popular, and their success is often due to their social implications. When Economic Man owns a business, he buys and sells at the most advantageous prices, and he competes in a free market with others. This is private enterprise. Correspondingly, his customers test the value of the goods or services he offers against that of his rivals. Whether producer or consumer, writes J.K. Galbraith, 'He never stops working on the excuse that he has enough money for the moment and would prefer idleness.' (The similarity to *Homo operans*, working relentlessly for clearly defined material rewards, is clear.) 'The essence' of this system, in Galbraith's words,

> is that individuals using income derived, in the main, from their own productive activities express their desires by the way they distribute this income for the various goods and services available to them in markets.

Hence the economic system of free enterprise is held to work automatically for the benefit of each consumer: 'satisfaction, even happiness, is maximized'.[6]

F.A. Hayek, a Nobel prize-winner in economics, takes the stand that any attempt to alter or to replace the system is perverse, even wicked. 'We have become a worldwide, peaceful, and prosperous society', he writes,

> by having learned to . . . follow certain abstract rules of honesty – rules establishing property and ultimately codified in the form of

private law. It is the rules of property and contract on which the growth of a worldwide, peaceful, prosperous society was based.[7]

The market economy, which – Hayek holds – is the only means of ensuring peace and prosperity, is said to be an outcome of unconscious social selection, analogous to natural selection. Attempts by liberals or socialists to create a better society based on reason are therefore regarded as destructive of social order. Hayek, however, sees little hope that people will obey his rules. This is hardly surprising: the market economy, the Economic Persons who operate it and Hayek's peaceful and prosperous world are myths. Such a utopia would have no place for gigantic corporations which smother competition, and still less for those that fall back on governments to bail them out of difficulties – both of which are commonplaces of the modern state. Nor would governments endow industries to enable them to export products at artificially lowered prices. The only advertisements would be catalogues of available goods. Public services would consist of little more than an army and police forces. Roads and sanitation would depend on private enterprise. Health services, schools and universities would be run for profit and for the benefit of those who could afford them.

Yet some governments have tried to base their policies strictly on the operation of 'market forces'. They are then often criticized for blindness to the realities of human action and human needs; but they are also inefficient. As I write, one obvious outcome is unemployment on a vast scale: while politicians advertise the virtues of personal industry and dutiful labor, their policies force millions to experience the calamity of idleness. But not only the workless suffer: the whole community loses the contributions of those excluded from productive work.

In practice, however, the economic model is only incompletely applied: the most doctrinaire advocates of a market economy do not (as far as I know) recommend reading the Sale of Goods Act as part of the marriage ceremony. More seriously, even in the countries most under the sway of monetary values, schools and universities depend largely on public funds; so do services for the health of the poor and the old, and for the welfare of mothers and children. These are among the essentials with which the market cannot cope effectively. Similarly, only legislation that interferes with market forces can protect people from industrial pollution or contaminated food, or from the effects of drugs such as nicotine. Hence current economic images of human society, like those derived from biology, have a prominent political role but fail to match the real world of human action. In any society, capitalist, socialist or hybrid, the part played by market forces depends on decisions concerning who is to benefit. The primary choices are moral, not technical.

Economic models are, of course, often much more elaborate than these comments convey. Among them are those that deal systematically with questions of choice: when an individual, a corporation or a government has to choose between alternative goods, what determines the decision finally made? In this context, 'goods' include much more than material objects – for instance, access to health services or radio programs, and the right to work or to enjoy leisure. When they try to answer, economists again make images or models of human behavior.

As an instance, H.A. Simon describes the theory of subjective expected utility (SEU), which is designed to analyze what happens when an individual makes a private choice: this may be minor, as between new shoes and new records, or major, as between a secure job and temporary but morally superior work abroad. The model, unlike Economic Man, recognizes that frivolous pursuits, such as chess and listening to music, and serious but non-economic activities, such as debating the nature of the good life, are valued by many people. But, as Simon writes: 'The SEU model assumes that the decision maker . . . understands the range of alternative choices open to him . . . over the whole panorama of the future. But . . . SEU theory has never been applied, and never can be applied – with or without the largest computers – in the real world.'[8]

This is an extreme case. Current theories of public choice are less ambitious but still deal with difficult questions. They examine decisions on how to share out scarce resources: among these are material goods, such as bread and vintage claret, and services, such as those provided by hairdressers and hospitals. As in biology, the problems are first classified. For instance, in any complex society, regardless of its political type, one may distinguish between 'inclusive' and 'exclusive' public goods. The former remain unaffected by an increase in the numbers of those who benefit from them: an example is an Act which regulates the hours worked by people in an industry. In contrast, an exclusive good deteriorates with crowding – for instance, access to a wilderness. In a democracy, citizens should, ideally, understand the various kinds of dilemma represented by these examples. Often, as obviously in the case of the wilderness, moral and aesthetic principles are of leading importance. These cannot usefully be given numerical values. Hence policies that disregard everything that cannot be measured are ill founded and unlikely to stand for long.

The concepts of public choice have several instructive resemblances to theories on the evolution of behavior. They use similar mathematics and, as a political scientist, I.S. McLean, remarks, they make a 'most impressive intellectual achievement'. They also illustrate yet again the fascination exerted by formal, simplified models or images of complex circumstances. McLean refers to the 'god-like status' assigned to public choice models by economists, and so reminds us of the deification of natural selection described in Chapter 6. Correspondingly, the theory of public choice has severe limitations, resulting in part from the 'meagreness' of its presumptions.[9]

These limitations can be exemplified without formulae. 'Chicken' games, for instance, illustrate the way in which a single logical structure can be applied to a great variety of situations which in practice are quite dissimilar. They are named after a contest in which two (one may think, remarkably stupid) people drive their cars straight at each other: the (less stupid) one who swerves first, or chickens out, loses. (He may be said to lose face.) A primary feature of a chicken game is the absence of any external, overriding power, or sovereign. A second requirement is that one or both parties must give way, and each hopes the other will do so. The most famous of 'chicken' encounters occurred in 1962, when the American president, J.F. Kennedy, and the head of the Russian government, Nikita Khruschev, clashed over missiles in Cuba. On this terrifying occasion, Khruschev withdrew ('swerved') and, to quote McLean again, 'the world's most desperate game of chicken ended co-operatively.'

McLean also describes a situation closer to everyday experience. A 'good and busy' man walks to work from Jericho to Jerusalem, and another commutes in the other direction. One day, as they approach, they see a traveller lying by the road. Both are concerned that the injured man should be helped; each hopes that the other will do the helping. But 'both men look at their watches in a preoccupied sort of way, and hurry past . . . , leaving the injured man to his fate.' Presumably, if this sort of thing happened regularly, the two would agree to help (that is, to 'swerve') on, say, alternate occasions. They would then achieve a common sense solution to a simple supergame, that is, one in which the situation recurs.

The limitations of formal models of human conduct are soon exposed when we turn from parables and ask questions about real events. Although people often help strangers unselfishly, on some well documented occasions persons obviously in urgent need have been ignored by others. Failing to help is said to be likely when others, who might also help, are present. It would be interesting to know whether hearing of such occurrences has led anyone, in similar circumstances, to take effective action. And it would be still more interesting to know what lessons politicians have learned from the Cuban affair. Such questions are, it seems, rarely asked, and still less often answered. Yet people do learn from experience, their own and others'; they can even learn to behave more morally than before. Such complexities of conduct are, as a rule, more usefully described colloquially than in terms of formal models: the Cuban crisis and the lessons learned from it, to be *understood*, require historical analysis.

The formal theories of economics often give us instructive *descriptions* of problems that face us in a democracy – descriptions more elaborate than those founded on Economic Man; but, when we look for solutions, theory usually has to give place to common sense. Our circumstances, and the ways in which we try to cope with them, are a resultant of more, and more complex, forces than can be put into a computer program; not all can be measured. Hence, to quote a political critic, 'economics is not equipped to deal with the wider political issues over which it has shown increasing domination.'[10] When they enter politics, economists navigating by algebra alone are evidently often in the same boat as sociobiologists – very much at sea.

Uneconomic persons

The difficulties are further shown if we return to the real world of human motives, and begin by asking what happens when one wants a supply of fresh human blood. R.M. Titmuss has described an enquiry into the morals and economics of giving blood for medical use. He contrasts the situation in the USA, where blood is paid for, with that in Britain, where donors receive only a cup of tea after they are bled; and he describes answers to questions put to unpaid blood donors. No donor gave an exclusively altruistic answer, for giving blood was rewarding in the sense that it evoked the interest and approval of others. But nearly all also gave moral grounds as the principal reasons for giving blood.

> For most of them the universe was not limited or confined to the family, the kinship, or to a defined social, ethnic or occupational group or class; it was the universal stranger.

The supply of blood in the USA, but not in Britain, entails much waste. Titmuss therefore asks whether a cost-benefit analysis would be useful; but he declines to attempt one because, he says,

> these economic wastes, immense as they are, are completely dwarfed by the social costs . . . of the market system and the social benefits of an alternative voluntary, gift-relationship system. . . . No money values can be attached to the presence or absence of a spirit of altruism in a society.[11]

Moreover, monetary payment deprives donors both of the freedom to choose whether to give altruistically, and of the reward of social approval.

Blood donors, paid or unpaid, are only a minority in any population. Economic theory is more usually applied to the world of employment and wages. Economic Man and his offspring developed during the industrial revolution. During this period the conditions of work of large populations were transformed. Until the nineteenth century, even in western Europe, most men, and many women, worked on the land or in individual crafts. Their work was diverse, and each person had to make frequent, if minor decisions on what to do next. In the new factories, workers were progressively deprived even of minor decision-making. 'By the later 1890s,' writes Norman Stone,

> tales of the 'mass-production' methods of America were amazing Europe. A Henry Ford, turning out motor cars by 'flow methods', . . . with semi-skilled workers at a moving belt perform-

ing the same operation, mindlessly, and then passing on the piece to the next worker for a different operation, was a portent.[12]

The theory of mass production was elaborately developed by an American engineer, F.W. Taylor (1856–1915). A nineteenth-century factory usually employed many skilled workers, each of whom was responsible for a finished item. 'Taylorism' entailed employing semi-skilled hands who performed only a single operation. The movements of the workers were laid down in detail, on the presumption that the most cost-effective procedure could be determined by what Taylor called scientific study. 'Under our system', he writes, 'the workman is told minutely just what he is to do and how he is to do it, and any improvement which he makes . . . is fatal to success.' No individual variation was allowed for. The workman, it was held, was and should be 'so stupid and so phlegmatic that he more nearly resembles in his mental makeup the ox than any other type.'[13] 'It was a formula for extraordinary tedium, which managers and unions in the older established industrial centres regarded with horror.'[14]

The anagram, therblig, which means a standardized pattern of movements, acknowledges the work of another American, L.M. Gilbreth. She describes how not only the workers' movements, but also their rewards, were precisely calculated:

> Under Scientific Management, with the ordinary type of worker on manual work, it has been found most satisfactory to pay the reward every day, or at the end of the week, and to announce the score of the output as often as every hour.[15]

Both Taylor and Gilbreth sound as though they had been influenced by neo-behaviorism, but in fact their proposals were published before the First World War. Hence the neo-behaviorists, in the 1930s and after, were in fact using methods like those of Taylorism to study habit formation by animals. An animal performing an act (an operant) 100 times for a fragment of food has something in common with a factory hand on an assembly line.

In the decades before the First World War, the metaphor of the machine was prominent. That was the period when a mechanical account of the conditional reflex first caught the popular imagination. (Today, we are more likely to think of ourselves, or at least of our brains, as computers.) In the same period behavioristic intelligence tests began to make their appeal in schooling. Infancy did not escape the influence of the machine age: authorities on child care laid down schedules for the feeding of babies from the bottle. An interval of four hours was strictly enjoined (by men) on mothers. Correspondingly, the cupboard-love theory of infant attachment was part of conventional wisdom: attachment to a parent, nurse or other adult was presumed to develop as a result of associating the person with material reward – usually, food.[16] (Today we know this to be incorrect even for monkeys.)

Such an outlook was not necessarily inhumane. The recommendations on feeding infants were designed to help inexperienced mothers to rear healthy children. Even factory managers sometimes paid attention to the workers' well being; but, on this attitudes were ambivalent. Gilbreth has a chapter on welfare work. This phrase covers building rest rooms, providing amusements and employing a welfare officer. Such measures are said to deserve 'nothing but praise'; but the author goes on at once to condemn them. She also states: 'Under scientific management, there should be no necessity for a special Department of Welfare Work.' If there are welfare personnel, they 'must go on the payroll as part of the efficiency equipment'.[17]

Later managerial developments include the well established Hawthorne effect, conveniently described by means of a parable. A humane management decides that working hours are too long, and reduces them, in the hope that nonetheless there will be no decline in output. To everyone's surprise, output *rises*. Later, in a slump, the firm gets into difficulties, and the management, with apologies, lengthens hours. Total output, *and output for each person-hour*, both rise. Later still, during a recovery, it is decided to improve the canteens and other amenities. Again, output rises. The common feature of all these changes is that the bosses are seen by the workers as taking an interest.[18] Such a subjective factor in production is not easily included in the calculations of a behaviorist or an economist.

Modern industry, run largely on principles developed from Taylorism, has produced societies of unprecedented wealth. Most readers of this book will be living far above the level of basic subsistence, and will own equipment that, a few decades ago, would hardly have appeared outside works of science fiction. But these goods have been acquired at a cost that cannot be measured in money, or by the methods of neo-behaviorism. Vast numbers of people have been deprived of personal involvement in their work or in what they produce; and the loss entails estrangement from the community, especially its government and the 'faceless' persons who manage public affairs. This is the alienation that results from living in a machine age. Moreover, in our society the greatest conventional rewards are earned by those who accumulate riches. Such people acquire not only wealth, but power; and they create communities in which the achievements of the opulent come to represent a moral principle. In 1983, an American president, Ronald Reagan, said:

> What I want to see *above all else* is that this country remains a country where someone can always get rich. That's the thing that we have and that must be preserved. [Emphasis added.][19]

In this way the young are taught to work only for material reward, and greed becomes a leading principle of conduct. Fortunately, they do not all learn this lesson very well.

Incentives

To those who do accept such teaching, asking what work means to people must seem naive: surely work is a means of earning rewards in the form of money, and the more one earns the more one can enjoy life in one's spare time. Such a view is convenient for neo-behaviorists, and also for economists. 'Modern economists', writes one of them, 'have nothing to say on whether work is pleasant or unpleasant.'[20]

Historians and social scientists have, however, much to say. A historian of labor, E.J. Hobsbawm, remarks that pleasure in work is commoner than is often supposed, and that among those who experience it there is a strong 'moral stigma against slacking': workers who feel some self-respect do not 'apply market criteria to the measurement of their efforts'.[21] Hobsbawm is referring especially to studies of European workers after the First World War. A majority of skilled workers are described as finding more pleasure than dislike for their work; and nearly half even of the unskilled workers did so too.

Workers are conventionally presumed to rebel against low wages: pay is what they value. This, the impoverishment theory, sometimes fits the facts, but the exceptions are equally significant. During the nineteenth century, and early in the twentieth, strikes and other disputes were commonly initiated by the skilled and higher paid. Some of the most important were by artisans 'incensed by the loss of dignity and status that had previously come with independence'.[22]

Another non-economic factor is loyalty to an organization, to a country or to a cause. Sen remarks on the importance of commitment. Every economic system has had to rely in part on attitudes to work which transcend 'the calculation of net gain from each unit of exertion'. It is a hopeless task, he says, to run an organization entirely on the incentive of personal gain.[23]

Dignity and commitment, unlike wages and hours, are difficult to fit into econometric equations; but the extent to which people like their work can be studied. In 1977, 5471 Finns answered questions about job satisfaction and exhaustion resulting from their work.[24] An important finding was the ill effect of pressure to complete an operation on time: the more a worker could control his or her own work, the greater the satisfaction and the less the fatigue. Both individual and social control had favorable effects. The resulting recommendations included 'active participation of workers' as 'the key element' in the reforms needed. Similarly, in West Germany, it is reported that millions would be happy to give up full-time, well-paid work that they dislike, and instead work part-time in conditions in which they would be 'more autonomous, and more dedicated to the pursuit of non-materialistic values.'[25] The Finnish authors also mention a difference in the attitudes of the sexes: the men tended to emphasize self-regulation and high wages; but 'women wanted their work to be socially meaningful to others'.

The Finnish and German findings show once again that workers need to express something of themselves in the work they do. 'Welfare' measures, however desirable in themselves, do not help. It has been hoped that new methods of production would erode 'Taylorism', and once again give workers more discretion. The old form of the assembly line is being replaced by procedures which could enable workers to see, and to take some responsibility for, what they produce. Modern technology can, rightly used, vastly improve the character of industrial work in all departments. But there is also an opposite process. M.J.E. Cooley quotes an English shop steward's comment on the widespread attempts to make modern methods of production acceptable without allowing workers any influence on decision-making: it is, he said, 'like keeping people in a cage and debating with them the colour of the bars.' Similarly, a foreman in an English factory comments on the effects of 'automating' a production line. 'Previously I had certain jobs to handle. I had certain headaches to take care of. That is what I'm paid for. Now I'm just pushing buttons. Doing nothing.' The same changes are taking place within computing itself. A computer operator has described himself as a tape-ape; and, with excusable incoherence, he also likens himself to one of Pavlov's dogs: 'give us a screen prompt and we leap into action with tongues hanging out.'

Cooley holds that the computer is the Trojan Horse which is bringing Taylorism even into intellectual work. He quotes the following statement, made as the outcome of discussions among the systems engineers of a large American company. (Cooley remarks that until he had checked that it was authentic, he 'could not believe that it could be serious'.)

> Our immediate concern . . . is the exploitation of the operating unit approach to systems design no matter what materials are used. . . . What we need is an inventory of the manner in which human behavior can be controlled, and a description of some of the instruments which will help us achieve that control. If this provides us with sufficient handles on human materials so that we can think of them as metal parts, electrical power or chemical reactions, then we have succeeded in placing human material on the same footing as any other . . . There are, however, many disadvantages in the use of those human operating units. They are . . . subject to fatigue, obsolescence, disease, and even death. They are frequently stupid, unreliable, and limited in memory capacity. But beyond all this, they sometimes seek to design their own circuitry [that is, they rebel?]. This in a material is unforgivable . . .[26]

This bizarre passage is not an isolated aberration: just as some neo-Darwinists present human beings, and therefore themselves, as unscrupulous egoists, so some systems theorists, like Nietzsche, mindlessly speak of people, and hence themselves, as though they were ill designed machines. One of them, G.E. Pugh, explicitly blames nature or, sometimes, evolution, for incompetent workmanship.[27]

If human beings are treated as persons, instead of 'operating units', they can be asked to state why they work. Daniel Yankelovich and others have published the results of asking this and similar questions in six industrial countries in the 1980s.[28] They list three kinds of answer. The first and most obvious, to survive, remains the objective of most of the world's population. The second is to increase one's wealth and other measures of material success. The third – with which the report is most concerned – is to express oneself and to achieve something that fully reflects one's interests and abilities. The inquiry therefore asked people to comment on the statement: 'I have an inner need to do the best I can, regardless of pay.' The answers measured the level of 'perceived work ethic'. In all countries this measure was at least moderate among a substantial majority of the population. Such a finding, though not surprising, does not match the presumptions of neo-behaviorism or of similar systems: according to them, we should work strictly for external, material rewards.

Countries differed greatly, especially in the proportion of people who reported a strong level of 'work ethic'. Israel, with 57 percent, scored highest (compare the account of kibbutzim in the previous chapter) and Britain, with 17 percent, the lowest. But even the British often emphasized the value of self-employment and of small economic units. Such findings illustrate the effect of local tradition – the cultural influence – on attitudes. They can be usefully discussed only in the context of people as social beings reared in communities which differ greatly from each other.

The inability to express oneself in work is a major source of alienation from society. The lack of any strong feeling that one belongs to a community is likely to go with acceptance of conventional 'materialist' values: at the extreme, work is seen as no more than a disagreeable way of getting money and perhaps status, so that life can be enjoyed in leisure time. The opposite attitude is most common among those who already experience intrinsically satisfying work or other activity. These 'post-materialists' look, not only for personal autonomy and freedom of expression, but also for the opportunity to take a constructive part in public affairs. Such people are as yet in a small minority – perhaps 12 per cent of the population of the USA.[29] Other minorities are represented by those who favor Gandhi's proposals for village life and by the inhabitants of kibbutzim. They are dismissed by practical men as dreamers; but it is often those who pride themselves on their hard-headedness who live in an imaginary world. This is the world of Economic Man and similar ideal constructs which exist only on paper or in computer programs.

The labor of study

The diversity of the incentives to work has been analyzed experimentally. The kind of work most accessible to psychologists is that done by students in school and college. Here too behavioral engineers have been active. By the 1960s mechanical methods derived from behaviorism were common in schools in the USA.[30] In an account of the new technology of teaching, J.G. Holland writes of the old, defunct concepts of knowledge, meaning, mind and symbolic processes which, he holds, have never allowed the necessary manipulation or control. Control is achieved with machines that present a student with questions and state whether the answer given is correct. Like pigeons, students learn a complex sequence best if it is presented in small steps.[31]

As an aid in some kinds of learning, machines are valuable. They allow study at the student's own pace and give teachers more time to encourage original thought, debate and spontaneity – 'defunct' aspects of education to which behavioristic principles and teaching machines can make little contribution. In the behaviorist's system, the development of correct responses depends, as usual, on reward. Punishment is not recommended. It is possible to suspect the intrusion here of a touch of non-behavioristic humanity; but the main reasons given for not using punishment are that it is ineffective, and that it has unfavorable long-term effects on behavior.

As before, the concept of reward presents difficulties. In machine learning, 'positive reinforcement' consists of being told that one's answer is correct. But B.F. Skinner, in his major work on teaching, also asks what schools have that 'will reinforce a child. . . . Children play for hours with mechanical toys, paints, scissors and paper, . . . with almost anything which feeds back significant changes in the environment. . . . The sheer control of nature is itself reinforcing.'[32] As before, we find that everything that is done, everything that is enjoyed, is classified as reinforcing because it is done or enjoyed.

If, however, we do not restrict ourselves to behaviorism, we can make a useful distinction between two kinds of reward. There are first actions carried out for external reasons (pay, praise. . .) and, second, those not driven in this way. The latter allow freedom to do what one chooses, and are said to reflect intrinsic motivation.[33]

Earlier I give examples of exploratory and other activities performed without evident reward. I now give others, from experiments on both children and adults. A simple experiment is to give nursery school children large felt-tipped pens and to record how they use them. They are next taken to another place and asked to make drawings. Some are rewarded for their drawings with a certificate; others are not. When they return to their school, those that have been rewarded play with the pens less than before, and less imaginatively.[34] So reward has an effect opposite to that expected. It is probably important that the activity studied is one that children do very readily, unprompted.

Similar findings come from experiments on undergraduates. E.L. Deci used mechanical puzzles that present problems in spatial relationships. His subjects are said to have found these tasks enjoyable and absorbing. The question was whether payment for working at the puzzles would influence behavior when the subjects were given free time to work on them alone; and interest (or intrinsic motivation) again proved to be reduced by payment. There are many other such examples.[35]

The experimenters explain their findings by a change in attitude: the subjects who receive rewards are less inclined to regard the activities as worth doing, *except* for reward. Yet the loss of interest is only apparent, for the subjects still need a challenge to their abilities. According to Deci, this need is often met by a shift of attention to the problem of getting the reward as quickly as possible. His argument may help to account for some kinds of delinquent behavior, for people often 'satisfy their intrinsic need to be creative and competent by devising ways to beat the system.' Crime or sharp practise (or cheating at school) then become less a means of earning a living (or high marks) than a way of avoiding boredom and of restoring lost freedom.[36]

It would, however, be misleading to suggest that external reward is always liable to interfere with efficiency, interest or enjoyment; or that its use always entails an authoritarian system. Reward may be anything from a substantial payment to somebody saying 'correct' or 'good': social approval is quite distinct from payment, and may have a quite different – sometimes, much greater – effect. As well, the effect of external encouragements (or threats) may depend on the task. In school, external reward, such as a token exchangeable for candy, seems to help with learning material by heart; it can also improve orderly behavior; it is perhaps especially helpful with emotionally disturbed children who can only with difficulty be induced to follow rules. But such rewards are counter-productive when one wishes to encourage a general interest in learning, innovative or independent activity, or lasting understanding of general principles. (By the same token, it is perhaps unlikely that this book will often be read for even indirect monetary reward, or for the sake of social approval. To say this, as we know, contradicts the views also of some sociobiologists. Whether I am right about this can be discovered only by investigation. Perhaps publishers should endow research on these lines.)

Another important distinction concerns the kind of use to which rewards (such as praise or certificates) are put. They may be used (in the behavioristic fashion) to control behavior. Or they may influence the way in which a person sees his or her performance: that is, they may give a person information about his or her competence. A student may, on the one hand, be rewarded for correctly going through a procedure laid down by the teacher; or, on the other, for showing ability in solving a problem or for displaying originality.

Should one disregard intrinsic motivation or emphasize it? How we answer this question depends on what kinds of learning we wish to foster, what aspects of personality we intend to encourage, and hence what sort of society we prefer. These are not merely academic questions devised in the quiet of an ivory tower: they are directly relevant to what goes on daily in schools and elsewhere. Deci and his colleagues describe the ways in which elementary school teachers in the USA use rewards. They find two extremes. 'Highly controlling' teachers 'make decisions about what is right and utilize highly controlling sanctions to produce the desired behavior.' In contrast, 'highly autonomous' teachers encourage children 'to consider the relevant elements of the situation, and to take responsibility for working out a solution' to each problem.[37]

The style of teaching quite quickly, at the beginning of the school year, influences the attitudes of the children toward school work. Those taught by teachers who emphasize autonomy develop more intrinsic motivation: they welcome a challenge to their abilities; they display curiosity, instead of a liking for the familiar and safe; they favor independent mastery, not help from the teacher; and they tend to work for their own satisfaction, not for praise or high marks. (To an authoritarian teacher such pupils are liable to be a source of anxiety.)

In most formal teaching, the tradition on the whole is authoritarian; hence the neo-behavioristic emphasis on external control is easily adopted in the schools. The result is further encouragement for children to work, not for knowledge or skill, but for rewards: the capacity for spontaneous learning is underestimated. Here is a passage from a famous review by the Canadian psychologist, D.O. Hebb:

> An experiment on teaching, performed many years ago, was impressive in indicating that the human liking for work is not a rare phenomenon, but general. All the six hundred pupils in a city school, aged six to fifteen, were suddenly told that they need do no work unless they wanted to; that the punishment for interrupting others' work was to be sent to the playground to play; and that the reward for being good was to be allowed to do more work. *All* the pupils discovered in a day or two that, within limits, they preferred work to no work (and incidentally learned more arithmetic and so forth than in previous years).
>
> The phenomenon of work for its own sake is familiar enough

to all of us, when the timing is controlled by the worker himself, when 'work' does not refer only to activity imposed from without.[38]

There are severe difficulties in applying this principle in the classroom, but they are being gradually overcome. Not before time: two-and-a-half millennia ago, in his *Metaphysics*, Aristotle wrote that a striving after knowledge is implanted in the nature of all mankind.

Work as play

At the beginning of the chapter, I imply that play is frivolous. But play covers a great variety of actions, whose functions include the promotion of physical fitness, the practise of skills and the learning of social roles. As the previous paragraphs show, even school work can, in favorable circumstances, be a kind of play. But, most often, play refers to unforced activity, undertaken for enjoyment. Yet the two categories, work and play, are not the same as disliked and enjoyed. In the dictionary, under play, we find not only 'amorous disport or dalliance' but also 'action, activity, operation, working; often implying the ideas of rapid change, variety . . . feeling, fancy, thought'; and so we are brought close to the mysterious achievements, in the arts and sciences, that represent the peaks of human creative ability and are especially clear instances of human action that cannot be reduced to biology. Michelangelo's four excruciating years while he painted the ceiling of the Sistine Chapel cannot usefully be analyzed statistically. Nor can Mozart's last three and greatest symphonies, which he wrote in less than two months when he was ill, in debt, and rejected by his patrons. Of them, Alfred Einstein writes:

> We know nothing about the occasion for writing these works. . . . It is possible that Mozart never conducted these symphonies and never heard them.
>
> But this is perhaps symbolic of their position in the history of music and of human endeavour, representing no occasion, no immediate purpose . . .[39]

Most readers will perhaps feel that such extremes of creation are wholly outside their actual or possible experience. Yet activities driven from within, sometimes performed with little or no external reward, are universal; and they can be studied systematically. In *Beyond Boredom and Anxiety*, Mihaly Csikszentmihalyi asks: What is enjoyment?[40] We are all familiar with its opposite: boredom can result from having to spend much time on something that is too easy, or from being forced to do something – common experiences for many who work for wages or carry out 'home duties'. An additional, painful experience, as a child will often say, is to have 'nothing to do'. At the other extreme, a task well beyond our powers may induce severe anxiety.

Csikszentmihalyi distinguishes sharply between leisure and comfort on the one hand, and enjoyment itself on the other. Enjoyment comes above all from unselfconscious absorption in an activity freely undertaken. Such an activity he calls autotelic, or self-driven. The activity should fully extend one's capacity. He selected a diverse group of occupations, and questioned people who performed them well. Some, such as playing basketball, rock climbing and dancing, require strenuous physical exertion; others, such as composing music, playing chess and performing surgical operations, do not. Some but not all the activities provided a livelihood. Despite the variety, they all had one thing in common: they were enjoyed. The enjoyment was independent of external reward, and was largely independent of outcome: the *doing* was crucial.

Absorption in what one is doing is called *flow*. Perhaps this does not convey the experience very well; but there is no accepted word for it in English – an interesting fact in itself. In flow there is a merging of action and awareness: one is intensely aware of what one is doing, but there is no *self*-consciousness, hence no anxiety.

The universality of the fine arts. Cast iron head of a queen mother from the west African kingdom of Benin (probably sixteenth century).

The reader may at this point expect a digression into some form of 'Eastern' mysticism; but in fact there is nothing especially mystical about flow. When people are asked about it, they liken it to experiences such as exploring a strange place, designing or discovering something new, or listening to good music. Still less mysterious are 'micro-flow' activities. These are things we do every day, often without awareness of their significance. Their role has been observed experimentally, by recording the results of deprivation. A number of students in a university in the USA were asked to make a minutely detailed record of all their 'playful, non-instrumental but rewarding behavior' during a period of 48 hours. Solitary activities included pacing the room and doodling; social interactions ranged from conversation to coitus. A week later, these people were asked to stop all such actions for two days. Most reported such disagreeable consequences that the deprivation had to be cut short. The symptoms included not only tension but also fatigue and sleepiness; headaches, irritability and depression; and loss of concentration and of the ability to do anything creative.

The universality of the fine arts. Australian aboriginal's traditional design in bark.

There was also a tendency to have minor accidents, such as blundering into a door or cutting a finger. Microflow activities are held to be an unconscious means of giving one a sense of freedom and of control over what one is doing.[41]

The symptoms that result from deprivation resemble those that accompany emotional upsets and psychoneurosis. During the twentieth century such states have been increasingly treated by a training in relaxation. The subject lies or sits in a restful position and concentrates on some aspect of bodily function such as breathing or the pulse that can be felt in the fingers. The emphasis on the subjective – on inner experience of consciousness – makes an interesting contrast with the objectivity of behavior therapy. The training, which is exacting if not strenuous, can lead to one kind of 'flow' experience. A similar procedure is given the less prosaic name, transcendental meditation.[42]

The beneficial effects, for some people, of such regimes match the findings of Csikszentmihalyi with one qualification: they are, if not antisocial, at least asocial. A person relaxing, concentrating on his breathing, or meditating (transcendentally or otherwise) is usually on his or her own; even when the subject is under training the only other person effectively present is the trainer, or *guru*. The same applies to some microflow activities: from the point of view of one's family, pacing up and down in one's study, or solitary walking, must seem to verge on the antisocial. But Csikszentmihalyi's subjects reported the most favorable effects from social rather than solitary activities. Individuals vary greatly but, on the whole, the findings on 'flow' do not fit the concept of human beings as solitary egoists.

Much remains to be learned. The 'death of school' has been predicted or advocated; and perennial attempts are made to reject industrial society. Nonetheless, schools and factories will probably continue to exist. We should therefore try to make them more effective, but not by treating human beings, of whatever age, as if they were laboratory animals in cages. Scientific methods have a role, as the preceding pages show; but the achievements of human behavioral science have depended on accepting its subjects as persons with feelings, desires and intentions which are just as significant, as causes of action, as are external 'reinforcements'. Central among them is the desire for stimulating and intrinsically satisfying work which also confers status in the community. Psychologists and economists also have to allow for the fact that we (and they) are sometimes frivolous: we play.

Part 5

Homo sapiens:
the human species

Austin felt that he had a lot of reading to catch up on – too much. His head was a buzzing hive of awakened but directionless ideas. There was Freud who said that we must acknowledge our own repressed desires, and Jung who said that we must recognize our archetypal patterning, and Marx who said we must join in the class struggle and Marshall McLuhan who said we must watch more television. There was Sartre who said that man was absurd though free and Skinner who said he was a bundle of conditioned reflexes and Chomsky who said he was a sentence-generating organism and Wilhelm Reich who said he was an orgasm-having organism. Each book that Austin read seemed to him totally persuasive at the time, but they couldn't all be right . . . Wittgenstein said, whereof we cannot speak, thereof we must remain silent – an aphorism in which Austin Brierly found great comfort.

David Lodge[1]

To end this book with a rival, supposedly complete, portrait of *Homo sapiens* would be pretentious and shallow; but there are questions, so far left open, to which it is possible to suggest answers. In the next chapter I describe, more fully than before, the consequences, logical and moral, of reducing human beings to genes or machines or, more generally, to nothing but physical systems. I also suggest a commonsense alternative. This is more than an academic issue: it has everyday, practical implications.

Equipped with common sense, in Chapters 13 and 14 we inspect some of the peculiarities of our species. These chapters, therefore, do attempt an incomplete sketch of humanity. We begin with our unique capacity for speech, on which there is much new knowledge. But we are not only loquacious: we are also musical and – a neglected feature – a teaching species, *Homo docens*. Tradition, in all its many forms, depends on teaching as well as imitation. And it supports both social stability and also rapid social change – even progress.

In Chapter 15 we finally reject pessimism and misanthropy, and ask: to what action does the argument lead?

12 The reductionist imperative

Einstein . . . was asked whether he thought everything could ultimately be expressed in scientific terms. Einstein replied: 'Yes, that is conceivable, but it would make no sense. It would be as if one were to reproduce Beethoven's ninth symphony in the form of an air pressure curve.'[1]

The preceding chapters give many examples of interpreting human action by a single principle, such as natural selection or conditional reflexes. Such explanations are commonly called reductionist. Some kind of explanatory reduction is a feature of many branches of science; yet 'reductionism' can arouse such ire that it becomes a term of abuse. One reason for the agitation we have already met: reduction sometimes seems to deprive us of freedom of action and to entail rejecting moral principles. 'A recurrent popular image', writes Austen Clark,

> is that of behavioural science gradually encroaching on the domain of free will, . . . as scientists find causal explanations of mental illness, deviance, and criminality; . . . the province of free will and action . . . must continually shrink under the impact of increased knowledge, finally to disappear with the success of physiological reductionism. . . . The door is [then] open to increasing technological reorganization of our social institutions. Our moral notions and our system of justice, which are based on the assumption of personal responsibility, would need radical overhaul.[2]

We see below that not only moral abstractions are at issue, but also, still more obviously, practical matters, such as what we regard as acceptable treatments for some kinds of illness.

In explanatory reduction, statements that belong to one theoretical system are used to account for findings or theories that belong to another: physiology explains psychology; chemistry explains physiology, and so on. Reduction in this sense is beneficial because the new explanation often covers more ground than the old: its theoretical statements have a higher generality. In the nineteenth century all the bewilderingly various substances and materials around us were found to consist of fewer than a hundred elements; these were arranged in a 'periodic' table which classified the elements in an orderly way. At first the atoms of which each element is composed were held to be indivisible (hence their name); but later, by a further reduction, the theory of fundamental particles (protons, electrons and so on) gave a still more general account of the elements. As a result, chemistry was, in a sense, reduced to physics. Much information about the properties of many kinds of atoms (that is, of the elements) could be related to their newly proposed inner structure.

This tremendous achievement, however, did not put chemists out of business: it did not relieve them of the task of studying chemical compounds. The properties of substances in which atoms are combined to form molecules still had to be investigated in their combined form; for the new physics did not, and does not, tell us what these properties are. Even water, seemingly a very simple substance made of hydrogen and oxygen, has many complex features – among them, the various crystalline structures of snow. The study of hydrogen and oxygen, each in isolation, does not tell us what all their properties will be when combined. The findings of physics do not *replace* those of chemistry.

Hence within the physical sciences reduction can lead to laws of high generality, but it does not dispose of the phenomena that have been reduced. The same applies, still more clearly, to attempts to reduce human action to biology or physiology or physics. Nonetheless, reductionist ideas continue to be influential, and not only those already described at length in this book. One of the most enduring and enticing is that of the human being as a machine – an image I have so far rather neglected. So now I turn aside from the logic of reduction to this image and its social role.

The human machine

The image of the human machine became prominent in the eighteenth century, early in the machine age, and is a notable instance of the power of metaphor.[3] In its primary usage, the word machine refers to an apparatus, made by human beings, to perform some function such as milling, pumping or indicating the time. In this sense, the human body is not a machine. But for some purposes, mechanical principles are usefully applied to it. Our bones act as levers, and our heart pumps blood. Like an automobile, we require fuel for producing mechanical energy and heat; and input and output balance, well enough, according to physical laws.

The machine analogy was influential in Pavlov's attempt to give a physiological account of humanity. Its most recent developments are the interpretations of human beings as neural mechanisms, or as robots driven by computers in the head. In the previous chapter I describe the use of this idea in industry.

Early in the twentieth century, Jacques Loeb (page 179), a contemporary of Pavlov, an arch-mechanist and a founder of the study of animal orientation, wrote this:

> We eat, drink and reproduce . . . because, machinelike, we are compelled to do so. We are active because we are compelled to be so by processes in our central nervous system. . . . The mother loves and cares for her children, not because metaphysicians had the idea that this was desirable, but because the instinct of taking care of the young is inherited just as distinctly as the morphological characters of the female body.[4]

Loeb was both an able experimenter and a successful popularizer with a large following. The popularizing has continued, and sometimes the reduction has gone to further extremes. In mid-century 'brain-waves' (the electroencephalogram or EEG) had their hour of glory. W.G. Walter, in a popular book (derived from radio talks) entitled *The Living Brain*, devoted his account of the brain almost entirely to these rhythmic changes of electrical potential.[5] Here was an extreme case of an enthusiast inflating a single phenomenon much beyond its usefulness. One difficulty is that we still do not know what part the rhythms play in brain function.

More important, two sets of radio lectures, put out by the British Broadcasting Corporation and heard world-wide, have been given by neuroscientists, both of whom have a broader vision than Loeb (or Walter). In lectures delivered in 1950, J.Z. Young asked us 'to organize *all* our talk about human powers and capacities around knowledge of what the brain does'.[6] Nearly 30 years later he similarly made the theme of his Gifford lectures 'that brains contain programs that regulate our lives'.

> Because of our brain organization we are able to love and to hate, to command and to obey, to create beautiful things and to enjoy them.

Moreover, happiness depends on 'the proper functioning of certain reward centres in the brain'.[7]

Young's broadcast lectures are entitled, *Doubt and Certainty in Science*. Colin Blakemore's, delivered in 1976, have a more reductionist title, *Mechanics of the Mind*. The cover of his book shows a man with his head opened to reveal a mass of wires and electronic equipment. But, unlike Young, Blakemore begins with no clearly stated theme; instead, in his last paragraphs, he writes of

> the promise that research on the brain will provide a genuine basis for the treatment of mental disease. But much more than that, it will give a greater understanding of man himself. . . . For without a description of the brain, without an account of the forces that mould human behaviour, there can never be a truly objective ethic based on the needs and rights of man.[8]

And he ends by stating that 'The brain struggling to understand the brain is society trying to explain itself.' Hence Blakemore's conclusion is, apparently, a thoroughgoing reduction of 'society' (and of the human beings that compose it) to neurophysiology. If so, it disregards a significant reservation made by Young:

> Biology can show some of the prerequisites for the development of human culture, but is quite unfitted to follow the full richness of its development.[9]

And elsewhere in his Gifford lectures Young enlarges on features of human existence that fall outside a neuromechanistic frame.

There are, of course, authentic grounds for putting the mind in the brain. They are certainly better than those for giving it to the liver or the heart (where it has sometimes been thought to be). A human being may have his or her heart operated on or even replaced, without any change of intellect or personality; and damage to the liver need not lead to any significant effect on one's intelligence or emotions. In contrast, there are the sometimes bizarre and often distressing results of brain damage. Among closely studied casualties, an American, 'H.M.', had an operation on his brain to treat severe epilepsy: his hippocampus (a structure that has long been a source of doubt and controversy) was removed on both sides. As a result H.M. lost his memory in an extreme sense: he remained able to talk intelligently, and to learn tasks that require manual dexterity; but he could remember nothing that had happened since his operation; his memory span was reduced to only a few minutes.[10]

There are also people, studied especially by R.W. Sperry, who have two separate memories.[11] The cerebral hemispheres are connected by a large band of nerve fibers, the corpus callosum. People with crippling and otherwise untreatable epilepsy have had this structure cut through. Astonishingly, at first there seemed to be no effect on personality or intelligence. But simple experiments revealed an effect on memory. If an unfamiliar object is out of sight, but is explored with one hand, one can recognize the object if it is later put in the other hand: it is remembered, and can be distinguished from other objects. 'Split-brain' patients cannot do this: to the second hand the object is new and strange. Similar observations were made with things seen. Suppose an object is exposed so that the patient can see it only in the left half of the field of vision. Later it is exposed in the right half-field. It is then not recognized. Hence in these conditions each hemisphere operates ('remembers') independently.

The cerebral hemispheres, especially the frontal lobes, of *Homo sapiens* are notoriously much larger than those even of chimpanzees or gorillas. The results of injury (for instance in accident or in war) correspond to the presumption that they are especially concerned with intelligence. Intellect may be seriously impaired by damage to the front part of the left hemisphere.

There is usually a deficiency of speech (aphasia), and also difficulty in coping with general ideas and in solving even simple problems in a new way. Thought is more restricted and standardized than before, less independent. Severe damage can also have a drastic effect on personality. The most famous case is that of an American, Phineas Gage, who at the age of 25 had an iron bar blown through his head from below his left eye. He had been an able and sociable foreman; but after the accident he became an irresponsible, foul-mouthed drifter.

There are many such histories. And other, less distressing findings carry a similar message. Wilder Penfield made famous observations by exposing the brains of conscious, epileptic patients, and stimulating them with a weak electric current. (The procedure was a diagnostic one, harmless and painless.) On stimulation of certain regions of the brain, the patient described the evocation of complete memories.[12]

Brain physiology can also contribute to methods for treating certain illnesses. In Parkinsonism, a region of the brain slowly degenerates: the result is progressive deterioration of normal movement of the limbs and of facial expression. Studies of brain chemistry, aided by experiments on animals, have led to rational and sometimes successful treatment.[13]

It is hardly surprising that neuroscientists – and not only neuroscientists – are impressed by such facts, despite the incompleteness of our knowledge of the brain. Certainly, a functioning brain is a necessary condition of a person's existence. Yet, when the brain of a living patient is exposed, one sees only a whitish, convoluted mass with a jelly-like consistency: one does not see *thoughts*. Whatever is done to it, one cannot, by observing only the brain itself, reveal a personality with feelings, intentions and memories. Penfield's observations were a result, not of recording nerve impulses or the EEG, but of asking the patient questions, and of listening to the answers.

Psychosurgery

The question whether the brain should be treated mechanistically, and equated with the mind or the person, is not merely a matter of philosophical debate. The importance, in some circumstances, of *avoiding* this form of reduction is illustrated by the practise of psychosurgery.[14] Above I quote Blakemore on studying the brain as a source of 'a truly objective ethic'. It is not clear what he has in mind; for he precedes this passage with a justly critical mention of frontal lobotomy (or prefrontal leucotomy). This operation was first used extensively by a Portuguese psychiatrist, Egas Moniz (1874–1955). Moniz adopted it, as a treatment for various forms of mental disorder, after hearing an account, at a scientific meeting, of observations on a chimpanzee. He was also influenced by the mechanistic implications of Pavlov's conditional reflexes.[15] He carried out 20 operations, but was then stopped by the psychiatric director of his hospital. He quickly published optimistic accounts of his work throughout the world, in several languages. Neither then nor since has the method had any valid support either from physiological theory or from properly designed clinical research; but Moniz received a Nobel prize for medicine in 1949.

The fame of Moniz is, however, surpassed by that of an American psychiatrist, W.J. Freeman (1895–1972), whose procedure was a kind of surgical mass production: in the 1940s and 1950s he performed more than four thousand lobotomies, usually by driving an ice pick up into the brain from above an eye. This occupied him for only a few seconds.[16] Many of his patients, and those of other psychosurgeons, were treated for 'aggressivity' (page 74); but, in recent decades, the indications for destroying parts of the brain have included depression, anxiety, drug addiction, and even the vague, 'emotional illness'. Most of the Americans who received psychosurgery have been women diagnosed simply as neurotic.[17] But some patients have been children, down to the age of five years; the grounds for injuring the brains of children have included 'mental retardation' and over-excitability ('hyperkinesis'). Freeman, however, considered black women to make the best patients: such people could be returned to their families in a passive condition but capable of simple housework.

The lack of a rational foundation for psychosurgery has allowed changes of fashion. After a time, attention was shifted from the frontal lobe to the amygdala, a small region, of complex and obscure function, at the base of the brain. The amygdala was chosen as a target because destroying it in animals such as rats can have a taming effect. It may remove (for example) the propensity for biting when handled. This, a defensive response, is of course called aggression. Unfortunately, in some experiments, amygdalotomy has had an opposite effect: it increased wildness.[18] Nonetheless, this operation has been carried out on large numbers of patients. Yet other structures have been destroyed. Individual surgeons have, or had, their favorite targets. 'It is as if each dentist specialized in a particular tooth, pulling that tooth as treatment for toothache, stomach ache, headache and heartache alike.'[19]

The commonest outcome of surgically damaging a frontal lobe, or of destroying an amygdala, has been called 'vegetabilization': the patient becomes sluggish, unresponsive and indifferent. As a result, she or he may be more easily managed, in hospital or at home, than before. For a minority of severely disturbed people, this outcome could just possibly be defended: many of the insane put a heavy emotional burden on those who look after them. R.W. Scheflin & E.M. Opton, in a meticulously documented survey, sum up this dilemma. Administratively, they say,

> a patient who is dull, apathetic, and so forth, is much easier to

control than a patient who is unhappy, demanding, angry, wilful, active, a patient whose soul is sick but alive.[20]

It could even be argued that not only their caretakers but some patients are themselves happier when their capacity for anguish has been abolished; but at best this would apply to only a few of those subjected to psychosurgery.

Psychosurgery was initially a product of two or three exceptionally energetic and ambitious men, but it has not remained merely an aberration of a few eccentrics: the orginators were honored; their methods were welcomed and adopted by many colleagues; and their work was supported by governments (pages 74–6). We can discern four elements in this response. First, physicians hoped that at last a cure had been found for intractable and distressing conditions. Second, communities and individuals wished for protection against misfits who, even if not dangerous, are at least liable to cause emotional disturbance in others. Third, authoritarian governments sought to silence critics by means that could be represented as humane. The fourth and most fundamental element is our current willingness to reduce the human personality to a mechanism available for tinkering, or to a biological preparation which could properly be manipulated in a laboratory.

The logic of reduction

Despite misapplications of mechanistic ideas, many must feel that there is still something to be said for explaining human action by scientific principles. We depend on the workings of our brains. For that matter, we are also products of evolution. The two last statements are hardly controversial, but they point to a dilemma: if we are bent on reduction, which is to have priority – neurophysiology, or neo-Darwinism? We might imagine a doctrinaire neuroscientist assuring a sociobiologist that his views on human evolution (and on everything else) are solely a result of processes in his brain; but his rival retorts that the neuroscientist believes this – or, at least, states it – only because it contributes to his Darwinian fitness. Such a dialogue, sterile from the beginning, could hardly be kept up for long.

Scientists, however, as we know, usually avoid this difficulty by adopting a simpler and more general form of reduction, from larger to smaller units: the activities of individuals are explained by their physiology, physiology by physics and chemistry . . . It is often taken for granted that all biological phenomena can, in the long run, be reduced to physics.[21] In Chapter 6 I quote Monod on this theme. It is possible to detect, among some biological scientists, a hopeless longing for the seeming certainties of physics and chemistry. We must now examine this idea further.

During the twentieth century biology has been transformed by studies of the inner workings of cells. Cells have been analyzed as chemical machines, equipped with parts of hitherto unimagined smallness. (The sequences of chemical changes have also proved to be of unsuspected intricacy: compare page 99.) These machines reliably produce heat, movement, growth and complexity. They seem to determine every accomplishment of every organism. By going small, molecular biologists have achieved, if not simplicity, at least predictability. True, predictions are subject to some error; but error seems to decline with advancing knowledge.

It does not, however, decline as we descend further from the molecules studied by cell biologists: at the level of the ultimate particles of physics, we face inescapable doubt. According to the principle of uncertainty (due to W.K. Heisenberg, 1901–1976), there is a limit to the precision with which certain very small quantities can be measured: the velocity and the position of an ultimate particle at a given moment cannot both be accurately stated. Whether this celebrated finding has profound philosophical implications has been much argued.[22] But at least it reminds us that reduction to the smallest can lead to doubt rather than certainty.

The limitations of explanatory reduction can be seen more clearly if we turn to more accessible things. The photograph opposite makes a statement at the level of biochemistry. From a biological point of view it is therefore an example of a 'reductionist' statement: it gives information only about the presence or absence of a certain enzyme in two liquids. The reader may find it useful to consider just what questions come to mind when it is presented without explanation. A biochemist might be led to think about the structure of the enzyme molecule, and its interactions with other molecules; such thoughts might even lead to reflexions concerning electrons – a further step in reduction. But most people – even biochemists – would be inclined to ask about the sources of the two solutions. If told that they were derived from red blood cells, they might reasonably ask, whose blood? And so on.

The enzyme (also called for short G-6-PD) is present in the red cells of most people, but in the African tropics about 20 per cent are without it.[23] The difference between normal people and the others is genetically determined. The high incidence is surprising, because the deficiency is liable to be accompanied by severe jaundice and early death. How can we *explain* G-6-PD deficiency? The probable answer is that possession of the gene responsible protects a person against malaria: the populations affected are all in regions in which malaria is endemic. (Compare the account of sickle-cell anemia in Chapter 6.) This explanation is not reductionist: it is at the level of ecology – the relationship of individuals and populations with their environments.

On the other hand, we may be interested in the difference between normal and jaundiced people. This we may *explain* genetically and biochemically by the presence or absence of G-6-PD; and *this* explanation takes us from the larger to the smaller. Hence any approach to a complete account of these phenomena requires statements at levels from the ecological to the physical. Some statements would be classified as reductionist, some not. What we accept as a satisfying explanation depends on what specific question we are asking.

It is possible to ask questions which lead to answers that interest hardly anybody: for example, the attempt might be made to relate the incidence of G-6-PD deficiency to body-build or skull shape. Anatol Rapoport illustrates the significance of asking the right questions in an elaborate parable. A scientist, completely ignorant of all such games, makes a study of chess. He sees that, while the players change, the board and pieces remain the same, and therefore concentrates his attention on the pieces. During a game the number of pieces on the board diminishes; so he carries out a population analysis, and works out how the number of pieces declines with the lapse of time. From this he derives an equation, and calculates the probability that a piece will be taken from the board at a given stage in the game. Later, he distinguishes the different shapes of the pieces; he also studies the geometry of the ways in which the pieces are arranged.[24]

Rapoport's scientist is evidently a specialist in the mathematics of populations; and so far his project reflects how research may be handicapped

A purely reductionist statement at the level of biochemistry. If one required an explanation, what questions would one ask?

by the narrowness of specialism, and by the limitations of a single person's knowledge. This is one source of excessive reductionism. But there is a more fundamental reason why his method is unsatisfactory. In the present chapter a central topic is the clash of those who pay attention only to genes or molecules or nerve cells, on the one hand, and those who recognize human beings as persons, on the other. Rapoport comments that his scientist's account of chess does not depict the game as we understand it, because the players, and what happens in their minds, are ignored: the players' strategies concern intentions and future action, and their actions are choices.

The clash can be avoided if it is realized that reduction is both inevitable and useful, but – by itself – always gives an incomplete picture.[25] If a physicist says that this book, though apparently a solid object, is *really only* ultimate particles in a lot of empty space, the word 'really' is empty or question-begging, and 'only' implies what is not true.

The notion that everything that can be usefully said can be said in terms of physics is indeed incoherent, because the existence of physics itself rules it out. Physics is one branch of knowledge; and the existence of knowledge implies a knower. Uncompromising physical reduction states that everything that exists is *nothing but* particles in motion. (I had better not try to say just what these particles are: the physicists themselves are, I understand, in some doubt on this point.) It is open to us to describe ourselves as complex mechanisms.[26] But, even if we do this, biologists will be deluded if they expect physics to exempt them from studying the diversity of species, or if they hope that physiologists will let them off investigating behavior. Similarly, social scientists and historians must continue to describe, classify and interpret individuals, populations and nations. No doubt, most of them are quite happy with this prospect.

Determinism

One form of reduction is the doctrine of determinism: that all events are decided in advance of their occurrence. Everything that happens is interpreted by this presumption. We are all, in a sense, determinists some of the time: we say that events are *caused*. If I put a lighted match to paper, 'inevitably' the paper will catch fire. But determinism seems to conflict with another presumption, also made in some contexts by nearly everyone: that human beings can choose between alternative actions.

The debate began, as far as we know, when the Greek philosopher, Democritus (about 460–370 BC), proposed that all matter consists of atoms whose motions accord with fixed laws. His apparent determinism was opposed by others, especially Epicurus (about 342–270 BC), who asserted that human beings can originate action; hence what we do is not fixed in advance, and we can choose between right and wrong.[27] The seeming opposites, determinism and freedom to choose, have had their counterparts in Christian theology: the Calvinist doctrine of predestination is set against the presumption of free will. The Greek determinists attributed our destiny to nature, but for Calvinists the determining agent is a deity.[28]

Controversy has continued to this day, but has led to no consensus. Perhaps then we should leave the philosophers to their bickering. If so, we should have to resign ourselves to a philosophical dissociation, or dual personality, in which we alternately believe, or at least act upon, two incompatible principles: while the admirable Dr Jekyll accepts freedom to choose, the sinister Mr Hyde espouses determinism. (Or should it be the other way around?) Scientists, who may be regarded as professional determinists, often in conversation adopt such agnosticism. Some echo Fitzgerald's Omar:

> Myself when young did eagerly frequent
> Doctor and Saint, and heard great Argument
> About it and about; but evermore
> Came out by the same Door as in I went.

Many scientists explicitly reject the debates of philosophers as pointless, and irrelevant to the progress of natural science. Yet, while they do this, they fall into logical difficulties. The clearest examples in this book concern natural selection (Chapter 6); but the reader will recall also the problem of defining aggression (Chapter 5) and the objection to saying that a characteristic is inherited (Chapter 7). And behaviorism is not a scientific theory but a metaphysical doctrine[29] (Chapter 10). Hence those who reject philosophy unwittingly adopt philosophical positions, especially that of determinism itself: all the biological images of humanity are determinist by implication, and therefore impose this teaching on a large audience.

To talk usefully about determinism, it is necessary to relate it to practice. 'A general thesis of determinism, applying to all events without restriction, is too general to be . . . falsified, or confirmed, or rendered probable, and is empty and uninteresting'.[30] We are now concerned with scientific determinism, of which the nature and implications have been examined by K.R. Popper.[31] In natural science, as in everyday life, we try to establish causes: tides are due to the combined influence of moon and sun; fermentation of wine requires the presence of micro-organisms. We also predict the results of human action in causal terms: if we organize a clean water supply for a town, the incidence of diseases such as typhoid declines; if we reduce the taxes paid by parents, the birth rate rises (perhaps). Knowledge of the 'certainties' of scientific findings can lead to an extreme position: all events in the universe are unalterable or predetermined; the future is as unchangeable as the past. Indeed, the idea of the future, as it is commonly understood, is no longer appropriate, and our feeling that time passes is an illusion. 'Billions of years ago the elementary particles' of which the universe was made 'contained the poetry of Homer, the philosophy of Plato, and the symphonies of Beethoven'.[32]

A slightly less extreme proposition acknowledges a difference between past and future, but states that every future event can in principle be calculated from what is already known, or at least knowable. This position is sometimes adopted by scientists when they make their views explicit. Such determinism is, however, not itself part of science: it is not a conclusion based on observation, still less on experiment. We are sometimes successful in our predictions, but often we are not; and there is much that we cannot predict and – as we see further below – cannot reasonably hope to predict. In science a statement about events is usually rejected if it is not testable; but the determinist doctrine cannot be tested. Hence determinism may be regarded as an attempt to reduce all phenomena to those that we believe we can predict with accuracy. Determinism is then a presumption or axiom, not a finding from the study of nature.

In a sense, however, experience regularly conflicts with thoroughgoing determinism: our predictions are subject to error. In biology, and above all in behavioral biology, errors are so large and so general that a large branch of mathematics has grown up to deal with them. As for physics, I mention the principle of uncertainty above.

That principle deals with events utterly remote from ordinary experience. We can learn more about determinism if we follow Popper, and compare clouds and clocks: his clouds represent physical systems which 'are highly irregular, disorderly, and more or less unpredictable'; while a precision clock is a system which is 'regular, orderly, and highly predictable' in its behavior.[33] Clouds also exemplify systems which have become more predictable as we learned more: weather forecasts have certainly become less inaccurate during recent decades. Does this signify that, after all, such systems are in the long run as predictable as anything else: that, as Popper puts it, all clouds are clocks? If so, our distinction between clouds and clocks merely reflects our present (and temporary?) state of ignorance.

But when we turn to scrutinize clocks, we find a difficulty: our measurements and predictions are always, to repeat, to some extent in error; and the more closely we investigate, the more clearly we find that our clocks contain some unreliability. The most general source of unreliability is that molecules are in continuous motion. Hence at this level, if no other, our measurements can be only approximate: they are always affected by chance variations. The phenomenon of radioactivity makes this especially clear: we can, with fair accuracy, say of a mass of a radioactive element how long it will take for half the atoms to decay; but we cannot say when a particular atom will lose (say) an electron, or what decides that *this* atom rather than *that* will be transformed. Unpredictability is not only at the level of atoms: a speck of dust can interfere with a clock's accuracy; and we have no good reason to think that we can predict the movements of dust particles.

The apparent operation of chance occurs indeed at many levels. In Chapter 6 I describe the evidence for the role of chance in evolution: mutations, that is, 'spontaneous' changes in the material of heredity, seem to provide the variation on which evolution depends. They are molecular changes. It is possible to establish the rate at which a particular type of mutation occurs, but not to identify, at any given moment, which individual genes are about to mutate. In this respect, genes are like the atoms of radioactive elements.

There is similar unpredictability in the conduct of whole organisms. In Chapter 10 I mention the exploratory movements of animals about their living space. We can make approximate predictions about such movements but we cannot say exactly where a particular animal will be at a specified time; and, if it is held that one day we shall be able to do so, that is a presumption: no valid ground exists for such optimism. We may put our exploring animal in a simple, artificial environment, and restrict its movements to a few channels – almost as if it were a tramcar. But even then we cannot say where it will be at any given moment. We are therefore led to conclude that

> to some degree *all clocks are clouds*; or in other words, that *only clouds exist*, though clouds of very different degrees of cloudiness.[34]

To stop at this point would present an indeterminism in which chance and randomness are emphasized; but such a bleak view of what T.H. Huxley called the 'cosmic process' is obviously inadequate. For, if we resign ourselves to the uncertainties of indeterminism (or, better, decide to enjoy them), we may recognize that the passage of time entails the emergence of novelty. In the living world this mysterious process has produced organisms as diverse as a mold on bread, a moss on a rock, a Californian redwood, a malarial parasite, an earthworm, a death watch beetle, and *Homo sapiens*. On an entirely different time scale, novelty arises from individual action. Animals improvize new sequences of movement by combining what they have learned on previous occasions (page 193). This ability may be regarded as a precursor of human creativity; but the human ability to create something new is of an order different from anything attained by animals. In everyday life it is reflected in the layout of a garden, the plan of a meal, the choice of clothes or of home decoration, the way we instruct our children, and much else. More obviously, it is shown in artistic and scientific achievement.

Popper regards physical determinism as a nightmare, because, he says, it represents the universe as a 'huge automaton' and human beings as 'nothing but . . . subautomata within it'. If so (for example), an author's belief that, in writing a book, he or she has produced something (slightly) new is an illusion: 'any physicist with sufficient detailed information' could have written it.[35] Of course, to find such a notion nightmarish does not show that it is false. But determinism results not only in nightmares but also in logical absurdity. It states or implies that any theory we believe to be true is accepted because we are constructed so that we accept it.

> Accordingly we are deceiving ourselves (and are physically so deter-
> mined as to deceive ourselves) whenever we believe that there are
> such things as arguments or reasons which make us accept deter-
> minism. Or in other words, physical determinism is a theory which,
> if it is true, is not arguable, since it must explain all our reactions,
> including what appear to us as beliefs based on arguments, as due
> to *purely physical conditions*.[36]   (Original emphasis.]

No doubt some scientists and others will continue to argue in favor of determinism. But, when they do so, they will be putting forward opinions which, on their own presumptions, are empty. So we may ask why they do not simply keep quiet. Their answer (assuming that they are willing to argue) would, no doubt, be that they are compelled . . .

Common sense

If the cases against determinism and extreme reductionism are valid, then the sciences of human behavior are partly undermined; for they are inevitably, to some degree, both determinist and reductionist. Science, says Popper, is the art of systematic oversimplification[37]; and the simplifications of the behavioral sciences include both determinist and other reductionist presumptions.

In an age dominated by electronics and computers the preceding argument is unlikely to be easily accepted: the influence of mechanistic metaphor is too great. But there are other reasons for dissatisfaction with the behavioral sciences. As we know, earlier in this century it was expected that behaviorism would lead to general laws that would give us new understanding of ourselves; and, as a result, many practical problems would be solved. But this has not happened: to quote a philosopher, 'there are no non-trivial deterministic laws of psychology or of any other behavioral science.'[38] Valiant attempts (described in Chapter 10) have been made to develop behavioristic treatments for mental illness. And, on similar grounds, behaviorism has been proposed as a massive source of help for teachers. But, in the outcome, success has usually come only when strict behaviorism has been discarded for common sense.[39]

Howard Gardner, in a review of human abilities, defines common sense as 'the ability to deal with problems in an intuitive, rapid, and perhaps unexpectedly accurate manner'.[40] It is, however, not necessary to insist on unexpectedness. Indeed, the quality I refer to as common sense is an expected and, in a sense, commonplace feature of human conduct. Its importance has, in this century, been rediscovered in a movement, in both philosophy and psychology, which puts much weight on our everyday, non-scientific ability to communicate among ourselves, and to account for our actions and those of others. Here is a philosopher, Gilbert Ryle:

> We know quite well what caused the farmer to return from the market with his pigs unsold. He found that the prices were lower than he had expected. We know quite well why John Doe scowled and slammed the door. He had been insulted. We know quite well why the heroine took one of her morning letters to read in solitude, for the novelist gives us the required causal explanation. The heroine recognized her lover's handwriting on the envelope. The schoolboy knows quite well what made him write down the answer '225' when asked for the square of 15. Each of the operations he performed had put him on the track to its successor.[41]

And here are the opening paragraphs of a book by a psychologist, R.B. Joynson:

> In one of G.K. Chesterton's stories, a man dreams of emulating the great explorers. One day he sets sail from the West Country and heads out into the Atlantic, confident that he is destined to discover an unknown land. For many weeks he wanders across the ocean, buffeted by storms and uncertain of his position. At last, a coastline comes in view; and, as he approaches, he sees the towers and domes and minarets of a strange civilization. Greatly excited, he makes his way ashore. To his astonishment, the natives speak English. He has landed at Brighton.
>
> The psychologists have dreamed of emulating the great natural scientists. They have noted the prodigious progress of experiment, from physics to physiology, and they have argued that the same methods should next be applied to human behaviour. We still know little more of human nature, they say, than once we knew of the solar system or the evolution of life. But the methods which have been so successful there can be equally successful here: we can attain the same powers of prediction and control over human behaviour that we have acquired over the world around us. . . . In earlier times, they claim, the ignorance of human behaviour was relatively unimportant. But now life is increasingly complex; problems of over-population, of economic and political unrest, become daily harder to resolve, and a technology of behaviour has become an urgent necessity. Fortunately, however, the science of human behaviour is at hand.
>
> A disillusionment awaits them. Human nature is not an unknown country, a *terra incognita* on the map of knowledge. It is our home

ground. Human beings are not, like the objects of natural science, things which do not understand themselves; and we can already predict and control our behaviour to a remarkable extent . . . Even people who are not psychologists understand each other very well. We speak English. The psychologists are landing at Brighton.[42]

When we say that human beings understand each other, we are, then, talking of experiences that we all have. Consider a reductionist's version of a simple movement: the right elbow is partly flexed, the fingers are extended, and the arm is rotated laterally so that the hand moves across in front of the face. This could be a description of how a gracious personage acknowledges a cheering crowd; but in Australia it is more likely to refer to somebody brushing away a bush fly. We understand both situations when we see them, or when they are described in plain words; that is, we know why they have been performed: we can *explain* them.

Our everyday explanations of human action are of several kinds. D.H. Hargreaves suggests four.[43] Here, based on his list, are possible answers to the question: why did Barnett write this book? First, there are answers which state *intentions* or goals: for instance, he wrote in order to earn royalties, or to persuade people to accept his views. Second are *reasons*, that is, states of mind during writing: he wanted something with which to distract himself, or he was anxious to provoke public attention. Third are *triggering events*, such as a letter from a publisher, or the appearance of an especially irritating and wrong-headed popular work. Last are *predispositions*: he enjoys the action of writing for its own sake, or he likes conveying information and ideas to others.

All these statements could be true: they are not alternatives. A full explanation of my writing this book requires answers of at least three of these kinds. (There might be no identifiable triggering stimulus.) And none can be replaced by reduction. A physiological analysis could, however, add further information: for example, the hypothetical irritating book might induce changes in the amount of certain hormones in my bloodstream, or a rise in blood pressure.

Of these kinds of explanation, only the last – the physiological – would colloquially be called scientific. (It is also reductionist.) The others represent personal knowledge; they are sometimes called hermeneutic: they are interpretations, and belong more to literature than to the natural sciences. To acknowledge more fully the value of such knowledge I choose a passage from a novel written as if it were the autobiography of the Roman emperor, Hadrian (AD 76–138), soldier, statesman and poet. 'It was to the liberty of submission', writes the fictional Hadrian,

> that I applied myself most strenuously. I determined to make the
> best of whatever situation I was in; during my years of dependence

my subjection lost its portion of bitterness, and even ignominy, if I learned to accept it as a useful exercise. Whatever I had I chose to have, obliging myself only to possess it totally, and to taste the experience to the full. Thus the most dreary tasks were accomplished with ease as long as I was willing to give myself to them. Whenever an object repelled me, I made it a subject of study, ingeniously compelling myself to extract from it a motive for enjoyment. If faced with something unforeseen or near cause for despair, like an ambush or a storm at sea, after all measures for the safety of others had been taken, I strove to welcome this hazard, to rejoice in whatever it brought me of the new and unexpected, and thus without shock the ambush or the tempest was incorporated into my plans, or my thoughts. Even in the throes of my worst disaster, I have seen a moment when sheer exhaustion reduced some part of the horror of the experience, and when I made the defeat a thing of my own in being willing to accept it. If ever I am to undergo torture (and illness will doubtless see to that) . . . I shall at least have the resource of resigning myself to my cries. And it is in such a way, with a mixture of reserve and of daring, of submission and revolt carefully concerted, of extreme demand and prudent concession, that I have finally learned to accept myself.[44]

Such a passage helps to present a portrait of a man who was remarkable in fact as well as in fiction. It also describes a personal policy which we can admire, and may even find helpful – if difficult – to imitate.

There are many other sources of this kind of knowledge. Solomon Schimmel, in an account of attitudes to anger and hostility in ancient Greece and Rome, suggests that current neglect of early moral philosophers has deprived us of a rich source of wisdom.[45] His authors include Seneca the younger (about 4 BC–AD 65) and Plutarch (about AD 46–120). Seneca's views on the upbringing of children sound like a summary of a commonsense text addressed to modern parents. Among the topics emphasized is what psychologists now name modelling, but used to be called setting an example. Similarly, Plutarch is quoted on the value of priding oneself on controlling one's own emotions; and we may recall how some modern treatments for mental illness have progressed from behaviorism to emphasizing self-awareness: as Schimmel writes, recently many psychotherapists have come to urge the importance of self-evaluation.

The ancients wrote in a style very different from ours. 'The forceful and dynamic images of Seneca . . . may be contrasted with the bland and weak self-evaluative statements' of modern therapists. When one reads the ancient authors, it is easy to see an important reason for the difference: they were moral philosophers in the traditional sense; they uninhibitedly mixed moral exhortation with 'clinical' advice. Today, psychiatric practise concentrates on benefiting the patient. References to a patient's obligations to others are not prominent. The ancient writers urged their readers to be virtuous. Moreover, they assumed that people, even those with severe emotional disorders, could sometimes respond to an appeal to reason; and that rational suggestions on what to do could be beneficial.

Has our understanding of ourselves advanced during the past 2000 years? The present century has seen a vast increase in our knowledge of the physiology of the emotions and of perception, and in our descriptions of childhood. But our useful new knowledge of how we manage our social lives is only piecemeal. (I give examples especially in Chapters 5, 10 and 11.) It is said that each generation has to write history anew for its own purposes. Perhaps we have to do much the same for the rules that regulate how we get on among ourselves.

One body of psychological theory and practise represents, at least for some people, a massive advance on ancient knowledge. Early in the twentieth century Sigmund Freud interpreted our behavior and since then many have interpreted Freud.[46] Psychoanalysis is not only a therapeutic method but also an account of what drives people to behave, think and feel as they do. In view of the obscurities and strangeness of psychoanalytic theories, it may seem incongruous to mention them under the heading of common sense. Nonetheless, they belong here, because they interpret human personality. A central feature of Freud's writings is his perception of character. Some of the extreme psychoanalytic types are familiar. The obsessional person is tight-lipped, excessively orderly, pig-headed, stingy, often distant in manner: if we know such a person we may expect that he or she will find it difficult or impossible to run the risk of making a mistake, or to accept an error when it is shown. An extreme obsessional can produce intolerable stress for a spouse or a family. Another is the narcissist, or egocentric person, whose self-admiration is at the expense of the ability to be concerned for, or interested in, others; such a person can make serious blunders by believing that his or her own desires or intentions are so valuable and important that their success is certain. Failure can then lead to uncontrolled rage or to depression.

These are extreme types. Obsessional behavior and narcissism are also character traits present to some degree in most people; sometimes, they are helpful. Meticulous tidiness and attention to detail are assets in some occupations, say, accountancy. In others, such as acting and politics, some degree of narcissism is almost obligatory.

A prominent feature of Freud's teaching is its emphasis on the symbolic significance of behavior and of our inner life. Some explanations based on symbolism are straightforward. A woman consults a physician because of severe bouts of vomiting. A full investigation reveals no cause, such as pregnancy, poisoning or intestinal obstruction. A psychiatrist is consulted. He learns that vomiting occurs only when her husband is at home; and he concludes that the woman is *sick of* her husband.[47] Other examples are more elaborate and, some would say, fanciful. If one dreams of a bridge, one is told, that is primarily a symbol of the phallus 'which connects the parents with each other during sexual intercourse'. But it also has

> a wider set of meanings . . . Since the male genital organ is responsible for the fact that one can emerge from the waters of birth into the world, the bridge depicts the passage from Yonder . . . to Here (life) . . . ; and finally, further removed from its original meaning, it indicates transition, or any change of condition whatever. That is why a woman who has not yet overcome her desire to be a man so frequently dreams of bridges which are too short to reach the other side.[48]

The impact of psychoanalysis comes largely from illuminating accounts of the lives of particular individuals. Many display traits such as narcissism or obsessional behavior in extreme and crippling forms. These case histories are not scientific in the modern sense; but they give insight into human conduct of the kind derived from the great writers of fiction. Behavior – often bizarre – is interpreted in the symbolic terms of the person's inner life. Successful treatment of psychoneurosis is held to depend on understanding fears and desires hitherto unacknowledged (unconscious). The extent to which psychoanalysis does cure such disorders has been inconclusively debated, but Freud and his followers have greatly influenced the ways in which we see ourselves. They have (for good or ill) returned to the ancient Greek advocacy of self-knowledge, and given it a new dimension. The poet, W.H. Auden, writes of Freud,

> if often he was wrong and, at times, absurd,
> to us he is no more a person
> now but a whole climate of opinion.

Classifying

The knowledge that we use in our everyday encounters with other people is – like the interpretations of psychoanalysis – not, in any ordinary sense, scientific: it is not founded on experiment; it cannot be summed up under clearly defined laws of human conduct; hence it cannot be formally tested or refuted by further observations. Howard Gardner puts such understanding of others and of ourselves as a major 'intelligence' or ability. In his system, it is distinguished from linguistic, scientific, spatial and muscular skills. He writes of 'the ability to notice and make distinctions among individuals and, in particular, among their moods, temperaments, motivations, and intentions' as something that we all have to some degree; but he suggests that it is especially well developed among psychotherapists, religious and political leaders, and witch-doctors.[49] Those virtually without it are called psychopaths.

As Gardner implies, our interpretations of ourselves (as of the world in general) are often *classifications*. Indeed, many explanations are accepted because they place something or somebody under a familiar heading. Like metaphor, classification enters all our intellectual activities, trivial or profound.[50] Much of it, though important, is so commonplace as to pass unnoticed. In a garden one may note that some plants are culinary herbs, others, quite similar, are not. The herbs, but not the others, if chopped up and added to a stew, will improve its flavor. Hence putting something in a particular category may enable us to make useful predictions about it. There is a story that Georges Cuvier (1769–1832), the famous anatomist, was confronted by the devil, who threatened to eat him. Cuvier looked the devil up and down and replied, "You have horns and hooves. Go and eat grass! You can't eat me!'

As we know from psychoanalytic theory, we use similar logic when we classify human beings. If, say, one 'explains' that an author has written a book because of a neurotic compulsion to write, the author is put in a recognizable category – that of people who seem impelled from within to work most of the time. Moreover, compulsive workers are exceptionally liable to coronary artery disease, and they can be usefully advised how to reduce this danger.[51]

We may also, by reclassifying an object, persuade oneself or others into a new attitude. A novelist tells how a writer sat at his window, and saw,

> as he supposed, a many coloured beetle of unusually hideous shape crawling across his paper. A second glance showed him that it was a dead leaf, moved by the breeze; and instantly the very curves and re-entrants which had made its ugliness turned into its beauties.[52]

In this book we are concerned not with how we see a leaf but with how we see ourselves. What are the consequences of classifying human beings as nothing but irascible apes, the vehicles of competitive genes, or simply mechanisms? We know the answer (Chapter 10). People are treated not as persons but as objects to be manipulated. 'To deny people', writes John Shotter,

> the opportunity to describe their experience, to explain and justify themselves, or . . . to participate in the effort to illuminate the reasons for their conduct is . . . to refuse to confer upon them their status as persons in their own moral world.[53]

Moreover, if people are regarded as only objects, then – to be consistent – those who hold such views must look on themselves too as some sort of machines, programmed or programmable to respond in set ways to outside influences. Some human beings do consistently believe themselves to be driven by external forces: they are paranoid schizophrenics.[54] Among the sane the feeling or belief that one is a mechanism may – Shotter suggests

– induce a failure of self-confidence; (and certainly many people are deficient in rational assertiveness).

> Viewing ourselves as merely the product of external causes, we may distrust and debunk even our own judgements; we become afraid to say what we think . . . feel, . . . want. . . . Unable to commit ourselves to a position, to something in which we really believe, we lose the capacity for sustained, self-directed, purposeful action.[55]

There is, however, an alternative position: those who look on human beings as mere objects may, by an unstated dispensation, exempt themselves. They often, after all, display independence and originality in their own work. Perhaps they consider themselves entitled to membership of a privileged group with power to control state affairs. If such a claim were accepted, the result could be rule by an elite answerable only to themselves. The concept of democracy (to which we return in Chapter 15) is founded on the opposite presumption: that every adult has both the capacity for self-expression and the right to exercise it.

Kinds of knowledge

The preceding pages may appear to put common sense in *opposition* to science. The seeming criticism of science is, however, in fact criticism of the metaphysical presumptions (usually unacknowledged) of some writers. The two kinds of knowledge are both valid. Common sense, of course, has priority: it came first; and, correspondingly, we cannot do without it. Experimental science hardly existed before the seventeenth century: it is a novel human creation, complementary to traditional knowledge but not replacing it. To sum up the argument, I ask what kinds of true statement can be made about these lines on the human condition, by a seventeenth-century poet, John Davies.

> I know my soul hathe power to know all things,
> Yet she is blind and ignorant in all:
> I know I'm one of Nature's little kings,
> Yet to the least and vilest things am thrall.
> I know my life's a pain and but a span;
> I know my sense is mocked in everything;
> And to conclude, I know myself a Man –
> Which is a proud and yet a wretched thing.

For some purposes, it would perhaps be interesting to know how the poet's cerebellum contributed to the minutely adjusted movements of his fingers as they guided pen across paper; and it would be still more interesting to know just what was happening in his cerebral cortex and associated parts of the brain at the time of writing. Or we might ask what was the effect (if any) of composing the poem on the poet's metabolic rate; or – a different but related question – did he have a hearty meal after completing the poem, or did he fast? Such questions could, of course, be asked about anybody, writing anything.

As we know, some people appear to assume that writing those lines must have contributed to the poet's biological fitness; but few, if any, would in fact be led, after reading the poem, to enquire about the poet's family. (Or, if they did, they would probably be interested in him as a person, not as an example of a law of population genetics.) Again, it might be asked whether writing the poem was rewarding. Such a question could perhaps be taken seriously if it were related to the economics of original writing in seventeenth-century England, that is, if it were raised in a sociological or historical context. (Alternatively, it could signify interest in how the poet *felt* when he had finished writing: did he experience relief of tension, fatigue, pleasure or desire for some other activity?)

All these questions, and many others, are legitimate. Each corresponds to some kind of authentic knowledge. And, to repeat, none can replace any other. But none is about the poem, considered as something unique. The poet's genes, metabolism, social status and much else combined to produce those eight lines. The lines themselves make something new. A reader may find them moving or tedious, significant or empty; but, in making any such judgment, one is considering the poem itself, not what caused it. One may also remark that many others have written on the same theme. (Compare Pope's lines quoted on page 59.) Though the poem is unique, it has something in common with other products of the human imagination.

In its most complete forms, reductionism implies that one final kind of knowledge can provide explanations, perhaps in terms of ultimate particles, that need no further addition or exposition. In contrast, when we look at all the knowledge we have, we find a 'universe of emergent novelty'.[56] Novelty emerges, at each stage, as we move from the smallest units to atoms; from them to molecules . . . to organisms, populations and communities. Novelty also issues with the lapse of time: in the living world the time scale may be that of evolutionary change or that of the life of a single individual. The production of novelty is an especial talent of human beings. Even if we are not great innovators, we each have our own ways of living and our own decisions to make.

13 Human communication

Words, from the earliest times of which we have historical records, have been objects of superstitious awe.

Bertrand Russell[1]

The everyday ability in which we most clearly express our individuality is speech. It also marks us off from other species. Yet even our wordless communication illustrates quite well how different we are from animals. I now return to the story begun in Chapter 4.

Communication without words

Some forms of human communication, especially the signals of infancy, resemble those of animals in being distinct patterns common to our whole species. They may be sounds, such as an infant's cry, sights such as a smile, or even odors, though we know less about them. The most fully studied of our non-verbal signals are those that we can see. We use facial expressions, postures and gestures to convey attitudes, emotions and intentions, often unconsciously. The whites of our eyes and our pink lips are distinctive features of human beings and help us to convey information to others: we can instantly tell in what direction a person is looking, and the understanding of speech and of facial expression is aided by watching lip movements. Still more important, the muscles of expression – those around our eyes, nose and lips – give the human face an exceptional mobility which is already evident at birth: a newborn infant can indicate disgust at an unpleasant taste.[2] By the age of 36 hours an infant can imitate the facial expressions of an adult.[3]

There has been much argument on the extent to which facial expressions are universal throughout the human species; and, if they are universal, on whether they are generally understood. Experimenters have sometimes used photographs, sometimes actors simulating various emotions in person. The grimaces that go with happiness, anger, disgust, melancholy and the combination of fear with surprise have proved to be largely independent of local custom: they are 'species-typical', that is, general among human beings. But even they are, by themselves, often misinterpreted.[4] Of course, in ordinary situations, facial expressions are accompanied by other clues, such as exclamations. Tone of voice can provide much information.

Most of our non-verbal signals, however, belong to the community in which we have been reared. We may distinguish two kinds.[5] The first, called emblems by Paul Ekman, have a well defined meaning in a given culture. An example is the head nod, which can mean yes; but the direction

Infant's imitation of mother's facial expressions.

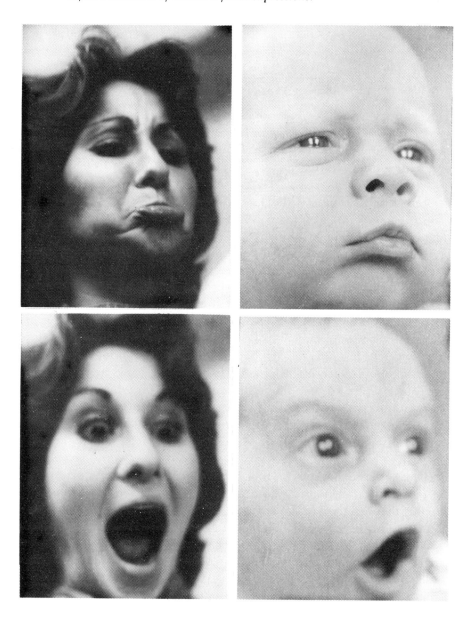

of the affirmative movement (vertical or horizontal) is not the same everywhere. This can lead to rather serious misunderstandings when two cultures meet. Moreover, some hundreds of millions of people in India (but not, as far as I know, anywhere else) have a lateral movement which conveys a muted assent. Other local emblems, familiar to many readers, include the shrug, which signals ignorance or indifference, and the wink, which signifies 'collusive agreement or flirtation'. In every community there are also typical emblems for welcome and farewell, and for insult.

The second kind, the illustrators, do not stand for words but reinforce speech. They include gestures that indicate direction, or the shape of an object. Such movements can be dispensed with (as when we telephone); but they accompany almost all speech, though their extent varies. Peter Farb quotes the Yiddish witticism, 'I talked my hands off', and the Italian, 'I held his arms to shut him up'.[6] Whether we are reared in a gesticulating culture or not, we learn, early in life, not only the speech of our community but also its wordless signals.

Non-verbal communication. The message conveyed by a facial expression often varies: in Tibet this gesture is a sign of polite deference.

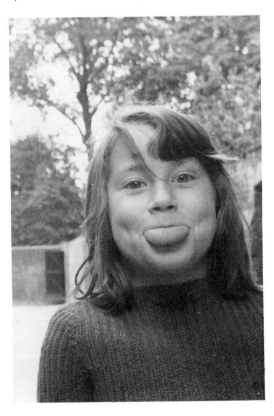

Primeval language?

Speech has sometimes been supposed to arise from gesture.[7] But no existing languages depend largely on wordless signs: all have vocabularies of thousands of words and an elaborate grammar. The notion that simple, 'primitive' languages still exist ('Me Tarzan, you Jane') is erroneous.

Some of the most elaborate tongues are those of the few remaining gatherer–hunters. The Arunta or Aranda, of the central Australian desert, have a grammar far more intricate than that of English, or even that of classical Greek. To quote Peter Farb:

> An aborigine . . . who can carry all of his material possessions on his back nevertheless carries in his head a language of unbelievable complexity and richness. Aranda verb stems, for example, can take about a thousand combinations of suffixes – as if, in English, the verb *go* could be converted into a thousand variants – like *going* and *gone*. Moreover, just one of these suffixes can so subtly change

Tradition in technology. Left, an Acheulian hand-ax, of a type used in Africa, Asia and Europe for about a million years. Eventually, such tools were replaced by elaborately flaked implements, such as the sickle blade on the right.

the original verb that often an entire sentence is needed in an English translation to express the different shades of meaning. . . . many archaic words, absent from everyday speech, have been preserved and are used on formal occasions, much as if a speaker of English today could switch suddenly to the Old English of *Beowulf* or to the Middle English of Chaucer.[8]

Existing languages, therefore, give us no guidance about the origin of speech. But they leave us free to exercise the imagination. One possibility is that the ability to converse arose slowly, over millennia, as a means of conveying to the next generation how to make and use tools. Alternatively, perhaps speech was invented by some towering paleolithic intellect, and adopted by her family on whose members it conferred a massive increase in Darwinian fitness.

But most people interested in linguistics concern themselves with what can be observed now, and with ideas that can be tested. If this rules out discussion of the origin of speech, it does not exclude studying the signalling abilities of chimpanzees and gorillas. Do these somehow represent the rudiments of linguistic ability? Chimpanzees seem to convey information about their surroundings by posture or gesture. Such signals are difficult to study, but E.W. Menzel describes a number of experiments with a group of eight chimpanzees which had been born in the wild.[9] While they were confined in a cage, food was concealed in one corner of their paddock, and shown to one of them. When they were all released together, some went ahead to the food, before the informant; but it was not clear exactly what signals they were responding to. In other experiments, communication about a 'dangerous' object, such as a model snake, was similarly observed. The gait of an ape moving toward or near such an object may convey information to other members of its group.

Apes can also be trained to communicate among themselves by quite unnatural methods. The signals in some experiments are plastic shapes. The chimpanzees first learn a 'meaning' for each shape. They are also taught to open boxes with various tools, such as a wrench, a key, or money (for a vending machine). They can then learn to signal to another for the tool needed to open a box; the box is available and visible only to the signaller. Success is rewarded with food, and the success rate can be better than 90 per cent.[10]

Such experiments illustrate the capacity of apes to use symbols, but are not themselves designed to reveal linguistic skills. The gestural abilities of chimpanzees have, however, been used in attempts to teach them human 'speech'. Apes cannot make the human range of noises: they lack a suitable larynx and other structures.[11] Some human beings are born deaf, and so never learn to speak, but communicate effectively by signs. Why should not chimpanzees do the same? B.T. & R.A. Gardner thought of asking this question, and so tried to teach a young female, Washoe, to use a sign language designed for deaf people.[12] The experimenters taught Washoe partly by moving her limbs into the desired positions, partly by encouraging her to imitate human beings. After four exacting years she had learned to use more than a hundred signs correctly. Since then, other apes have reached similar standards. Some have learned to use colored plastic shapes in place of words.

Initially, the animals name objects, such as banana, dog, bird and so on. Learning to sign banana when a banana is seen (or desired) is, however, not a major linguistic achievement. But apes have also learned verbs, including give-me, go and put, and adverbs such as up and down. Moreover, they put strings of words together, such as 'Please give piece-of cabbage'. It has therefore seemed that apes can, not only sign separate words, but also make up sentences.

This assertion has led to a lively and sometimes rather odd controversy.[13] At one extreme are experimenters who are indignant at any derogation of the linguistic abilities of their chimpanzee protégés, for whom they may feel great affection; at the other are those who seem to resent the notion that apes can in some sense rival the human species. A notable intervention has been that of H.S. Terrace and of his chimpanzee, Nim (or Neam) Chimpsky. When Terrace began his experiments with the infant Nim, he assumed that chimpanzees had the capacity to make sentences, but wished to know more about it. Nim was intensively studied to the age of four years and four months. During the two final years, the average length of his utterances fluctuated between 1·1 and 1·6 signs. When he did make long utterances, they were repetitions: 'give orange me give eat orange me eat orange give me eat orange give me you.'[14]

Terrace came to doubt the then current conclusions on chimpanzee abilities, not only from observation of his one subject but also from critical study of the reports and films of others. When one observes the gestures of apes with their human captors, it is difficult to distinguish spontaneous utterances from those prompted (unintentionally) by the experimenter. There is the possibility, even probability, that the sentences produced by chimpanzees are a result of aping their industrious human teachers, not of creating novel sequences of words. Yet children can make new statements even at the age of two; moreover, they do so without being systematically taught to speak. The rather disappointing conclusion is summed up by C.A. Ristau & Donald Robbins: 'there is no unequivocal evidence that apes can master conversational, semantic, and syntactic aspects of language.'[15]

The facts of anatomy match those of behavior. Spoken language requires a special kind of memory, for it consists of units – sounds, words and phrases – which last only a short time and have to be put in order. A distinctive region of the human brain, Broca's area, is especially concerned with this ability: destroy it on one side (usually the left) and the ability is destroyed too. Apes do not possess such a structure. Another such region in our brains, Wernicke's area, is essential for understanding speech.[16] Hence, to quote Noam Chomsky, 'there is no evidence of continuity, in an evolutionary sense, between the grammars of human languages and animal communication systems.'[17] The evolution of humanity is mysterious; and among the mysteries that of the origin of language has a leading place.

Words about words

This conclusion leads to the question: just what are the special forms of human languages? A prominent feature is that one writes of languages in the plural. Human speech, unlike the signals of an animal species, is immensely diverse: learning one language does not enable us to understand another. Nor is there one ideal language. Instead, a general property of our speech and writing is the capacity for change. The flexibility of language is essential when we introduce new techniques or new ideas. Novel expressions turn up spontaneously in new situations. Here is a remark, perhaps obscure to some readers, overheard in a Royal Air Force mess in the 1940s: 'Let's get the gen on that wizard new kite before some operational type goes and prangs it.' In translation it loses force: 'Let's get the (printed) information on that excellent new aircraft before somebody on flying duties takes it.' Language is then a means by which a new tradition can be rapidly created and maintained. But changes in our speech may also reflect losses of obsolete custom and ways of thinking.

In Chapter 11 of *Genesis*, language is 'confounded' by divine intervention, so that people 'may not understand one another's speech'. But, despite the tower of Babel, all languages can be learned by almost any person, especially in childhood. And all languages have certain design features in common.[18] Here I select the most important.

Nearly all the many thousands of words of each language are *arbitrary*: their form rarely corresponds to what they stand for. (Words such as shriek and whistle are exceptions.) And there is no limit to the introduction of new words or phrases. The scope of human languages is, in this respect, of an entirely different order from that of any animal species. Most animal communication, as we know, reflects a specific internal state of the signaller: it is exclamatory and, when it is evoked by an external event, it refers to the situation at the moment of utterance. In contrast, much of our speech is descriptive and has the quality known as *displacement*: it refers to past or distant events, or even to possible happenings in the future. And descriptions may be fictions. Indeed, fiction and fantasy have a role in all societies.

The wide scope of our utterances leads to a feature obvious when pointed out but still surprising. Our languages are *productive*: we daily speak or write sentences that we have never uttered before, though they are constructed (as a rule) of often used words or phrases. Certain expressions, such as greetings ('How are you?'), are frequently repeated, but these are the exceptions: we both create and understand, 'without a thought', novel sentences, whenever we have a conversation. Correspondingly, what a person says is far short of what he or she could say. We have at our command a vast number of possible sentences most of which we shall never utter, or even hear.

It may seem trite or naive to say that our sentences usually have a meaning. But the last sentence (one, I believe, that I have never written before) has itself an important meaning. I am now referring to a property of language called *semanticity*. When we hear or read a sentence, what we attend to, as a rule, is its significance, but not its individual words or its grammatical form. To quote Noam Chomsky, 'specific sound-meaning correspondence' is at the level of the sentence, not of individual words.[19] Here is an example. 'Hebe said, "Jill gave Robyn a shirt".' This simple report can equally be rendered as: '"Robyn", Hebe said, "was given a shirt by Jill".' Yet other versions are easily thought of: for example, 'Hebe said that Jill had given Robyn a shirt.' In ordinary circumstances, a hearer of any of these statements would register the nature of Jill's generosity, and that Hebe had reported it; but the precise form of words would probably not be recalled.

The role of the sentence as a unit is brought out by P.N. Johnson-Laird, in his description of what he calls an odd but familiar phenomenon rarely discussed by linguists.[20] We choose the words we need in conversation, with little or no hesitation; and we usually understand what is said to us. Yet we may be in severe difficulty if we are asked for the meaning of an individual word. His example is the sentence, 'Have you ever spent any time in Sussex?' What, in that sentence, is the meaning of 'time'? (He adds: 'Children have the happy knack of stumping us with such questions.')

Meaning (or semanticity) is also illustrated by ambiguous sentences. 'They are boring students' could refer to the effect of professors on students *or* that of students on professors. (The context should give a clue: who is speaking, a professor or a student?) And here is a sentence with four possible meanings: 'The police were ordered to stop drinking after midnight.' Ambiguity can also be fun. In a novel, a glamorous young woman remarks, 'I like eating oysters without dressing.'[21] A sentence heard or read calls up an image in our thoughts: it is not merely a standard stimulus evoking a standard response. We continually interpret what we hear or read. Hence the image may not be the one intended by the speaker. A listener to the sentence about Jill, if biased by an inappropriate assumption, might believe that Jill had given Robyn a *skirt*.

The productive (or creative) quality of language combines with the phenomena of meaning to give our speech features which do not allow any simple behavioral or psychological description, let alone explanation. The ability to create new sentences and to understand them goes far beyond the creation of new jargon or of jokes. It can go with the appearance of novel and influential ideas. John Passmore points to the imperfections, or lack of exactness, in ordinary language as an aid to originality. He gives the example of Freud's expression, unconscious mind. In the past, he remarks, mind 'had commonly been *defined* as consciousness'.[22] (Such creative uses of metaphor and of words – compare Chapter 3 – do not, of course, excuse slovenly or deceitful use of language, or neglect of logic.)

The ability to create and to understand novelty in our speech stands by itself as something that requires its own methods of study, just as the human species stands alone, apart from the animal kingdom. It is accompanied by an additional design feature of language, *reflexiveness*, or the ability, exemplified in the preceding paragraphs, to use language to scrutinize language.

Language and machines

A problem, to which much logical effort has been directed, is that of designing machines that translate from one language to another. Despite the ingenuity of those who have tried it, the outlook for this venture is not promising. Machine translation has, however, generated some good stories which, like the difficulties of machine translation themselves, illustrate the complexities of our languages. My favorite concerns the reported result of translating English proverbial sayings into Japanese and back again. One emerged from the double process as: 'Blind men are mad.' With a little effort (I found that free association helps) one can derive the original.[23]

Computers cannot use free association. Yet attempts at machine translation have been founded on the presumption that computers can be programmed to simulate human thinking (or cognition). This idea, and indeed the equation of brains with computers, has now become popular. Here is an incident from a thriller, in which (after an earthquake) the protagonist is suddenly faced with a murderer bearing a knife.

> There was no room for manoevre, and from the breadth of his shoulders she knew the man was very strong. . . . That part of her mind which was a fighting computer assessed a dozen permutations of possibilities in milliseconds, and as the blade leapt at her she took it two inches below her raised forearm, through the sleeve of her robe . . . [24]

The most advanced computers are of the digital type: each unit of information ('bit' or binary digit) represents a decision on which of two equally probable alternatives has been chosen. When one uses such a system, information is redefined as anything that removes uncertainty. A coin falling tails provides information; and a die falling with (say) four uppermost provides more. But information in this sense need have nothing to do with meaning: two messages, one nonsensical and the other full of significance, may each carry the same amount of information.

A computer is effective because it can perform long sequences of operations at a hardly imaginable speed. The procedure, however, is cumbrous: for instance, if it is to identify even so simple a shape as a hexagon, it must compare the shape with a series of alternatives, one by one; has it three sides, four sides . . . ? This is not how a human being identifies such a simple plane figure; and it certainly is not how we recognize, say, a face in a crowd – or an assassin with a knife; nor is it the method used by the Eskimo or the Puluwatans in identifying the features of their surroundings (pages 195–6); still less is it how we understand a sentence. Yet some computer scientists who attempt machine translation have assumed that, whenever we respond appropriately to a word or a sentence, we unconsciously run almost instantaneously through a series of comparisons.

In Chapter 3 I illustrate the always present phenomenon of metaphor. Above, in the present chapter, I give examples of the related phenomenon of ambiguity. H.L. Dreyfus illustrates what computers cannot do when he cites the sentence, 'The idea is in the pen.'[25] Taken literally, this is nonsense; but, as he says, it would be comprehensible in the context of a discussion about a promising writer who does not quite achieve what is hoped for. Context is all important in many less obscure statements. At first, 'the box was in the pen' sounds odd. But if a small boy were writing with a pen, then playing near his pen, then looking for his box and found it in his pen, there would be no problem. A playpen is more than commodious enough to contain a box. Can a digital computer be programmed to cope with contextual problems of this sort? Even very simple colloquial utterances present difficulties. It would be irritating if a friend, when asked 'Can you pass the salt?', merely replied, 'Yes!' Yet, taken literally, the question is merely a request for information. We have no satisfactory account of how we understand sentences, and no reason so far to think that study of computers will provide one.

Functions of language

The properties of languages are not the same as the purposes for which we use them. We may conveniently classify the types of message conveyed by language under the headings of expressive, descriptive and argumentative.

In Chapter 4 I list the principal meanings of animal signals: most convey what in ourselves we should call an inner state, an attitude or possibly an intention ('I love you!'), or a command ('Go away!'). These may be called expressive; but to equate the cries or postures of animals with human expression of the emotions is obviously a mistake. Recall, for example, the lines by John Davies, quoted on page 249. Expressive utterances may reflect only a momentary feeling; but, at the other extreme, they may be an attempt, such as that of the poet, to state a universal truth about the human condition. Sometimes they are efforts, such as the exhortations of priests or politicians, to persuade people to adopt some belief or policy. Whatever their role, all except the simplest are highly individual: each represents not only a general message but the personality of the speaker.

Individuality is found even in our attempts to describe accurately what we observe. Here is an example from Samuel Johnson's *Journey to the Western Islands of Scotland*, published in 1775.

> A hut is constructed with loose stones, ranged for the most part with some tendency to circularity. It must be placed where the wind cannot act upon it with violence, because it has no cement; and where the water will run easily away, because it has no floor but the naked ground. The wall, which is commonly about six feet high, declines from the perpendicular a little inward. Such rafters as can be procured are then raised for a roof, and covered with heath, which makes a strong and warm thatch, kept from flying off by ropes of twisted heath, of which the ends, reaching from the centre of the thatch to the top of the wall, are held firm by the weight of a large stone. No light is admitted but at the entrance, and through a hole in the thatch, which gives vent to the smoke. This hole is not directly over the fire, lest the rain should extinguish it; and the smoke therefore naturally fills the place before it escapes. Such is the general structure of the houses in which one of the nations of this opulent and powerful island has been hitherto content to live. Huts however are not more uniform than palaces; and this which we were inspecting was very far from one of the meanest, for it was divided into several apartments; and its inhabitants possessed such property as a pastoral poet might exalt into riches.[26]

And here is a passage from another master of description, Rudyard Kipling, also an Englishman, written rather more than a century later about a house in India.

> There were the usual blue-and-white striped jail-made rugs on the uneven floor; the usual glass-studded Amritsar *phulkaris* draped to nails driven into the flaking whitewash of the walls; the usual half-dozen chairs that did not match, picked up at sales of dead men's effects; and the usual streaks of black grease where the leather

punka-thong ran through the wall. It was as though everything had
been unpacked the night before to be repacked next morning. Not
a door in the house was true on its hinges. The little windows,
fifteen feet up, were darkened with wasp-nests, and lizards hunted
flies between the beams of the wood-ceiled roof.[27]

Johnson and Kipling illustrate both the changes in usage that can take place
in a few decades and also personal style.

The use of language for description is quite unlike any animal communication; and the same applies, even more clearly, when description leads to
debate. 'The two most important higher functions of human languages',
writes K.R. Popper,

> are the *descriptive function* and the *argumentative function*.
>
> With the descriptive function of human language, the regulative
> idea of *truth* emerges, that is, of a description that fits the facts. . . .
>
> The argumentative function of human language presupposes the
> descriptive function: arguments are, fundamentally, about descrip-
> tions: they criticize descriptions from the point of view of the reg-
> ulative ideas of truth; content; and verisimilitude.

And he adds that we owe 'our humanity, our reason' to the development
of these features of language.[28]

The passage from Popper is both a statement about language and an
example of the use of language by a philosopher. We use words, not only
to express our feelings and to describe events, but also to argue, to criticize,
to debate questions. Much of this book illustrates the argumentative role
of language.

The genesis of speech

The intricacies of adult speech are matched by the development of
language in childhood. The earliest sounds regularly made by infants, those
of crying, are analogous to the expressive cries of other species. But, after
a few weeks, babbling begins. These variable but agreeable sounds have
no counterpart in the sounds made by infant apes. They are sometimes
said to be random; but, though unintelligible, they are generally supposed
to be precursors of speech. Like exploration and play, babbling is spontaneous: it does not depend on any special stimulation from outside, and it is
not, in any ordinary sense, rewarded. An infant can hear the noises he or
she makes; but on whether the noises give pleasure we can only guess.

Babbling may, however, be to some extent imitative.[29] The noises made
by infants with Chinese-speaking parents are said to differ from those of
infants with parents who speak English. The pitch of an infant's babbling
also reflects that of the voice most recently heard. Imitation is often mutual.
A mother may respond to babbling by copying it, or at least by repeating
the tone of the utterance.

Between babbling and talking a child may utter word-like noises not heard in the language used around him. One boy I know, early in his second year, frequently made a noise which I represent as scheez; later he preferred something like thingay or thingur. (Eventually he became highly articulate in English.) Of course, at all stages, such efforts evoke a parental response; but we have no exact knowledge of how either babbling or 'non-standard words' contribute to a child's development of speech.

Our first unambiguous words usually occur in our second year. It might be expected that they would express feelings of pleasure, pain and so on; and that descriptions (which are distinctively human) would come later. Moreover children, like apes, are very imitative; hence early speech might be supposed to depend on simple imitation of older people.

Both presumptions are wrong.[30] During the early years emotions are conveyed non-verbally, by smiles, cries and postures. The first sentences concern not feelings but what is happening around the child: 'car coming'; 'cat on chair'. They also include statements of desires: 'want apple'; but most speech is unrelated to any reward except that derived from speaking itself, or from the verbal response of another person. The expression of the emotions in words arises independently, at three years or later: it reflects development of awareness of the self.

The role of imitation is complex. It seems to play an important part in the learning of words, but the words and sentences, even of a young child, are not merely a result of aping adults: they are often new. Invented words are sometimes the result of misapplying rules: examples include foots, instead of feet, and throwed, instead of threw. Errors of word order may also occur: 'What this is?' These quickly disappear even if they are not corrected. Again, neither reward nor punishment, in the usual senses of these terms, is needed. Nor, while they are learning to speak grammatically, do children learn the rules of grammar. As G.A. Miller remarks, neither children nor their parents *understand* language.[32]

Grammatical speech, as we know from hearing our children, develops gradually. The process can illustrate, once again, the phenomenon of semanticity. J.S. Bruner refers to the case of a child who wants to ask, 'Have you looked?', and has not yet mastered the use of the perfect tense. An alternative is available: 'Did you look yet?'[33] For the child, as for an adult, the precise form of the language used is not important, but the meaning is. And to convey the meaning needed, the child improvises a new (and, in this case, slightly ungrammatical) sentence. Like adult speech, that of children falls outside any other, familiar kinds of behavior. Although the very complex phenomenon of imitation plays a part, it does not fully explain even the earliest sentences. From the early years, a child constructs novel sequences of words, by adapting grammatical rules to need and to his or her incomplete vocabulary.

Hence the more we learn about language, the wider we see the gulf between our own ways of communicating and those of other species. 'The normal use of language', writes Chomsky, 'is not only innovative and potentially infinite in scope, but also free from the control of detectable stimuli, either external or internal.' Language is 'an instrument of thought and self-expression' for all human beings.[34]

The musical species

Grammatical speech is universal in human communities; it is an exceedingly complex and distinctively individual form of behavior, developed spontaneously in early life; it is regulated by rules which differ between cultures; it is often a source of pleasure, but is in many ways mysterious; and it is wholly distinct from anything done by animals. The possible variations it offers are infinite.

Exactly the same statements apply to musical behavior. We are, in an important sense, the only musical species. The animals we are most likely to call musical are birds, such as nightingales, whose songs are music to *our* ears. One species has been described, by zoologists, as singing as if 'charged by passion'.[35] Moreover, the individuals of some species have not just one typical song but a repertoire; and the members of yet others sing

The musical species. Polynesian with mouth bow.

duets. Often, the typical song is learned by imitation early in life. Yet the differences of bird song from human music are more important. Bird song, as we know (Chapter 4), has, for each species, one or two well defined functions: it may help the singer to hold a territory, or it may be part of courtship. The precise role depends on the species. Similarly, duetting among birds is species-typical: either all pairs of the species sing duets, or none.

Among human beings there is no uniformity. Martial music has been used in the defense of invaded ground, but there is no rule about it. A youth may serenade a maiden, but human courtship can succeed without a musical accompaniment. Different cultures have their own types of music and uses for it: among many examples, compare the melodies of the bagpipes of northern Britain with those of the sitar in India. There is also change – sometimes rapid – in historical time: compare the orchestral works of J.S. Bach (1687–1750) with those of Johannes Brahms (1833–1897); both wrote their music in central Europe.

The changes sometimes reflect social transformations. Much of the earliest written European music was influenced by the Christian churches; but the religious component faded, and reappears today, as a rule, only in unexpected ways. We may contrast the *Litanies for the Virgin Mary* of Claudio Monteverdi (1567–1643) with the *Lady Madonna* of the Beatles (twentieth century). A devotee of the one might be disconcerted by the sound (as well as the meaning) of the other.

Music is distinguished by the precision of its sequence of notes, each of which has a specified pitch, duration and loudness. As a rule, emphasis, produced by increased loudness, provides a distinct rhythm. Within this framework is much variation, even if we disregard the diversity of musical instruments. Western music has a chromatic scale with intervals of a semitone; but most of such music is written in a single key which uses only seven of the twelve notes in the chromatic octave; this is the diatonic scale. (Changes of key nearly always occur in a single piece.) But the sitar music of India has intervals of less than a semitone. And in the music of Australian aboriginals the octave is less important than is usual.[36] The kind of music a person enjoys or sings or plays depends on upbringing. As usual in human affairs, what is accepted as 'natural' is what happens in the familiar environment.

Probably, all human groups, however 'primitive', are musical. John Blacking describes the music and dances of simple, especially African, societies. He quotes the opinion of the Venda of northern Transvaal that music is essential for human survival.[37] Accounts of the music of such societies also illustrate its variety. Colin McPhee describes a community in Bali.[38] There as elsewhere music and dancing depend on oral tradition. The orchestra consists of tuned bronze gongs, gong-chimes, metallophones, drums and cymbals. McPhee gave a set of instruments to some children and provided a teacher. The children responded enthusiastically, and in six months learned to perform well. Their music was traditional in form but also allowed individuality.

An orchestra is, of course, not essential. L.D. Koranda has studied the traditional songs of American Indians in Alaska. They refer to every facet of life, but especially to hunting. On the coast women play games with rocks and sing 'indelicate' songs about coitus or about a woman's lack of attraction for her husband.[39]

Musical ability everywhere develops early. In Chapter 9 I mention the response of infants of a few months to a change in a sequence of notes. This indicates not only a liking for variety but also the capacity to distinguish quite small differences. By similar tests infants of five months have been found to detect a change of rhythm; but they treat a familiar melody transposed to a new key as the same, just as adults do.[40] Infants can also imitate the pitch of the voices of older people and other sounds; and they improve with training even before they reach six months.[41] The spontaneous singing that first develops in the second year does not, however, follow any formal musical system: there are no distinct notes, but instead there is a wavering, atonal sound. Identifiable notes are heard later in the second year. Successive notes are then commonly close together in pitch (seconds), but soon major and minor thirds occur in undulating rhythm. Longer intervals, such as fourths, become prominent still later. Soon after eight

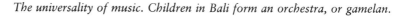

The universality of music. Children in Bali form an orchestra, or gamelan.

years, a child can recognise melodies and rhythms and can sing accordingly. Musical ability now equals that of untrained adults. These statements apply to children who hear music of the 'western' type in their early years. Experience of other kinds of melody may have a different effect.

Composing sad songs with a bush piano, in the Kalahari.

Music has been called a universal language. This is an attractive idea, but it does not fit the fact that the music of one culture sounds strange to ears from another. Nor is music a language except in a remotely metaphorical way. It can be used in conveying a message, as when a bugle sounds 'Lights Out' or when words are sung instead of spoken; but usually music consists of sounds which do not correspond to any other experience: they are on their own. Music can of course imitate events, such as a thunderstorm, bird song, or a battle; but the 'program music' that does this is not characteristic of music in general. I break off writing this paragraph to listen to a *Gloria* by Antonio Vivaldi. To a Christian listener (as no doubt it did to the composer) this piece conveys an important message that could be expressed in plain words. But for me no image is evoked: it is a sequence of lovely and moving sounds which cannot be translated; the role of the human voices resembles that of the instruments.

The scope of music in representing or evoking emotion is familiar.[42] A funeral march may signify respectful grief, a jigging tune may stand for bucolic jollity. In experiments, people with no special training have listened to passages from conventional western music, and have largely agreed on the moods represented. If a composer is writing for, say, movies or television, and wishes to convey sadness, he is likely to choose a minor key and a slow tempo. It does not, however, follow that music portraying melancholy induces depression in the listeners. Even grand opera, in which music is reinforced with words, and which commonly ends in tragedy, does not leave members of the audience staggering out bent on suicide. On the contrary, tragedy – which is traditionally accompanied by music – purges the emotions through the experience of pity and terror.[43] Music has, indeed, long been held to have a healing influence.

Language is an essential means of maintaining our complex traditons, that is, beliefs and practises transmitted from generation to generation; it is therefore conservative. Yet it is constantly changing; hence it is a medium of our adaptability to changing demands. The demands may come from our environment or from ourselves. Similarly, the language we learn in early life is a means by which we conform with the customs and attitudes of the group in which we are reared. But what we say also reflects our individuality: we each have some freedom in our choice of words. Much the same applies to music, except that few of us can give musical expression to our personalities.

Tradition is often maintained by imitation. Our children do imitate us, even when they also rebel against their elders. In the conduct of each generation, as in language and music, there is a blend of conformity with a rejection of the past. But imitation is aided by yet another ability that belongs to us alone. We energetically and, on the whole successfully, teach our young skills. Teaching as we practise it occurs in no other species; yet, despite its importance, it has been astonishingly neglected as a distinctively human trait. There is no well-developed anthropology of teaching. It requires a chapter to itself.

14 Teaching and tradition

Hast thou comprehended the breadth of the earth?
Declare if thou knowest it all.

Job: 38:18

Mathematics, music, morality and speech are easily seen as novelties that emerged with the human species. But they would hardly exist without yet another novelty, teaching. We could justly call ourselves *Homo docens*, for teaching is a distinctively human trait.[1] And it presents the same kinds of problems as our other peculiarities.

The modern neglect of teaching in the lists of human characteristics is surprising. The classical Greek word (παιδεία), usually translated as teaching or education, came also to mean culture. In the great age of Latin literature it was translated as *humanitas*, and it was further used to mean civilization.[2] But, equally significant, it could also mean discipline or correction. In a traditional textbook, from twelfth-century France, Grammar is illustrated as a woman holding both writing materials and a whip in her right hand. Her pupils, all boys, are also shown. Two have been stripped to the waist for castigation.[3] Perhaps the notion of ourselves as a teaching species is rarely found because we associate teaching with punishment.

Teaching, as I use the word here, has two features: first, it induces an intended change in the behavior or understanding of another person; and, second, it tends to be persisted in and adapted until the pupil reaches a certain level of achievement. That is how 'teaching' is to be understood in this chapter. The definition is chosen to force attention on an activity which is central to all we do under the heading of education. The definition, however, excludes much of what is commonly called teaching, for instance, lecturing: the intention is to concentrate on the mutual interactions of teacher and pupil.

It may be objected that the definition is too restrictive. The abilities we learn from others are of two main kinds, closed and open.[4] Examples of closed abilities are making a chair of a standard design, drawing a solid object in accurate perspective and performing routine mathematical calculations. All may require exertion by both expert and apprentice; but there is little difficulty in deciding when the pupil has achieved the required skill. In contrast, open abilities can be developed without end: examples are designing and making novel kinds of furniture; painting original pictures; applying mathematics to new problems; teaching; and the kind of capacity required for critically analyzing the evidence or logic used in a book such as this. Success in teaching open abilities is not always easy to identify (and is notoriously difficult to examine). Nonetheless, some teachers try to develop originality, independence and critical skill in their pupils, and my definition is intended to cover their efforts.

Teaching is not only a central feature of human social life: with the related phenomena of tradition it gives us a cross bearing on fundamental principles of human biology. We are reminded once again of how heredity and environment interact during individual development; and so we return to the crucial importance of studying ontogeny – the growth of the individual, from fertilization onward, in a complex and varying environment. We also see more clearly how social transmission of skills, ideas and attitudes differs from the transmission of genetical information.

Social learning by animals

My definition of teaching refers only to human beings, but we may ask whether animals do anything analogous. Learning as a result of instruction belongs in a wider category, social learning; and this occurs even among birds. During the 1940s, in England, blue tits took to tearing the tops off milk bottles, and drinking the cream. The sudden spread of this habit must have suggested, to some indignant householders, that the birds were spreading the news by word of mouth; but in fact the habit probably began as a chance occurrence, and became general because in winter such birds feed in company. Another kind of social learning among birds, as we know, depends on imitation: to sing the typical song of the species a bird may need, early in life, to hear it sung by adults.

Imitation is important for many mammals, and is usually of something seen. Cats imitate other cats, and so are helped to develop new habits: that is, they can learn by observation. A domestic cat, that has seen another cat learning to pull a string to get food, quickly acquires the string-pulling habit when given the opportunity: its learning has been speeded up by watching the demonstrator cat. In natural conditions such an ability probably helps young carnivores to learn how to hunt. Other kinds of mammals that readily imitate include the dolphins. These animals even copy the sounds they hear, and also the actions, not only of other dolphins, but also of human beings: one saw a diver scraping algae from the side of its aquarium; it used a scraper for the same purpose, and ate the algae.

Among the monkeys and apes, imitation is so important that the development of skills and new habits of all kinds is usually, perhaps always, a social process. Many experiments have confirmed their notorious capacity to *ape* the movements of another. Moreover, it leads to the development of local traditions. One troop of monkeys may select, from the foods available, a diet different from that of another troop with access to the same resources. Baboons are plant eaters; but the members of some groups have become occasional flesh-eaters during a period of a few years. There may even be local habits in the ways in which foods are treated: the most famous example is of a troop of Japanese monkeys that took to washing sweet potatoes before eating them. Such innovations are attributed to the originality of young individuals – independent spirits who break away from the entrenched customs of their group.

Another similarity to human conduct is in the use of tools by chimpanzees. Sticks may be selected and stripped, then used to get edible insects out of their nests. Smaller twigs are used for tooth cleaning. Young apes develop these habits by imitation: there is no evidence that adults help their young to develop skills by anything analogous to teaching.

Imitation, clearly, can occur when the model is quite unaware of it. There is, however, a kind of social learning in which an older animal does take an active part. The large cats, such as lions, evidently 'encourage' their young to hunt: they pull down prey and leave it for the cubs to kill. Smaller carnivores, such as meercats (a kind of mongoose), eat insects; when the young begin to take solids, the mother catches insects and holds them in her teeth; the young then snatch them from her. But even 'encouragement' is not quite analogous to teaching in my sense, for there is no evidence that the parents systematically adapt their behavior to the needs of the young. Perhaps, however, there is an approach to it in the 'discouragement' sometimes observed among chimpanzees. Young apes are adventurous feeders; an elder may take an unfamiliar food from a young one, and so restrict its feeding.[5]

There is only one group of activities in which animals do display behavior clearly analogous to teaching. The conduct 'taught' is always *keeping away*, and the method is always punishment. The most familiar case is weaning. A bitch bats off her now independent puppies; a macaque or a langur drives away her year-old offspring, sometimes with cries from the victim. Similarly, some kinds of status system require individuals to keep their distance: one member of a group of, say, baboons approaches another; the second animal gives a standard signal, and the first withdraws. The signal may be violent, such as a bite, but may be quite mild, such as blinking. It may seem to be stretching a point to call all such acts punishment, but, as so often, we are obliged to extend a commonplace notion beyond its everyday use: we are dealing with signals that tend to cause withdrawal; these include harmless acts, such as blinking or slapping the ground, as well as assault with battery. Lastly, in territorial conduct, too, deterrent signals are used. The justification for likening all such conduct to teaching (in my sense) is that in each the deterrent activity is persisted in until the target animal does keep away.

Such facts provide an opportunity for yet another myth about the human animal. It could run like this. The findings in the preceding paragraph show us that punishment is natural. Parents should force independence on their children at an early age, but strict discipline is needed for proper social behavior. The young learn only by the experience of punishment. Transgressors should be 'taught a lesson' by rigorous chastisement. And so on. But in fact these minor analogs of teaching only bring out, once again, the gulf between other species and ourselves. Our teaching need not depend upon punishment. Very little is concerned with spacing people out. We teach skills, ideas and principles of action; and we also teach others *about* what goes on in the world. Animals do not do any of these things.

Teaching in simple societies

The idea of humanity as the teaching species is, however, valid only if teaching is a general feature of human communities. In technically advanced societies schools are taken for granted, and some of what goes on in them is teaching in my sense. What of 'primitive' societies? Systematic

*The **universality** of teaching. Children of the Kalahari and their instructor.*

studies, by anthropologists, of teaching as a distinct (and important) social phenomenon, are rare, but there is much information on how children are induced to become sociable. Among people with simple technology, socializing children is a prominent feature of their lives; and it usually depends on praise, not punishment. W.N. Stephens remarks on how often anthropologists (from 'western' countries) 'seem amazed at the amount of affection, care, attention, indulgence and general fuss lavished on infants and young children.'[6]

As an instance, the survival of the Aleut people depends on their highly developed skill at hunting. Abilities such as throwing a spear from a kayak, and a detailed knowledge of animal behavior, are taught with great thoroughness. These hunters are notably tolerant of youthful conduct and misconduct. Among the Eskimo, a child may be applauded for wearing out a pair of shoes. At another climatic extreme, among the Pitjandjara of the western Australian desert, the children are encouraged from their second or third year to imitate adults, to dig with sticks and to track edible insects and lizards; success is rewarded with praise. Similarly, the San people of the Kalahari spend much time encouraging their young, even at two years, to pursue, catch and bite large insects.

In a pre-literate Melanesian society in New Guinea, the adults go further. They regard educating their young as a sacred duty. Each generation is obliged to transmit the tribal customs to the next. These customs are said to have been originated by mythical heroes of the past. A child's wilfulness is regarded as an offense against these heroes; but blame is usually attached to the parents, not the children.

Such attitudes are, however, not universal. In equatorial Africa, the agricultural Nyansongo of Kenya remind us of the diversity of human customs. They, too, train their children in the skills they need, between a rather late weaning and initiation at 10 to 12 years of age; but in doing so the adults adopt a contemptuous attitude, and use punishments such as deprivation of food, beating or exposure to cold.

Teaching is not always recognized as a necessary activity: in some groups children are expected to become both skilled and sociable simply by imitation. The desert-living Papago of Arizona assume that children will grow up without intervention by adults. Among the Africans of the northern Transvaal, the children have to acquire the skills of herding and of gathering food, some of which depend on an intimate knowledge of the local natural history. The adults say that they learn merely by doing; they are enthusiastic imitators, and the adults exert no compulsion. Similarly, among the Tallensi of northern Ghana, there is no systematic training: adults assume that skills will grow, just as children become bigger without being taught to do so. In Malaysia, the Semai who, like the Tallensi, practise a simple agriculture, positively deny that they teach their children: learning is by imitative play. A similar attitude is found among the rice-growing people of Okinawa:

when Okinawans are asked how children learn agricultural methods, they say that children are not taught but learn by themselves. When an elder stops work, a child picks up a tool, such as a hoe, and copies its use.

Children as teachers

From the preceding paragraph it might be supposed that teaching is, after all, not universal in human communities; but this is because one usually thinks of a teacher as an adult. There is much evidence of children as teachers.

In simple societies it is normal for children to look after young children. In a survey of information scattered in anthropological reports, it was found that even infants of less than one year are cared for more than half the time by other children.[7] In another review, of 50 communities, the age at which children became responsible for babies averaged about six years.[8] In such communities, older children and adolescents not only care for the young but, to quote Meyer Fortes, 'are always transmitting what they know of the cultural heritage to their younger brothers and sisters and cousins.'[9] Often children learn more easily from older children than from adults. This may well apply to the Nyansongo, mentioned above, among whom older children not only care for and feed younger ones, but also correct their conduct. In Okinawa, where adults say they regard teaching as unnecessary, teaching by children is usual. Young children not only follow and imitate older ones while they are collecting seaweed for pig food: the older children also show, by throwing out unsuitable pieces, what should be gathered, until the novice learns. Similarly, games may be used to teach small children how to gather food for chickens. In the Philippines, the Tarongans say that children have more time to spend on teaching than have adults; and older children support juniors in activities such as sweeping yards, preparing pig food and caring for infants. These are taught partly through play but partly by more formal methods, including mild punishment.

Hence the accounts of pre-agricultural or simple farming communities suggest that, whether adults say they teach or not, there is generally some teaching by the young. In technically advanced societies, especially in cities, children have fewer opportunities to join in adult activities, and can rarely instruct juniors in them. Yet children teach even in rich countries. Jean Piaget has given a minutely detailed account of how children in Geneva play marbles.[10] His interest is principally in stages of development as children grow up and respond to the local rules and customs; but he emphasizes that the code which regulates their play had been made up by the children, on their own, and is transmitted by them from one generation to the next. There is intense preoccupation with elaborate rules, imposed by older children on their juniors.

Other examples of juvenile tradition come from a famous account of the customs, rhymes and other lore of English and European children.[11] Iona

& Peter Opie trace current sayings and doggerel back, through the centuries, to conditions and events of which modern children know nothing. During a fight a child may call for a truce by exclaiming 'fains' – a word, derived from Old French, long discarded by adults. Another kind of moral rule is that a present, once given, must not be taken back. Some English children recite:

> Give a thing, take a thing,
> Never go to God again.

In the seventeenth century a children's saying was, 'To giue a thing, and take a thing, is fit for the Deuil's darling.' Perhaps some of the rules of

Children care for, and teach, younger children. Below, girl of seven years in central Java. *Opposite,* young girl in Togo, and her baby sister.

social behavior are instilled in us by a form of teaching we now barely remember.

Accounts of care-giving and teaching often emphasize the children's desire to help younger ones. This nurturant behavior is sometimes said to be altruistic (in its primary sense).[12] But there seems to be no comprehensive account of what impels children to help others. Sometimes there are clear practical advantages, if not in care-giving at least in teaching. In a peasant family, more hands make light work. For games, extra players may be needed. Moreover in some societies children are required by adults to look after young ones: in effect, they are taught to do so. Even where this does not happen (as in technically advanced communities) children imitate the care-giving conduct of parents, if not with other children then with dolls.

None of these influences rules out a truly altruistic attitude. Concern for others arises readily in young children. Perhaps there are few, if any, societies of which it is not a feature. Certainly, in most societies, whether technically primitive or advanced, children form communities of their own, according to age; and the customs of these often unacknowledged groups can be passed on, generation after generation, without adult intervention.[13]

Tradition

The passing on of rules and rhymes by children is a special case of tradition – the 'transmission of statements, beliefs, rules, customs, or the like' (*OED*). Most of what is taught, in schools or elsewhere, is traditional. Teaching has a strongly conservative influence. Correspondingly, to say that something is traditional often implies that it should be respected and followed. Language, teaching and tradition are all aspects of social stability.

Today, it is easy to underestimate the conservatism of tradition. Here, therefore, are two examples, on very different time scales. The earliest traditions we know of are paleolithic – those of the making and use of stone tools. Our ancestors depended on such tools for several million years, but the methods of making them altered (or 'evolved') with time. One type is the Acheulian hand-ax (page 254), elaborately chipped into a conical shape, with a base that fits into the palm of the hand. We have many examples of this kind of equipment, for it was in use for about one million years. For many thousands of generations, in Africa, Europe and Asia, the one technology was meticulously followed. No doubt its practitioners regarded it as natural, normal and inevitable. If innovators did arise, their proposals must have been ignored or rejected. Perhaps all eccentrics and

The tradition of making stone tools has persisted into our own time. Aboriginals strike flakes from cores in New South Wales.

nonconformists were turned away, to wander and die alone. But eventually, in circumstances of which we know little, the hand ax was succeeded by new kinds of stone tool, produced by new methods.[14]

The duration of recorded history, compared with that of the paleolithic, is only a moment. Yet within it are epochs of prolonged stability, contrasted with our own period, when both technology and fashions in the arts change in a decade. Here is a comment by E.H. Gombrich on the marvellous paintings and sculptures of ancient Egypt.

> The Egyptian style comprised a set of very strict laws which every artist had to learn from his earliest youth. Seated statues had to have their hands on their knees; men had to be painted with darker skin than women. . . . No one wanted anything different, no one asked him to be 'original'. So . . . in three thousand years or more Egyptian art changed very little.

But there is also 'a direct tradition, handed down from master to pupil, and from pupil to admirer or copyist, which links the art of our own days, any house or any poster, with the art of the Nile Valley of some five thousand years ago.' The Greeks learned from the Egyptians, 'and we are all pupils of the Greeks'.[15]

After the Egyptians, change is as prominent as stability, for the Greeks, of course, did much more than copy from Egypt. The Egyptians showed a person's feet from the side – an aspect that is easy to draw, as many of us have learned from experience; and they liked to show the big toe. So in an Egyptian portrait a man may seem to have two left feet. The Greeks were not content with this. 'It was a tremendous moment in the history of art when, perhaps a little before 500 BC, artists dared for the first time . . . to paint a foot as seen from in front.'[16] This was also a time of technical innovation. In the sixth century BC a period of invention began quite unlike anything that had happened before. Farmers acquired spades and scythes, metal-workers devised tongs and carpenters adopted planes. In the same period, domestic animals were put to crushing ore, grinding corn and other tasks which had hitherto required human muscle power.

Since then, changes have continued at an accelerating rate. Those in the arts exemplify further some of the peculiarities of tradition. Paintings of the European Middle Ages are often without perspective, and so have a flat or even top-heavy appearance. The invention of scientific perspective is usually attributed to Florentine painters of the fifteenth century. Later still, painters developed new techniques which finally allowed them to paint a landscape as a 'faithful record of what the artist saw in front of him'.[17] These methods were owed to people of genius, such as John Constable (1776–1837), and the results were not always instantly accepted. When Constable was a member of the selection committee of the English Royal

Academy, one of his own paintings, now regarded as a masterpiece, was (accidentally?) brought in to be judged, and was greeted by the exclamation, 'Take that nasty green thing away!' Yet, as Kenneth Clark remarks, 'Constable, like all revolutionaries, was an eager student of tradition.'[18]

Constable's achievements included marvellous representations of the water of rivers and lakes. Late in the same century, another artistic revolution, that of impressionism, caused a greater upheaval; and, to quote Clark again, 'it also came from ripples – the sun sparkling on water and the quavering reflection of masts. . . . All one could do was to give an impression – an impression of what? Of light, because that is all that we see.'[19] The works of Claude Monet (1840–1926), a leader of this movement, were first greeted with ridicule: they were called disasters, and he and his allies, lunatics. The early impressionist works were presented, in Paris, in a special show, the 'Salon of the Rejected'. Yet, in a few years, impressionism was accepted. During the last three decades of the nineteenth century, it dominated European art. But, after that, painters discarded it. It has, however, left us with many masterpieces, and also new ways of looking at our surroundings. People learned, perhaps for the first time, that shadows have colors.

We are accustomed to think of historical changes as progress; and in the fine arts innovators have taught others how to do things not done before, and those who enjoy paintings to see in new ways. But changing traditions do not necessarily make a forward movement. Sometimes, innovation is, paradoxically, backward looking. The flat appearance of pictures painted in the Middle Ages, filled from top to bottom with figures of people, flowers and trees, was revived by Henri (Douanier) Rousseau (1844–1910), whose jungle scenes are now famous and familiar. Similarly, the portraits of ancient Egypt were usually profiles, but with the eye shown as seen from in front; and Pablo Picasso (1881–1973) painted many of his faces in rather the same way: they are shown in both profile and full face.

We benefit greatly from the fact that modern paintings and music do not make those of the past obsolete. We can enjoy the art of all ages from the paleolithic to the present, and the music of several centuries. The fine arts and music change with time, but not along a straight line of progress or improvement. In this they differ from the natural sciences, for in them it is possible to define progress as something measurable. The accuracy with which we can predict physical processes has steadily increased: eclipses and other celestial events are among the most obvious. In ancient Rome, the extreme of accuracy was represented as *ad unguem* – to a finger nail; today, we measure small fractions of a millimeter. At the other end of the scientific spectrum, the number of species named and accurately described is rising steeply. Between these extremes are successes such as those of genetics or geology. As a result, only recent scientific writings are much read, except by historians: modern science is not only built on past achievements but also replaces them.

Heredity and inheritance

In Chapter 6 I make the distinction between 'inherited' in the legal or social sense, and 'genetically determined' – an expression referring to differences between individuals that correspond to differences in their genes. When we speak of our inheritance from the past, we usually refer to the first – to what is passed on as tradition, including artistic and scientific knowledge. We can now attempt an exact comparison of genetical transmission with what happens when information is transmitted, by non-genetical means, from generation to generation. There are at least five important differences.

First, genes are transmitted only at conception, but cultural effects act gradually: a child slowly absorbs ideas, attitudes and information during development, and the ways in which children (and adults) are influenced by their environment vary greatly among societies. General rules are therefore difficult to establish. A beginning has, however, been made. L.L. Cavalli-Sforza and others have recorded traits, ranging from religious belief to fondness for television, and from belief in extra-sensory perception to the kinds of person consulted during illness, among Americans and Taiwanese. Unsurprisingly, religious and political attitudes, as well as minor habits, proved to reflect the example, and perhaps the teaching, of parents; older brothers and sisters also have much influence. A hereditarian would suspect a genetical influence, but unrelated friends have a similar effect. Moreover, the populations included a number of twin pairs. As we know, if genetical variation has a significant influence, uniovular ('identical') twins should be more similar than are binovular pairs; but this was not found. Evidently genetical variation is not important in the transmission of the traits studied: the resemblances of parents and offspring in these features represent tradition, not heredity in the biological sense.[20]

The second difference is obvious: we acquire an equivalent set of genes from each parent, while the transfer of traditional knowledge gives scope for very unequal contributions. A child may be culturally influenced by only one parent or, if adopted, by neither. In Chapter 7 I describe how adoption can profoundly influence intellectual development.

A third difference follows from the second, and can be expressed in terms of natural selection. In Chapter 6 I describe the idea of kin selection: according to the modern theory of natural selection, animals must be expected to behave in ways that favor, not only their own survival and breeding, but also that of their kin. Kin in this context refers to close biological relatives. But in the transmission of items of knowledge, especially new ones, a person's conduct toward biological kin may be irrelevant. The artists who developed the use of perspective, the chemists who taught the atomic theory, and the politicians who (almost) abolished slavery did not ensure the survival of their ideas by aiding their kin. Their 'cultural kinsmen'[21] were those who shared or accepted their innovations. Correspondingly, conduct that favors the survival of one's genes may be quite different from activities that propagate one's beliefs. This book will perhaps persuade a few people to accept some of my opinions, but its influence on my grandchildren may be nil. More generally, there is no good reason to think that either stability or change in our customs and attitudes depend on uniformity or variation in the genes available.

Fourth, and more obvious, genes are distinct entities: their positions in the visible chromosomes of the cell nucleus can be mapped; when they are transmitted, actual material passes from parent to offspring. The two sets of genes, one from each parent, make up our genotype. This, our genetical constitution, is distinct from our actual characteristics (phenotype): as we know, a person with a given set of genes can be normal or imbecilic, pink or brown, vigorous or feeble, according to the environment. There is no parallel to this in the transfer of customs, skills, ideas, attitudes or moral principles.

Lastly, there is no detailed account of what, in the past, has caused the appearance of new (mutant) genes. We know only that some things, such as radioactivity, increase mutation rates; and that mutation, and hence genetical variation, is universal. But on changes in human practises and ideas we can say a great deal. They are part of the subject matter of history, and it is one of the tasks of historians to explain them. We can also perhaps learn from history what to do and what to avoid.

Reiteration

We are accustomed to analogies between the transmission of genes and the transmission of customs, for both processes cater for stability as well as change. But the analogies are misleading. When genetical change accompanies adaptation to new conditions, it affects whole populations: over generations, one genetical type replaces another. But cultural changes influence human communities in much more complex ways. Different, 'rival' attitudes, beliefs and practises may coexist without competing; and, when they do compete, they may do so by rational argument or other social processes without analogy in genetics. Moreover, cultural change can occur during the development of a single individual. Today the elderly can often look back on a lifetime of adapting, sometimes rather desperately, to novelty.

Social and technical upheavals reveal a conflict between the desire for stability and the willingness to accept change. Resentment of novelty is not confined to those who derive wealth and power from the existing order:

it is to some extent present in all of us. Those who oppose change may appeal to tradition – to the 'goodly usage of those antique times'; but today they may also resort to imperfectly understood biology. In its modern version instinct becomes something that we are driven to by our genes; and our genes in turn are an outcome of a struggle for existence which has imposed on us a fixed nature. Human nature is then a set of standardized impulses; these we cannot resist; or, if we do resist them, we do so at our peril. The emphasis is always on what is believed to be changeless.

In the social life of animals we find uniformity within each species, but the search for similar constancy in human action is a wild goose chase. The human species, like others, is presumed to be a product of natural selection, but it is one in which selection is superseded. No doubt changes are still going on in the genetical make-up of human populations; but nearly all the significant innovations that we know of are cultural, and independent of genetical variation. They are transmitted by tradition, hence by teaching: that is, by the exertions of each generation to provide for the next.

A combination of unique qualities enables us to bequeath to our successors not only custom but also the opportunity to change it. We are conscious both of ourselves and of others. Our empathy for others aids us in our efforts as teachers. Our awareness of ourselves allows us to plan ahead: we can imagine what we, and our descendants, will be doing. We not only make, but also state, intentions; and we argue about them. We are restlessly exploratory; and we have the intelligence to benefit from satisfying our curiosity. Our explorations sometimes lead to new proposals. The persistence of innovation depends on teaching by the innovator as well as imitation by followers.

Alas, this is an ideal picture. Let Alexander Pope restore the balance. 'Man', he says, reasons 'but to err';

> Chaos of thought and passion, all confused;
> Still by himself abused, or disabused;
> Created half to rise, and half to fall;
> Great Lord of all things, yet a prey to all;
> Sole judge of truth, in endless error hurld;
> The glory, jest, and riddle of the world!

But error, even if endless, is not always a handicap. The question is whether we learn from it. More generally, is the human species capable of continuing to improve on its past record? And, if it is, in what sense of improvement? We cannot afford to wait for genetical change: we must, for the most part, accept our genes as given. We can better our lot only if we change our ways by conscious intention.

15 The question

I believe it will be easier for the generations to come; our experience will be at their service . . . One wants to make history so that those generations may not have the right to say of us that we were nonentities or worse.

Anton Chekhov

A historian has described how 'In 1895 the novelist Henry James acquired electric lighting; in 1896 he rode a bicycle; in 1897 he wrote on a typewriter; in 1898 he saw a cinematograph. Within a very few years he could have had a Freudian analysis, travelled in an aircraft . . .'.[1] But it is not divulged whether Henry James looked on all these innovations with approval. Did he regard them as examples of progress? And, if he did, in what sense? The question we face in this final chapter is what further social change is both possible and desirable.

Perfectibility

According to K.R. Popper, 'In science (and only in science) can we say that we have made genuine progress: that we know more than we did before.'[2] (Not all Popper's fellow-philosophers agree with him.) Popper here equates progress with increasing knowledge – an unusual definition. The primary meaning of progress is forward motion: the movement may be toward a desired destination or over a cliff. A wider meaning occurs in expressions such as the progress of a disease, where the meaning is close to that of growth or development. But usually, when we speak of progress, we imply improvement, and so approval. Even evolutionary change, as we know, is sometimes regarded as commendable. Darwin, at the end of the *Origin*, wrote: 'as natural selection works solely by and for the good of each being, all corporeal and mental endowments will tend toward perfection.'[3]

Correspondingly, our idea of social progress usually reflects optimism about the future: we presume that societies can be made better, and that the human lot – and human beings themselves – can reach higher standards than before. This is the much debated notion of human perfectibility.[4] Perfectibility, in this context, does not imply that human beings can eventually reach a state beyond improvement: on the contrary, it signifies that any person, or group, however advanced, can learn more, or think of some further achievement. In the western world, belief in perfectibility, in this sense, has developed only gradually, mainly in modern times. It has been much encouraged by the open-ended growth of the sciences and of their social applications. To that extent we are justified in saying that today science and the idea of progress go together.

But, in the present century, technological changes have accelerated far beyond the capacity of traditional human institutions to absorb them. Hence today the confident prediction of change for the better is giving place to a questioning attitude. There is, after all, no necessary connexion between, say, rapid travel or communication, on the one hand, and a full life, or even contentment, on the other. John Passmore, in his *Perfectibility of Man*, quotes from a buoyant popular work by H.G. Wells (1866–1946), published in 1922.

> 'Can we doubt that presently our race . . . will achieve unity and peace, that it will live, the children of our blood and lives will live, in a world made more splendid and lovely than any palace or garden that we know, going on from strength to strength in an ever widening circle of adventure and achievement?'. Only too easily, [Passmore continues] many of Wells's readers would now reply, can we doubt whether our children will live in a world more splendid and more lovely . . . At best they will live in an air-conditioned box, at worst they will not live at all, or will live in a hovel in a devastated world.[5]

Edward Gibbon, in his history of the Roman Empire, suggests that people have always 'a strong propensity to depreciate the advantages, and to magnify the evils, of the present times.' Our period too is in many respects one of decline and fall, notably of recently powerful empires. More important, despite the progress they have made possible, science and technology are today increasingly associated with appalling weapons of war and world-wide destruction of forests, waters and other features of our environment. It would be easy, therefore, to turn this chapter into a jeremiad:

> And I brought you into a plentiful country, to eat the fruit thereof, and the goodness thereof; but when ye entered ye defiled my land, and made mine heritage an abomination.[6]

Freedom

Yet awareness of present evils need not blind us to the promise of our times. Of present examples of progress, the least arguable concern health and longevity. Large populations have been freed from hunger and the worst infectious diseases. These measures have sometimes been opposed, but there is no significant movement to put the clock back to a regime of intermittent famine and of high death rates from infection at all ages. Freedom from most illness, and surviving to old age, are therefore imaginable for people of all nations. They may be regarded as good in themselves, but they raise the question: for what kind of life do we wish to be preserved? To answer, a first step is to identify past or present changes welcomed as enlarging human freedom. These include the abolition of slavery; the retreat of racism; the improvement of the lot of wage earners; the beginning liberation of women; and the enhanced awareness of the needs of children.

Slavery has, until recently, been a feature of most civilizations; in many it was held to be a natural feature of human existence, like the use of domestic animals (Chapter 4). Yet today it is not only illegal: to enslave a person is almost universally regarded as wicked. The change in outlook began in eighteenth-century England – the headquarters of the slave trade. Black Africans were being seized, treated worse than cattle, and sold in North America.[7] It was part of proverbial lore, among slave traders and owners, that to beat a slave was to nourish him. Such assertions were repeated approvingly by Christian missionaries. D.B. Davis also quotes a French missionary who 'assured his readers that Negro slaves, when well fed and gently treated, were the happiest people in the world.' And Roger Anstey shows how those who took part in the slave trade regarded it as an honorable and even genteel occupation.[8]

The slaves did not agree. Modern history could reasonably be written as a period of struggles for freedom; and revolts by slaves make a central feature of these struggles.

We also find evangelical Christians, notably the Englishman, William Wilberforce, leading the opposition to slavery.[9] For some historians, abolition therefore represents a triumph of moral principle over greed. Others, however, point also to the economic disadvantages of slave-owning: employing people for wages is often more profitable, especially when slaves are likely to rebel. No doubt the motives of abolitionists were mixed and are often undiscoverable. What we do know is that by the end of the nineteenth century property in human beings was generally abhorred; and this attitude was strongly developed even among Europeans who retained a firm belief in the superiority of their own race. Here therefore is an example of moral progress which helped to better the lot of large numbers of people.

Like the presumption of slavery as natural, the doctrine of inferior races has been and is used to justify economic exploitation. Today the extreme case is in the Republic of South Africa. It can also help a government to persuade people to support a war against human beings of a different physical type – for instance, that of the USA in Vietnam. But racism is not only a matter of profits and power. Eric Hobsbawm, in an account of the mid-nineteenth century, writes:

> Racism pervades the thought of [this] period to an extent hard to appreciate today, and not always easy to understand. (Why, for instance, the widespread horror of miscegenation and the almost universal belief among whites that 'half-breeds' inherited precisely the *worst* features of their parents' races?) Apart from its . . . legitimation of the rule of white over coloured, rich over poor, it is perhaps best explained as a . . . means by which a fundamentally inegalitarian society . . . rationalised its inequalities.[10]

In the twentieth century, as we know, racism has steadily lost support. Its retreat has been in part due to vigorous resistance by its victims, but help has come also from history, anthropology and genetics. These have provided relevant facts. Historians tell us how northern Europeans, who have recently thought themselves so superior, were, in an earlier period, barbarians to the Chinese, to black Africans and to others; and earlier still to Greeks and Italians. Here is a passage from a standard historical work.

> Standing on the Pyrenees, the dauntless Viceroy conceived the project of conquering the whole of Europe; and in all human probability had he been allowed to carry his plan into execution he would have succeeded. . . . The cautious and hesitating policy of the . . . Court lost the glorious opportunity, with the consequence that Europe remained enveloped in intellectual darkness for the next eight centuries.[11]

Some readers may be in doubt about the occasion to which this refers. The conqueror was the Saracen, Mûsa, and the pusillanimous court, that of Damascus in the eighth century.

Anthropologists have similarly combated racism by teaching their compatriots in rich countries to respect the morals, knowledge and languages of 'primitive' peoples. And from genetics we are beginning to learn (with a struggle) to understand the interactions of nature and nurture: when many members of an unprivileged group (darkly pigmented, poor or female) are stunted or seemingly stupid, we look for and soon find environmental causes of their deficiencies (Chapters 7, 9).

The change in attitude also has a moral element: to claim special privileges or superiority for one's own type is increasingly held to be barbarous. In 1917, in the depths of the First World War, a moral philosopher wrote: 'The doctrine has been gradually gaining ground that our duties towards our fellow man are universal duties, not restricted by the limits of country or race.'[12] Social Darwinism and sociobiology suggest that our 'nature' enforces an opposite policy: but their message is founded on biological errors. Today, the universality of our obligations is acknowledged in international organizations. The constitution of the United Nations Educational, Scientific and Cultural Organization (UNESCO) states that the Second

World War was 'made possible by the denial of the democratic principles of the dignity, equality and mutual respect of men'; and in 1952 UNESCO expressed these principles in a document, *The Race Concept.*[13]

The movement for racial equality has still a long way to go. During the last quarter of the twentieth century there have been setbacks. In the USA, France and Britain in particular, there have been outbreaks of violence against people with dark skins; and governments have sometimes failed to check, or have even condoned, this barbarity.

The period of opposition to racism has also been one, appropriately, of colonial upheaval. Within living memory the British, Netherlanders, French and others took pride in their imperial status: school children, in their history lessons, learned an attitude of superiority over others of different type or custom, even if this sometimes also imposed an honorable duty (the 'white man's burden'). Today, the word colonialism is almost as derogatory as slavery. This is a moral advance, even though the former colonies are still struggling to free themselves from internal tyranny and from external economic and military domination.

The campaign against slavery occurred during the beginning of the age of organized labor, at a time of major struggles against 'wage slavery'. Early in the nineteenth century, Wordsworth wrote these lines on workers arriving for the night shift.

> Men, maidens, youths,
> Mothers and little children, boys and girls,
> Enter, and each the wonted task resumes
> Within this temple, where is offered up
> To Gain, the master idol of the realm,
> Perpetual sacrifice.

Resistance by the victims was gradually organized. And so, two decades later, we have Thomas Arnold, a famous Christian educational reformer, writing to a friend: 'you have heard, I doubt not, of the Trades Unions; a fearful engine of mischief, ready to riot or assassinate; and I see no counteracting power.'[14] The 'counteracting power' was, however, visible enough: during much of the nineteenth century, in Britain, the USA and elsewhere, trade unions were illegal, and men who tried to form them were likely to be imprisoned, deported or shot. That the trade unionists nonetheless persisted is not surprising if one studies the conditions of the wage earners in the richest countries. In Britain, in 1842, it was necessary to prohibit, by Act of Parliament, employment in the mines of children under the age of seven. Five years later, another Act forbade employers to require women or children to work for more than ten hours a day. In the USA, struggles for shorter working hours have been going on for nearly two centuries. The ten-hour day was introduced for some kinds of work early in the nineteenth century, but 100 years later a 60-hour week was still common.[15]

Today, we are accustomed to trade unions as part of the social scene. They protect the lives and standard of living of their members, and often of non-members also. Occasionally, they bring down governments, but for the most part they are props of the existing order. Of course, they interfere with individual freedoms – especially that of employers to fix wages and conditions of work but also, in some industries, that of workers to accept lower wages or longer hours. In the ideal worlds of Economic Man and of *Homo egoisticus* trade unions are indeed against nature; and for extreme devotees of the market economy they signify economic ruin. If, however, we look at facts, instead of ideal images, we find that strong labor organizations can raise productivity. When unions have a recognized influence on management, and employees are treated as more than pairs of hands, workers put more into their work and stay longer with the same firm; management can then improve efficiency by learning from the workers (Chapter 11). The result is economic benefit, and a more agreeable life for both workers and managers.

Such conclusions accord with common sense, but they arouse many kinds of opposition. Progress has been far from smooth. Independent labor organizations were destroyed, between the two great wars, in fascist Italy and Spain and in Nazi Germany. And countries with socialist economies have no unions of the 'western' type. Nonetheless, throughout much of the world, wage earners are now protected from levels of ruthless exploitation which, a century ago, were regarded as normal.[16] This change has gone not with the ruin inferred from simple images of humanity but with an unprecedented increase in material wealth.

During the same period, battles to enlarge the contributions made by women to society have been fought with increasing success (Chapter 9). The accompanying social upheaval entails changes in family structure which show how absurd it is to represent humanity as bound to fixed ways by genes or by instincts. Our lack of 'instinctive knowledge' obliges us, in the short run, to rely on tradition (Chapter 14). In the longer term we have, however painfully, to use our intelligence in deciding on the sort of lives we want to lead.

Some people fear a decline in traditional family life as an outcome of freedom for women. There may indeed be circumstances in which young children suffer if both parents go out to work. But perhaps children suffer also in a conventional family, in which their mother, regardless of her wishes and abilities, is isolated, restricted to domesticity and treated as an inferior. Such opposed views are often put forward without evidence; but actual findings do exist. In an Australian study children of primary school age with working mothers were found to have as much parental care as those whose mothers remained at home; and both young children and adolescents with working mothers were more independent and superior in a range of skills. Other studies have reported similar conclusions. Evidently, this outcome of women's liberation need not affect children adversely.[17]

One presumption, that goes with dislike of freedom for women, is of the nuclear family as natural and inevitable. Yet the unit of parents and children only, as the usual family group, appeared gradually in western Europe after the fifteenth century. It provides another example of the variety of ideas of kinship found in different societies (Chapter 8): as the nuclear family grew in importance, the significance of cousinhood declined. At the same time, the dominance of the father, as head of the family, became more prominent. His role was forcefully supported. In seventeenth-century France a royal pronouncement equated loyalty to one's father with allegiance to the monarch. In England the commandment to honor one's father (but not, evidently, one's mother) was held to signify respectful subservience to all one's superiors. This doctrine was reaffirmed in the American colonies, especially New England. Many, it seems, felt that the stability of the state rested on an unchanging family ruled by the father. Today, neo-conservatives have revived support for patriarchy as both natural and desirable (Chapter 9).

The recent changes in the family are usually seen as releasing women from bondage; but they have also gone with new attitudes to children: we are justified in speaking of children's liberation. An account of the family in Europe of the sixteenth and seventeenth centuries describes a 'determination to break the will of the child'.[18] In that period the number of boys (not girls) sent to school instead of apprenticeship greatly increased; there they were flogged so ruthlessly that blood was often drawn. Even undergraduates, notably in the University of Oxford, were similarly chastised. A sixteenth-century Netherlander, named Batty, whose educational works were translated into English, had

> developed the theory that the providence and wisdom of God had especially formed the human buttocks so that they could be severely beaten without incurring serious bodily injury.[19]

(A modern Batty would perhaps attribute such a beneficent effect to natural selection.) Philip Ariès, in a history principally of French childhood, suggests that, by the eighteenth century, 'the schoolboy had been more or less tamed'.[20]

The rise in the numbers of schools and of schoolchildren was owed partly to the efforts of educational reformers. Apprenticeship at an early age provided training only in special skills. The reformers wished, not to inflict more violence on the young, but to introduce children to learning through a general education, and so to enable them to become more versatile and adaptable. We still have not achieved what the reformers wanted; but, gradually, schooling has become more widespread, more humane and more efficient. This is one of the essential features of the growth of freedom.

The idea of democracy

All the reforms mentioned in the previous paragraphs have been strenuously resisted, and not only by those who feared loss of wealth or power. Most are still resisted. Opposition is regularly supported by spurious biology: the last resort of those who object to social changes is often to say that the changes are against human nature; we are told of genetically fixed features which make reforms impracticable or dangerous. More important, the biological stories help to create a climate of opinion in which attempts at reform are felt to be futile. The relevant biology (presented in earlier chapters) gives no support to such arguments. The genetical makeup of individuals cannot be predicted (or, in important respects, even identified): child-bearing, like marriage, is a lottery, hence the notion of a genetically superior class is invalid. Moreover, even the idea of genetically fixed features is wrong: one's complement of genes is fixed at conception, but not how we develop thereafter; hence genetical determinism of socially important traits is untenable. By environmental measures we can enable children to develop good health, abilities and other valuable qualities; hence preoccupation with genes (which are often hypothetical) interferes with effective action. Lastly, we have much to learn about the ways in which we should rear our children; hence we can put no limit to the range of environmental action.

These conclusions point the way toward further enlargement of human freedom. They are part of the foundations of democracy. But democracy, like property, is a most ambiguous category; and it is not enough to imply that it is a 'Good Thing' and to leave it at that. Two kinds of modern state are called democratic by their supporters. Both had a beginning in violence.[21] One emerged from the English revolution, which began in 1640, and the American and French revolutions a century and a half later. From them came liberal states. Monarchies and aristocracies were overthrown or lost power, and governments, in these and other countries, increasingly represented not the needs of landowners but the demands of commerce and banking. Initially, only men of property could vote. Later, all men were enfranchised, and eventually all women also. In such societies, the word democracy, when not a mere catch-phrase, always has some implication of equality. It may be identified with universal adult suffrage; but the right to vote, though necessary, is not sufficient for a free society: despite the popular vote, in liberal states power to influence the course of events remains largely in the hands of owners of industries and banks.[22]

The leading alternative to the liberal state is some form of socialist economy. The Russian revolution of 1917 and the Chinese revolution of the 1940s were each intended to found a democractic state in something like the original Greek sense – one in which a formerly oppressed majority forms a government. The socialist movement originated as rejection of capitalism and the unregulated market. The dominant principle of liberal society was held to be the accumulation of material wealth: this had replaced the authentic satisfactions of work and play. In such a society, human relationships are devalued in favor of those based on money and impersonal administration. Human beings are treated as machines, or as domestic animals with uniform needs. (Compare Chapters 10 and 11.) Friendship, fellowship and even civility deteriorate, and are replaced by the type of relationship we now experience when we draw money from an automatic teller. The feeling that one belongs to a community, and confidence that one can influence public affairs, both disappear. Such estrangement, or alienation, is incompatible with any current concept of democracy.

In a socialist society there were to be no divisive economic classes: every person could live a full life and take an individual part in communal affairs. Early socialist writings include anticipations of modern findings on the need for self-expression in work (Chapter 11). 'Self-realization through creative work', writes an American political scientist, 'is the essence of Marx's communism. I believe this is the most valuable and enduring element of Marx's thought.'[23] Perhaps one day our descendants will see this principle effectively applied on a large scale.

Traditionally, liberals and socialists are at daggers drawn; but liberalism has never been a unified doctrine and two sharply opposed systems have developed from it. One concentrates on the market economy and on authority (pages 208–9), the other, that of liberal-progressives, advocates much wider freedoms. When we examine their fundamental objectives, liberal-progressives and socialists have much in common. The deepest gulf lies between those who regard greater freedom as unattainable or undesirable, on the one hand, and those seriously concerned to advance individual freedoms, on the other. It is the members of the second group who can draw strength from authentic human biology.

This book is addressed principally to readers in the liberal, that is, capitalist democracies. The expression, the free world, often applied to them, though euphemistic, acknowledges genuine advances – the relief of much poverty, the growth of education, the provision of health services, and the existence of civil liberties. Such achievements have been made possible only by refusing to accept that, in Hobbes' words, 'the business of the world consisteth almost in nothing else but a perpetual contention for honour, riches and authority.'[24] To build on them further, we need a portrait of ourselves compatible with what we know of human potentialities and needs; and, even if the portrait is only an outline, we must also allow for the sheer complexity of human existence. In doing so we have to identify, with the greatest possible clarity, the practises which seem to confirm the pessimism of Hobbes and his modern followers. These restrict our liberties, and are the obstacles to further advance.

Among the most important obstacles is what the famous liberal economist, J.M. Keynes, called the 'avarice and usury' entailed in a liberal economy. Keynes, in 1932, held that these ills would continue to be 'an economic necessity' for at least another century. But he looked forward to a better world than that of 1932 and wrote:

> When the accumulation of wealth is no longer of high social importance, there will be great changes in the code of morals. We shall be able to rid ourselves of many of the pseudo-moral principles which have hag-ridden us for two hundred years, by which we have exalted some of the most distasteful of human qualities into the position of the highest virtues. We shall be able to afford to dare to assess the money-motive at its true value.[25]

The accumulation of riches then becomes not an end in itself but a means toward increasing human freedom.

The 'pseudo-moral principles' to which Keynes refers have been analyzed by another famous liberal, J.K. Galbraith, who – like Keynes – has had much experience of the corridors of power. He writes of the inadequate open discussion of the use of public resources, and of the resulting limitations of 'what is often called democracy in the United States'.[26] He describes the difference between two kinds of expenditure – a difference common to all rich countries. On the one hand are 'frivolous' items, such as 'automobiles, television, cosmetics, intoxicants'; the economics of producing these goods allows the formation of large, powerful firms that can afford vast outlays on advertising and on lobbying politicians. On the other hand are housing, medical services, public transport and education, which are politically weak and inadequately funded. As a country's wealth increases, the contrast between the two categories 'becomes first striking and then obscene.'[27] The success of such an economy depends on 'organized public bamboozlement'. To sell soap, its incidental and trivial properties, such as its smell, 'are held to be of the highest moment. Housewives are imagined to discuss such matters with an intensity otherwise reserved for unwanted pregnancy and nuclear war.'[28]

The bamboozlement goes with the close ties, outside any democratic process, of commercial interests with government. These interests range from the arms industry to the manufacture of cigarettes. In 1961, a United States president and military leader, Dwight D. Eisenhower, warned of the 'military-industrial complex' and spoke of the dangers of 'an immense military establishment and a large arms industry. . . . The potential for the disastrous use of misplaced power exists and will persist.'[29] Eisenhower's warning was only one of many, spread over half a century, against the merchants of death.[30] Two decades later, a United States vice-president, Walter Mondale, stated that 'America is no longer an arsenal of democracy;

it is quite simply an arsenal.'[31] By the 1980s, the USA was more of an arsenal than ever before, and the USSR was close behind.[32] The consequences are an ever present threat to the lives of everyone, economic chaos and, for many, impoverishment.

The manufacture of armaments, whether by private enterprise or by the state, is an international activity. Other concerns with a powerful but unobtrusive influence on governments are the transnational consortia. In an account of these organizations, Anthony Sampson writes of 'how easily the doors of government open to big money lobbies, and how closed they are to anyone else'.[33] The power of commercial interests, outside democratic control, to act against our everyday needs is exemplified by the industries that produce food and drugs.[34] Sugary foods and soft drinks are advertised with great skill, and at vast expense, to the detriment of the health of whole communities. Among such activities is the sale of foods for infants. When legislation forbade the sale of some of these 'formulas' in rich countries, they were energetically promoted among uninformed parents in poor ones. A similar lack of scruple among manufacturers of medicines and other pharmaceuticals has often been exposed, and occasionally, as in the cases of thalidomide and the Dalkon shield, becomes widely known.[35]

The extreme instance of a vast industry dedicated to the destruction of its customers' health is the tobacco industry. The tobacco companies have taken advantage of the fact that nicotine, like heroin, is addictive: it causes dependence in those who use it. In the 1980s the annual turnover of the world trade in tobacco is 22 billion dollars, of which two billion is spent on advertising and lobbying. In each year, the direct effects of smoking kill about a million people, many in vigorous middle age. Publication of the truth about smoking, especially on television where it would have the greatest effect, is actively and often successfully obstructed. The most prominent response of governments has been to make tobacco a source of revenue: for decades their support for the warnings of physicians has been weak and grudging. The most effective resistance to being poisoned, as to being bombed, has come from below, and in rich countries is now increasingly successful. The companies are therefore turning to the nations of the 'Third World' where ignorance and corruption will, they hope, enable them to maintain their profits.[36]

The policies of these enterprises have been called corporate crime; but, even when their crimes are undeniable, penalties are light or negligible. Correspondingly, those who manage or profit from the corporations hold

> that business ethics are morally superior to mere formal legalism. Thus free enterprise – the pursuit of fair profit, the generator of wealth and employment, the backbone on which social welfare is possible – can be viewed, at least by corporate officials, as *the* primary ethic . . . of an industrial society.[37]

The result is not a free society, but one in which much effort is put into proving that we are a hybrid between *Homo egoisticus* and *H. operans*, moved principally by selfishness, greed and nepotism.

In such communities, writes Galbraith, 'the purposes of the producer are dominant'; and the story that the interests of the producers and public are the same is a 'disguising myth'. Similarly, the belief that economic growth will automatically reduce inequality does not correspond to the facts of recent history. The governments of present-day liberal democracies can hardly be expected to remedy this state of affairs. To ask them to do so resembles going to a doctor for treatment when 'he serves even more devotedly as the local undertaker.'[38] Hence a necessary step toward enlarging freedom in such countries is to liberate government, and the management of affairs generally, from undemocratic control: the interests of consumers must become paramount. Galbraith and other liberal-progressives do not, as Marx did, emphasize conflict between economic classes as a source of progress; but their concern for consumers may be regarded as no less revolutionary than any proposal made by Marx.

Among the problems shared by liberal-progressives and socialists is the precise scope of state regulation. Socialist doctrine is often associated with advocacy of a powerful state; many forms of liberalism urge the opposite. But all modern, technically advanced communities, whatever their economy or government, accept 'paternalistic' regulations. Among the least controversial are certain compulsory immunizations, measures that ensure clean water and other safeguards against infection and pollution: without these, calamities such as typhoid, cholera, smallpox and tuberculosis would still plague us. To be free from such infections we readily accept restrictions on our liberty. The methods used depend on the technical knowledge of specialists, and can rarely be decided by votes in a parliament or local council. The same applies to quality control of food and drugs and to much else. To this extent, 'paternalism' is unavoidable.

But how far should regulation go? Adding fluoride to tap water greatly reduces the incidence of dental caries; yet it was opposed by some on the ground that it should not be forced on people, however few, who dislike the idea. More important, what are the proper limits of legislation to prevent people from ruining themselves and their families by smoking or excessive drinking? Neither wholesale elimination of controls, nor unrestrained control of all we do, is justified or practicable: decisions have to be taken piecemeal.

We need also to consider the influence of state action on the readiness of people to act vigorously on their own behalf and to improvize. Perhaps too much protection leads to supine dependence, social inertia, reliance on 'them' to do everything needed. But, if so, what is too much? Unless such questions are exposed and debated, freedom, like democracy, becomes a catch-phrase without substance. Even the eternal vigilance required to keep the advances already made is lacking.

No large community has yet achieved the level of liberty proposed either by liberal-progressives or by socialists. The foremost obstacle to enlarging our freedoms is war. We do not know what a modern liberal-progressive society, or a socialist one, would look like in a disarmed world. Since 1914, the world has been unremittingly at war or, if formally peaceful, has been arming itself with ever more destructive and expensive weapons. The world of the late-twentieth century is therefore in a continuing state of economic incoherence. The most prominent politicians command gigantic military forces and, whether of the left or of the right, whether voted for or not, in the absence of authentic democracy use this power to support the narrow interests of their own class or clique. Hence we are continually reminded of a famous, and usually misquoted, remark by the Catholic historian, Lord Acton (1834–1902). 'I cannot accept', he said, 'that we are to judge Pope and King unlike other men' with the favorable presumption that they do no wrong. 'Power tends to corrupt and absolute power corrupts absolutely. Great men are almost always bad men.'[39] Today, to Acton's Pope and King, we may add President, Prime Minister and Chairman, of both sexes. A task for future democracies is to find ways of electing managers of the state who are not corrupted by power.

Equality again

Power in the hands of a few goes with the attempt to control how we think and what we believe. One source of agreement between liberal-progressives and traditional socialists is the role of freedom of thought. In 1913 J.B. Bury wrote:

> A long time was needed to arrive at the conclusion that coercion of opinion is a mistake, and only part of the world is yet convinced. That conclusion, so far as I can judge, is the most important ever reached.[40]

Bury wrote at a time when methods of manipulating attitudes were still primitive. Today many new methods, not all obviously coercive, are used by authority (and by advertisers) to control opinion (Chapter 5). These are usually in the hands either of private and very rich persons or of governments. Hence democratic appearances, such as voting, need not threaten the arbitrary power of the rich or of self-appointed oligarchs. No society has yet achieved a democratic press or broadcasting system. Better access to information is one of the changes needed for a free society. This is recognized in laws such as the Freedom of Information Act of the USA. Without this Act, several passages in this book could not have been written.

General availability of information on the management of the state is one aspect of equality; and the growth of equality in this sense is one measure of human progress. But, on the face of it, to assert the 'equality of man' is unbiological and absurd: of all species, our own is the most diverse. Among the features in which we vary greatly is the ability to understand and to make use of information. In the present context, however, equality refers to the concept summed up as equality of opportunity. Yet, taken literally, this notion too is absurd: it implies equal access, for every child, to every possible occupation. Whatever resources we devote to education, such uniformity could hardly be achieved. Moreover, each family differs from every other, and provides its children with its own distinctive environment.

The tidy and effective slogan of equality of opportunity refers, in practise, to meeting the need of every person to develop his or her abilities, and to live a socially useful existence. A prerequisite is education, adapted to their needs and talents, for all young people, regardless of wealth or parentage. Like the adult franchise, the right to literacy and numeracy is now recognized in both socialist and liberal states, where it makes a beginning of general access to knowledge, understanding and enjoyment. But education for all at the higher levels is still resisted by some of the privileged (Chapter 9). It is certainly expensive – as expensive, perhaps, as preparation for modern war. The fundamental objection urged against it is, however, again biological: that most of the children of the poor, or of other lowly groups, are genetically unfit to benefit from any but the minimum of schooling. This argument could lead to a ludicrous implication: that, for most of the world's population, *no substantial improvements in learning or teaching are possible*. If this had been presumed in the past, literacy and numeracy would remain the privileges of a small minority in each country. We know, however, that education in every country could be much better than it now is (Chapters 7, 9).

At present, China and India have large but diminishing illiterate minorities, and some countries are even worse off. Nor is large-scale illiteracy confined to peasant populations. In the 1980s about 16 per cent of American adults cannot read, and a much higher proportion are functionally illiterate; and these figures are rising. Jonathan Kozol has detailed the evidence and has put forward a program of remedy; but, when he was writing his book, the United States government was proposing a large reduction in federal funds for education.[41]

These hardly credible, but fully established, findings represent only part of the current undermining of democracy. In the nineteenth century, it was said that religion was the opium of the people; today, the opium of the people is either opium itself or, much more often, television. If this were only a matter of entertainment, it would represent a drastic change in our social lives. But it is more: for the factual information formerly conveyed by speech or in print is increasingly replaced by colored moving pictures. Advertisements represent the process in an extreme form. A few decades ago, it was taken for granted that goods or services should usually be advertised by describing them: the descriptions, true or false, were factual statements on what was offered. Today, in rich countries, the most important kind of advertisement is a brief drama or other entertainment in which the descriptions, if any, are likely to be of the prospective buyers – as they are or as they wish to be.

Despite the lack of authentic information, advertisements have a considerable impact on the viewer. The advertisers' methods have therefore been extended to politics. Advertising agents are hired to tell politicians how to sell themselves and their policies. Rational debate on matters of life and death is replaced by reports that last less than a minute. Hence the values of show business increasingly determine the ways in which we see both public figures and public problems: the bomb comes to be equated with beer, terrorism with tennis, famine with face powder. An American educator, Neil Postman, has analyzed this process. His country, where it is most advanced, is – he holds – the best entertained and the least informed in the 'western' world. Since public affairs are regarded in the same way as advertising they are no longer treated as a matter of serious debate. Postman cites a report, published in 1983, on how the American president's staff had ceased to be concerned when the president made incorrect or misleading public statements, for the public had become accustomed to incoherence.[42] Other countries are going the same way. To counter these trends is a formidable task that requires, once again, that we put more resources into education. More can then be discovered on how to bring out the diverse abilities of our children. Among the most important are those needed for being an effective citizen in a democracy.

Health services, education and access to information are rights. So is the opportunity to work, but one not recognized in practise in liberal states: in them, unemployment is a regular feature, except in wartime. Its destructive effect on the personality may be difficult to analyze statistically, but is not in doubt. Being out of work goes with poor mental health and a high suicide rate; and the ill effects extend to the families of the workless.[43] A modern society which fails to offer every healthy adult productive employment, in a socially accepted activity, is deficient in a fundamental sense. As we know, human beings have an imperative need for stimulation and exertion provided by productive work (Chapter 10).

Emphasis on rights should not be at the expense of acknowledging duties. Every healthy adult, if not prevented, may properly be *required* to work productively and to take some responsibility for what is going on in the community. The maintenance of health, too, increasingly imposes duties, of which some are so familiar as to be usually unnoticed. In rich countries it is not only illegal to defecate in the street or by the roadside: it is also regarded as disgusting. Yet the idea of cleanliness as a duty is in most places only recent, and in many countries has still to be accepted. In Samuel Butler's fictional country, *Erewhon*, ill health is a criminal offence, punishable by imprisonment. I do not suggest that our descendants will adopt this policy in full, but further movement in that direction is likely. There is already, in some communities, a tendency to regard smoking not merely as foolish but as morally wrong. On the positive side, perhaps exertions that maintain athletic fitness will come to be socially demanded to the same extent as cleanliness.

As the preceding paragraphs show, rights and duties are not always separate categories: productive work, and taking part in communal affairs, come under both headings. Rights and duties are also alike in providing opportunities for self-expression (Chapter 11). These statements are not biological: the idea that we are entitled to make demands on our fellows, and that we also have obligations to them, is solely in the human dimension. And it is the acceptance of these rights and obligations that make what is valuable in the modern concept of democracy.

A beginning

To some readers, however, even the modest proposals in the previous pages may seem utopian, that is, 'of impossible and visionary perfection' (*OED*).[44] They may appear quite incompatible with the state of the world in the last quarter of the twentieth century. As I write, large numbers of Africans are dying of starvation. Much larger numbers, in the Americas, Africa and Asia, continue to depend precariously, from year to year, on the crops grown around their villages. Their labor is demanded by the need to survive: for them fussing about self-expression in work, or about the use of leisure, would seem ludicrous. Yet the temple at Segesta, with which this book begins, reminds us that even poor communities, in which there is a continual struggle for survival, are rarely satisfied only with material goods. Moreover, even for the most indigent nations, a different and richer life is accessible. The biggest obstacle is the squandering of the world's resources on arms. It is not utopian for the people of poor countries to work for advances similar to (or better than) those achieved, during the past two centuries, by the richest nations. Among their immediate objectives must be an increase in income and an improvement in the distribution of wealth. But these are not ends in themselves: they are means by which people can be relieved from the urgent demands of mere survival, and so can become more nearly free to choose what lives they want to lead.

To plan for such progress would be utopian if the pessimistic, biological portraits of humanity were accurate and complete. But if we see the images as caricatures they become useful: instead of stating what is preordained, they tell us what to avoid in our own conduct. They also (when we examine them) point to kinds of authentic knowledge that we might otherwise ignore.

To a cynic, optimism may, however, seem utopian in another sense – that of imagining an ideal state, in which the right, the good and the beautiful reign supreme. But saying that social progress is possible does not signify that we can devise one kind of social organization superior to all others. In the twentieth century, if one favors human freedom, liberal-progressivism and socialism are the principal systems on offer. Both, in their most familiar (and conspicuously imperfect) forms, depend on large organizations for the production of most goods. But, for some purposes, 'small is beautiful'.[45] In most modern advanced states, authentic private enterprise, such as that of a small business providing a local service, has little scope. Similarly, the satisfactions of life in a small, intimate community, such as a village, are lost. In Chapter 10 I give examples of perhaps successful reversals of this trend – the kibbutzim of Israel and the Ghandian villages of India. We do not know to what extent these foreshadow important features of the twenty-first century. In any case, we must expect them to have different roles in different countries.

In such questions, a formidable set of unknown quantities is presented by China, the world's largest nation and ostensibly 'Marxist'. The political upheaval through which we are living is well expressed in a remark attributed to a leading Chinese politician, Hu Yaobang: 'Marx never saw a light bulb, Engels never saw an aircraft and neither visited China.' A special feature of modern Chinese society, which does not fit any conventional political theory, is its source in a revolution, not of industrial workers, but of peasants. Correspondingly, in the 1980s, the Chinese have emphasized the importance of local, small-scale enterprise.[46] To what extent this experience will influence other peasant populations, such as those of Latin America and Africa, has yet to be seen.

If peace and disarmament are achieved, the twenty-first century could become a period of social and economic experiments on a vast scale. If so, every society will be much influenced by applications of the biological and other sciences, but the experiments will not be in the biological dimension; nor will they have the tidy logic of laboratory studies. The experimenters will be part of the systems they explore; and the experiments, as they progress, will themselves change the systems under study. But, as in the objective sciences, no finality will be expected: novelty will repeatedly emerge. Also as in science, social action should be founded on freedom of opinion and respect for dissent; and agreement should be reached, not because it is announced by authority, but on evidence fully debated.[47]

It would be utopian to hope that any community will soon reach such an extreme state of rationality, but movement toward it is possible. Admittedly, even to achieve global relief from wretchedness must seem a daunting task, liable to induce a feeling of helplessness. Many of those who could work for reform are sheltered, for the present, from the severest ills; and, as Bertrand Russell has written, 'Powerlessness makes us feel that nothing is worth doing and comfort makes the painfulness of this feeling just endurable.'[48]

The pain due to knowledge of human distress can be relieved in another way – by altruistic behavior: instead of making oneself more comfortable, one can act against the distress. Unlike mere comfort, such action can even be exhilarating. The most obvious goal is disarmament; but the objective need not be some grand undertaking. There are plenty of urgent tasks to choose from. On the international plane one may contribute to famine relief; one may help refugees from poverty or oppression; or one may support organizations such as Amnesty International. More parochially, one may attempt improvement in the health services or schools of a neighborhood, expose local corruption, or help to preserve a woodland from destruction. Perhaps even writing a book on biology and freedom could be included in this list. Such small operations have a cumulative effect and, on the historical time-scale, sometimes a rapid one.

Often, the aim must include restrictions on liberty – especially the liberty of the powerful, violent and selfish to kill, to coerce, to deceive or to rob others. But the prime test of such action is whether it helps to enlarge the proper freedoms of our human kindred. These are the freedoms, familiar enough, outlined in earlier pages.

We are not doomed to repeat, 'instinctively' or compulsively, the errors of our predecessors. All human communities can learn from their own or others' errors and successes, and can transmit what they have learned. When we try to do this, simple images of our species obscure understanding and obstruct action. Human beings and societies are infinitely diverse and changeable. We continually create novelty, and enjoy doing so. Freedom requires the recognition of complexity.

Glossary

Words are the tokens current and accepted for conceits, as moneys are for values.

Francis Bacon

Most of the words examined below occur in several places in this book; and most make difficulties for both reader and writer. Other special terms are defined or explained where they occur in the text. All special terms, whether listed below or not, are in the index. Italicized words have their own entries in this glossary. Raymond Williams (1976) discusses some of the words more fully. Most of them have many more uses than those I mention.

adaptation   In biology, has two useful meanings; each refers to a process of change. (A third use, in sensory physiology, is not relevant to this book.) First, an organ or other system, such as a muscle, may become more effective with use; this is individual, physiological or ontogenetic adaptation. Second, a population may change genetically: for instance, a characteristic, such as immunity to a disease or to a poison, initially rare, may become general; when the difference between immune and other individuals is genetically determined, this is genetical or evolutionary adaptation. For the relationship of this kind of adaptation with the idea of *natural selection*, see Chapter 6. The expression, 'an adaptation', often used in biological writing when a trait or a characteristic is meant, should be avoided: the obvious question is adaptation for what? It is, however, all right to say that a feature is an adaptation for, say, life in a desert.

aggression   A word of many meanings, and a source of much confusion to biologists, social scientists, politicians, and others. Often an extreme example of the misuse of the flexibility of language. There is scope here for a study in social linguistics. (Chapter 5, especially page 57 and accompanying text.)

alienation (anomie)   For this book, an important concept concerning a person's feelings about other people, about work, and about society in general. The nearest to a familiar equivalent is estrangement. In an extreme case, in a modern rich society, a person may feel (i) powerless to influence what happens (sometimes despite an urgent perception of the need to change society); (ii) that work and most other activities are unsatisfying; (iii) that laws and conventional customs are unacceptable, and should be changed, if necessary by illegal means; (iv) isolated from other people (Seeman, 1959; Israel, 1971). See *play*. (Chapters 11,15.)

altruism   The primary meaning is a moral one: the principle of acting on behalf of others regardless of the cost to oneself. Has been grossly misused by some biological writers in a manner that suggests indifference to moral codes or an incapacity to follow them. See *bioaltruism*. (Chapter 8.)

analogy   In this book we are especially concerned with 'reasoning by analogy': if one sees two kinds of object, both of them similar in some important way, one may conclude that they are therefore similar in other ways. For example, two kinds of animals are seen to have wings, and to fly; hence both are also assumed to have feathers. (But one may be a bat, which has fur.) This type of 'reasoning' can lead usefully to testable hypotheses. It is sometimes said that there are no false analogies; and it is correct that the term 'false' is not properly applicable to analogy. But to say this is irrelevant when an analogy is criticized as misleading (because it leads to an unjustifiable or incorrect conclusion). See *metaphor*. (Chapter 3.)

anomie   See *alienation*.

anthropomorphism   Describing or explaining animal behavior as if the animals were human. This can lead to misleading and even absurd *analogy*. See also *metaphor*. (Chapter 3 and many other places.)

behaviorism (behaviorist) The doctrine that the proper subject matter of (scientific) psychology is overt behavior only. (Behaviorist: somebody who holds this view.) 'Mentalistic' information and concepts, such as emotion, intention, thinking, wishing, are rejected or, at least, are not accepted as causes of human action. This is a metaphysical principle, not a finding from scientific research. See *metaphysics*; *neo-behaviorism*. (Chapters 10, 12.)

bioaltruism   Behavior that lowers the 'Darwinian' *fitness* (chances of survival and reproduction) of the actor, but increases the fitness of a member of the same species. (Sociobiologists use *altruism* in this sense.) The behavior (unlike altruistic conduct in its primary meaning) is defined by its effects on others: nothing is stated or implied about the intentions of the actor. See *sociobiology*. (Chapter 8.)

biogenetic law   Ernst Haeckel used this phrase for the statement that the stages of individual development (from the fertilized egg) correspond to the stages of the organism's evolution. (They often do not.) Also referred to as the recapitulation of ancestral characters.

capitalism   An economic system in which the means of production (factories, mines . . .), exchange (such as banks) and communication (including the 'media') are predominantly in private hands. Often equated with a *market economy* and, misleadingly, with free enterprise. Contrast *socialism* (Chapter 15.)

conditional ('conditioned') reflex   A response, or act, elicited by a previously indifferent stimulus, as a result of the repeated application of the indifferent stimulus at about the same time as an existing stimulus for a similar act. But it is easier to understand CRs from examples. (Chapter 10.)

conscious, consciousness   Formerly signified the awareness one has of oneself. (The Latin conscientia and the Greek συγείδησις had this meaning.) Compare Hamlet's 'conscience doth make cowards of us all' – meaning that self-knowledge does so. But today, when a physician says a patient is conscious, the term may mean only 'responsive': the patient flinches from pain or answers questions. This is a *behaviorist's* definition. The movement for a return to common sense is now allowing the traditional concept back into psychology. (Chapter 12 and elsewhere.)

democracy   Now a term of almost universal approval, but until recently quite the reverse: from Plato to the nineteenth century, it had abusive implications like those that communism often has today. Originally contained two important ideas, opposition to despotism and expression of the popular will. Today, it may signify universal suffrage, freedom of speech and assembly, and equality before the law (of which only universal suffrage has been fully achieved anywhere); or it may signify equality of opportunity (also still only an aspiration); or some combination of freedom and equality. See *equality*; *socialism*. (Chapters 9, 11, 15.)

determinism   See *metaphysics*.

dominance   In *ethology*, usually signifies an animal's having, within its group, prior access to food, place or a mate. More emphasis is now sometimes put on an animal's role in the group: for example, leading on the march, defending against predators and so on. Should not be confused with relationships based on *territory*. More important, equating animal dominance with human rank leads only to absurdity. See *status*. (Chapter 4.)

drift   In genetics, change in the genetical composition of a (small) group as a result of the accidents of survival or mortality. Contrast *natural selection*. The extent to which drift, or random change, has been important in evolution is controversial. (Chapter 6.)

drive   An internal state that causes a particular activity; the word drive is often accompanied by an epithet, as in aggressive drive. A residue of vitalistic ideas, such as that of vital forces (élan vital, primus impetus), popular in earlier periods, and a source of much confusion even today. As it is commonly used, psychology and ethology would be better off without this term. Authentic knowledge on internal states that influence human action is of two kinds. (i) Some phenomena can be described physiologically, for example, the level of male hormone in the blood. (ii) There are 'states of mind' – intentions, desires . . . See *instinct*. (Chapter 4.)

dulosis Ants of one species (called dulotic) enter the nests of another and take the brood; the latter are reared in the nest of the dulotic species, where they contribute to the survival of the colony. Dulotic ants are called slave-makers, although human slaves are of the same species as their owners. Also, if a species of ant is dulotic, this behavior is general throughout the species. Hence the term slave-maker represents a weak and misleading *analogy*. See also *anthropomorphism*. (Chapter 4.)

economic man See *market economy*.

equality In this book, used in a social and moral context. Like *democracy*, the political demand for equality has stood for rejection of privileges, such as those of the rich or of a ruling group (feudal lords, for instance); positively, it represents the aims of (i) equality before the law; (ii) access to education; (iii) the opportunity to perform satisfying work; and much else. The corresponding slogan is equality of opportunity, which implies opportunity to develop in one's own way and access, in early life, to learning a wide range of skills. It has nothing to do with reducing all people to the same level – a nonsensical notion; but it does imply that all kinds of people are entitled to respect. None of this has yet been fully achieved anywhere. (Chapters 9, 15.)

ethical naturalism See *naturalism*.

ethology As used in this book (and in many others): the science of animal behavior. Then refers to a division of biology: compare ecology, genetics. Unfortunately, sometimes used in a narrower sense, to refer (for instance) to the study of animals in their natural surroundings. Worse still, some people write of 'the ethologists' when they mean the authors of popular nonsense about animal or human behavior – an exasperating usage.

eugenics Maintenance or improvement of the quality of a human population by controlled breeding. Galton's original definition was wider; but this has always been the significance of the word in practise. (Chapters 7, 9.)

eusocial insect Species of insects which have co-operation in care of young, reproductive division of labor (queens, workers. . .) and at least two generations living together in the same colony. Termites, ants, wasps and bees. If one is concerned primarily with the human species, the main interest of these insects is in their differences from human beings. (Chapter 4.)

evolution The descent – or ascent – of organisms from very different organisms in the past; implies that such changes will continue in the future. Detectable evolutionary change usually takes millions of years, but can be traced in the fossil record. The fact that evolution has occurred should not be confused with theories of how it is brought about. See *natural selection*. (Chapter 6.)

exploration In ethology, apparently unrewarded movements of an animal about its living space: a means by which an animal acquires information about its surroundings and, especially early in life, develops more general abilities. The tendency to approach novel objects and places (equivalent to human curiosity). Also highly developed in the human species. See *play*. (Chapter 10.)

Fascism During the twentieth century political movements have arisen, in many countries, with some or all of the following features: (i) fervent nationalism; (ii) racism, xenophobia and persecution of minorities (especially Jews and persons with dark skins); (iii) militarism; (iv) hostility to civil rights and democratic procedures; (v) promotion of police power; (vi) misogyny; (vii) anti-intellectualism; (viii) attachment to a conspicuous leader. Political parties or other organizations, whose policies are founded on several of these attitudes, are conveniently called Fascist, after the first of such movements to become prominent, that of Italy (under Mussolini). The most successful, for a brief time, was that of the German 'National Socialists' (Nazis, led by Hitler).

feudalism In Europe, a social order in which land was held by a vassal as fee from a lord or a monarch; accompanied by *serfdom*. A complex system in which lords, vassals and serfs all had both rights and duties. Not at all like the *status* systems of animals. (Chapter 4.)

fitness In biology, the fitness (also known as Darwinian fitness) of an organism is some measure of its contribution to later generations. Similarly, the fitness of a gene is some measure of its incidence in later generations. See *inclusive fitness*; *natural selection*. (Chapter 6.)

gatherer–hunters People without agriculture or domesticated herds, who live by gathering and hunting food. The condition of all human beings until about 10 000 years ago. Judged by surviving groups, gathering was more important than hunting. (Chapters 5, 8.)

heritability In genetics, a confusing term: a measure of the extent to which the resemblance of organisms to their parents, in a specified population at a specified time, is genetically caused. (Resemblances can be due solely to developing in a similar environment: a human example is the language we speak.) The heritability of a trait varies with the environment. A trait, such as stature, can have a high heritability, yet can be greatly altered by environmental change. (Chapters 7, 9.)

homeostasis Maintaining a steady internal state. Examples are deep body temperature, blood sugar, and other aspects of blood composition.

homology   A special kind of *analogy*. Similarity in structural relationships, especially during embryonic development. The structures of the human fore-limb, a whale's flipper and a bird's wing are all homologous. The similarities are attributed to descent from a remote common ancestor. To try to 'homologize' patterns of behavior is not useful. (Chapter 3.)

hunter–gatherers   See *gatherer-hunters*.

inclusive fitness   A measure of an individual's *fitness*, plus that individual's contribution to the fitness of relatives other than direct descendants. *Natural selection* is held to produce behavior that favors not merely an individual's own descendants, but also others that have the most genes in common with that individual; these are likely to be close relatives. Such an effect is also called kin selection. (Chapters 6, 8.)

instinct   An interesting but tiresome word. It is convenient here to distinguish three meanings. (i) Intuition, or unconscious skill, is acquired by experience. It is exemplified when a person is said to perform a movement instinctively, as in catching a ball or warding off a blow. More important, somebody may be said to know, by instinct, that another person is, for example, sad or hostile. This is an essential kind of knowledge. Unlike the other meanings of instinct, it does not come into ethology (chapter 12). (ii) Animals often perform acts, sometimes complex, without instruction or learning how to do them. A striking contrast with human action is the making of elaborate structures, such as nests or webs, without practise. Among animals, much complex social behavior, including courtship, care of young and territorial displays, also seems to develop regardless of experience (again, unlike human social conduct). But this appearance may be deceptive: special early experience (learning), as for the song of many birds, may be crucial. The factors that influence the development of behavior can be revealed only by experiment. The social behavior of each animal species, however it develops, is usually very uniform. It is therefore often said to be species-typical – a term that does not commit oneself to any statement about how it develops. (iii) The third meaning is allied to that of *drive* – an impulsion to behave in certain ways or to achieve certain ends such as survival. (Chapter 4.)

intelligence quotient (IQ)   A measure of scholastic achievement. Each set of tests of 'intelligence' is designed for a particular population, such as that of Paris, at a particular time. There are tests for each age, say, from 5 to 15 years, usually of reading, writing and arithmetic. The population provides the norm, or average: a child may achieve the typical score for his or her age, or above or below it. Hence the IQ is not a test of an absolute feature, measured on an independent scale, as (for example) stature is: it relates the child to the population to which the tests apply. Nor is the IQ a measure of 'innate' or 'genetical' ability: a child's IQ can change substantially during the school years; and there is now massive evidence of the effects of schooling and other conditions of rearing on IQ. But the information we have on the environmental factors that influence the development of the intellect is very incomplete: the IQ tells us nothing of how people learn. Nor does it test originality, creativity or other factors related to success or to social worth. (Chapter 9.)

kin selection   See *inclusive fitness.*

Lamarckism   Colloquially, the proposition that acquired characters are inherited. The popular presumption has been, and remains, that the effects of use and disuse of organs are passed on to the offspring. On this view, the apparent skills of, for example, a predator in tracking and killing prey, or of bees in constructing a comb, are inherited memories, derived from ancestors that had to learn these skills. Attempts to demonstrate such effects experimentally have all failed. This is to be expected. Heredity depends on the genes contained in the chromosomes of the cell nucleus. Lamarckism requires that when I learned the methods of microscopy or the theories of genetics, my genes were altered so that my children either did not need to learn them, or learned them more easily than I did. There is no known way in which this could happen. The passage of traditions, from one generation to the next, is sometimes said to have a 'Lamarckian' character; but this is misleading, for nothing corresponding to genes is involved. Tradition depends on social interactions – imitation and *teaching*. (Chapters 6, 14.)

liberalism   A good example of a nomadic political term – one that wanders from meaning to meaning (Arblaster, 1984). In Australia the 'Liberal Party' has a conventional conservative policy: much emphasis is put on encouraging privately owned industry; even quite mild forms of public enterprise, such as a national health service, are resisted; privately owned schools are strongly supported; there is little enthusiasm for trade unions. In Britain the 'Liberal Party' is quite distinct from the Conservative Party: it emphasizes individual freedoms, including women's liberation, and the value of small, independent business. In an earlier period, a person said to be liberal was open-minded, perhaps unorthodox; but in politics early liberalism was associated with personal liberty (including the freedom, in a *market economy*, to employ others and to enrich oneself as a result), and with the weakening of the authority of monarchs, aristocracies and churches. Today, we have a movement in politics, which may be called liberal-progressivism, which accepts an economy based on private ownership, but emphasizes universal rights, including equality before the law and universal access to education and health services; it rejects racism; and favors equal rights for women. Liberalism, in this sense, is a term of abuse in some influential quarters, notably in the USA; and the ignorant equate it with *socialism*. See also *democracy*. (Chapter 15.)

liberal-progressivism   See *liberalism*.

market economy   An imaginary or model economic system in which production of goods and services is carried out by firms or individuals competing freely with each other, and wages are determined strictly by the value of the goods produced. Consumers (who are Economic Persons – also imaginary) always buy the cheapest goods appropriate for their needs and invest wisely. See *capitalism; liberalism*. (Chapters 11, 15.)

mechanistic interpretation   ('physicalism'; 'materialism') In several places in this book, I refer to attitudes to human beings, apparently founded on the presumption (often unacknowledged, even unconscious) that we are nothing more than mechanisms, even machines. The corresponding philosophical doctrine, conveniently called physicalism, states the following: there is nothing beyond the entities described in physical science; a person is no more than a complex physical mechanism; the mind is nothing over and above the brain, and our experiences are identical with brain processes (Smart, 1963, 1981). This is a metaphysical principle, not a finding from experiments. It is, however, in one aspect, founded on experience: behaving, thinking and feeling all depend on the brain; the functioning of the brain is a necessary condition for the existence of a person. This is a truism (unless one believes in the existence of a bodiless soul). Like the question of determinism, physicalism is nonetheless a source of inexhaustible controversy among philosophers. (Physicalism does not, however, necessarily entail determinism.) In this book, I am cool toward mechanistic interpretations, because – in the hands of non-philosophers – they often go with a thoughtless *reductionism*. They distract attention from the emergence of new properties as one ascends from small objects, such as the ultimate particles of physics, to large ones, such as organisms and – above all – persons. Scientists and others who take mechanistic interpretations for granted should examine the implications of their position. See *metaphysics*. (Chapters 11, 12.)

metaphysics   Principles, presumptions or beliefs that fall outside, and are prior to, what can be found out by, or inferred from, observation of the external world; may influence moral attitudes, scientific research policy and much else, sometimes unconsciously. Examples of metaphysical statements: (i) there are two modes of being, mind and matter; (ii) there is only one mode of being; (iii) everything, including human action and *consciousness*, can be explained by the physical laws concerning ultimate particles. Other such principles are normative: they state what ought to be done. Such 'metaphysical directives' include: (iv) all scientific statements should be refutable by observation; (v) psychologists should study only overt behavior. For us the most important metaphysical position is determinism: this states (vi) either that, if we knew enough, we could predict everything that will happen; or that everything that happens, or will happen, is (was) contained in and inescapably implied by, earlier conditions of the universe. See *behaviorism*; *mechanistic interpretation*; *naturalism*; *reductionism*. (Chapter 12, and many other places.)

naturalism (ethical naturalism) The presumption that moral principles (or 'laws') are, or should be, determined by fixed regularities (or 'laws') in nature. Hence, if it is held that, by a law of nature, organisms are persistently and uninhibitedly competitive, then human beings, too, must be (should be?) in constant competition among themselves. Alternatively, if animals are said to be naturally co-operative, then human beings must or should be so too. Both these views have been held. Both confuse natural laws with norms or principles of conduct which depend on local convention and vary widely from one human society to another. (Chapters 6, 8 and elsewhere.)

natural selection Shorn of logical errors (including expressions such as 'the survival of the fittest') the concept of natural selection is founded on the following true statements. (i) There is individual variation among organisms. (ii) Some of this variation is genetically determined. (iii) Some organisms contribute more to the next generation (have greater *fitness*), than others. (iv) Differences in fitness, like other differences, are partly determined genetically. 'Natural selection' further states or implies two (at first sight opposed) inferences or conclusions. (a) In a constant environment, all the characteristics of an organism (at least, those typical of the group) are the best possible: that is, departure from them would reduce *fitness*. This is the principle of optimality. The process involved is called stabilizing selection: the characteristics of a population are kept in a steady state. Environments are, however, never constant, though (for a given period) some are less inconstant than others. The principle of optimality is, indeed, an abstraction: but it can lead to testable hypotheses in particular instances: for example, that the typical thickness of the fur of an arctic mammal is the best compromise between too little insulation and too much weight. (b) Differences in fitness lead to changes in populations over generations; and such changes have been responsible for *evolution*. This is the central Darwinian proposition. See *tautology*. (Chapter 6.)

neo-behaviorism In this book, refers to the doctrine that all human action can be explained by the effects of *reward* and *punishment*; and that a scientific psychology must be based on *behaviorism*. See also *operant*. (Chapter 10.)

neo-conservatism See *New Right*.

neo-Darwinism In this book, refers to the doctrine that all the features of all organisms, including the human species, reflect the action of *natural selection*. (Chapter 8.)

New Right In this book refers to a rather mixed bag of politicians, economists and writers, including 'neo-conservatives'. The policies and attitudes they advocate include support for a *market economy*; an assertive nationalism; respect for authority; strong military and police forces independent of local control. The virtues of the nuclear family headed by a man (patriarchy) are emphasized. The welfare state, trade unions and liberation of women are opposed. The ownership of property of all kinds is held to be natural or instinctive; so is prejudice against persons of different color or nationality. *Instinct* and intuition are held in higher regard than reason. Only some of these views are held by each movement or individual. See *democracy, fascism, liberalism, social Darwinism, socialism.*

normative principles See *metaphysics.*

operant Any action that is accompanied by, or consists of, *reward or punishment* of the actor. The reference, in practise, is to experiments in which an animal performs some act, such as pressing a lever, and as a result receives or avoids a pellet of food, a shock or some other stimulus. *Neo-behaviorism* is based on the results of experiments of this kind. (Chapter 10.)

optimality principle See *natural selection.*

play A word with such a diversity of meanings, at first sight it defies analysis. In this book, a leading theme is play in the sense of an activity performed without reward, often spontaneously: the only pleasure is that derived from the activity itself. In both human and animal behavior, we associate play with the young. Among them, the functions of the activities called play include exercising the muscles and nervous system, and learning skills, including social skills. But adult human beings 'play' games, and 'work' at hobbies. Above all, if they are fortunate, they may so much enjoy the work that gives them a livelihood, that they exert themselves at it for the sake of doing it, or in order to do it well, regardless of the resulting conventional rewards (pay or status). Hence play and work are not always opposites. Contrast *reward* in the *behaviorist* sense. See *exploration.* (Chapters 10, 11, 12, 15.)

punishment (negative reinforcement) Technically, a response-contingent aversive stimulus. Less formally: an animal performs some act and, as a result, experiences pain or discomfort (an aversive stimulus); thereafter, performance of the act (which includes approach to the place where the stimulus is experienced) is avoided. Contrast *reward.* See *neo-behaviorism.* (Chapter 10.)

race Another word with many uses, some purely propagandist. Formally, a population, within a species, that is genetically distinct and geographically separated from other populations. The separation may have been in the past. The human 'racial' features most commonly attended to are only skin deep: color, hair form, nose shape; there is no evidence that they are genetically correlated with socially important qualities, such as intelligence (however measured) or moral worth. If correlations are found, they are likely to be the result of consigning people of a certain type to special conditions, such as slums, or to inferior or ill paid work. (Chaper 7.)

recapitulation See *biogenetic law*.

reductionism (explanatory reduction) In its most general sense: the findings or theories that belong to one study or discipline (for example, psychology) are explained by those of another (for example, physiology). Such reduction is sometimes supposed to lead to a simpler account than before, but this is an illusion. In this book I describe attempts to explain human action by the theories of *evolution* or of *behaviorism*. I also discuss the common presumption (among scientists) that biological phenomena can be fully explained by the principles of physical science. As usually stated, all these proposals contain serious errors. See *mechanistic interpretation*. (Chapters 8, 10, 12.)

reinforcement See *reward*; *punishment*.

reward (positive reinforcement) May be formally defined as a stimulus that strengthens or makes more probable an act that is followed by that stimulus. Contrast *punishment*. See *play*. (Chapter 10.)

serfdom An alternative to *slavery*. A serf (or villein) and his family worked a plot of land held from a lord, paid the lord in goods, money and labor, and, with his family, was legally obliged to remain on his land unless released by the lord. A feature of *feudalism*. (Chapter 4.)

slavery Human beings as property. A slave-owner may demand services from the slave, control his or her family life, and sell the slave. For a slave to run away carries severe penalties. (Chapters 4, 15.)

social Darwinism An ill-defined body of opinions founded on the belief that the theory of *evolution* can explain human society; in particular, the use of 'Darwinian' theory to justify conservative social policies: promotion of competitive 'free enterprise'; opposition to women's liberation and to public enterprise such as state schools and a national health service. A notable combination of *naturalism* and *reduction*. See also *neo-Darwinism*. (Chapter 9.)

socialism May be formally defined as an economic system in which the major industries, the banks, and the means of communication, including the 'media', are predominantly under public or co-operative control in a state with a democratically elected government. Contrast *capitalism*. Traditionally, socialists emphasize the demand for *equality*. They criticize those aspects of *liberalism* which lead to advocating freedom of employers to exploit their employees. But socialism and liberal-progressivism have much in common. Communism, in one usage, is a synonym of socialism; but today it is associated with violence (*'aggression'*!) and with Soviet tyranny. See also *democracy*. (Chapter 15.)

sociobiology   The interpretation of animal and human social interactions in terms of the presumed action of *natural selection*. Sometimes used, confusingly, as a synonym of social *ethology*. (Chapters 6, 8, 9.)

species-typical behavior   See *instinct*.

stabilizing selection   See *natural selection*.

status   In *ethology*, relationships of *dominance* and *subordinacy*. Status system is conveniently used in place of the polysyllabic and etymologically inappropriate expression, dominance hierarchy. In human affairs, status, like rank, has much more complex meanings. (Chapter 4.)

subordinacy   The obverse of *dominance*.

tautology   In a literary context, signifies repetition or redundancy: saying the same thing twice. In this book, we sometimes meet expressions which seem to inform or explain, but because they are merely repetitive, do not. In biology, the most notorious is 'the survival of the fittest'. This and similar phrases, carelessly used, purport to account for evolutionary change by saying that the fittest individuals (tend to) survive, while the less fit do not; but when we ask what is the criterion of *fitness*, the answer is survival! Hence we have the survival of those that survive: the adjective 'fittest' itself contains the information contained in 'surviving'. A more fundamental comment is that the theory of natural selection tells us nothing about the real world: it consists only of analytic statements like those of pure logic and mathematics. These systems begin with a set of definitions, and deduce consequences such as those of the geometrical theorems familiar to every schoolchild. All the statements, or propositions, in such a system have been called tautologous because they are contained in, implied by or follow from the initial statements: they are not about the world of observed events. In evolutionary theory, mathematics is used to infer what would happen to evolving organisms, if the properties (behavior and so on) of the organisms were of a certain imagined sort. Such sets of presumptions are called models. The conclusions are still not about the world of real organisms; but they can be used to suggest phenomena to look for in nature: that is, they can be a source of testable hypotheses. When this method is used, the real world usually proves to be more complicated than the model. See *adaptation*; *natural selection*. (Chapters 6, 8 and elsewhere.)

teaching   An objective definition, devised to make possible comparison of human beings and animals, is: behavior that tends to alter the behavior of a member of the same species (the pupil), and tends to be persisted in until the pupil achieves a certain standard of performance. In this book, 'teaching' is applied only to human action; but animal analogs of teaching are described. There is a sense in which teaching is a distinctively human trait, of central importance in all human communities. (Chapter 14.)

teleology An organism is sometimes said to have evolved a trait 'so that' it can survive in certain conditions: for instance, a long tongue so that it can feed on ants. Such a statement seems to imply that the end served, or the function to be performed in the future, is a cause of the acquisition of the trait; hence it may seem that, in some sense, foresight is involved. This is an example of a teleological statement, in which an outcome is said to be a cause (a 'final cause'). Final causes are not explicitly accepted as responsible for evolutionary change. There is, however, a sense in which we may properly speak of the function of a trait. We may, for instance, say that color vision has a function for a human being: a person with defective color vision, at least in primitive conditions, may be supposed to be at a disadvantage (have impaired Darwinian *fitness*). The causes of the development (*evolution*) of color vision are then assigned to the past. Such an account of biological function is sometimes called teleonomic. In contrast, a physicist does not say that electrons have the function of joining their atoms with other atoms (Nagel, 1961). See *adaptation*.

territory In *ethology*, a region, occupied by an individual or a group of animals, from which other members of the same species are excluded. Often reserved for a region defended from others. The similarity to human occupation or ownership of land, or to human property, is remote; in some writings, it leads to a misleading *analogy*. Distinct from *status* within a group. (Chapter 4.)

work See *play*.

Notes

Preface

1 Popper (1966, vol. 1, p.ix).
2 Mellow (1974, p.468).
3 For a recent survey, Prins (1984). See also Ford (1985).
4 Holton (1973).
5 There is correspondingly an urgent need for an education, at least for some people, that crosses disciplinary boundaries. See Barnett & Brown (1981, 1983); Barnett *et al.* (1983).

Chapter 1 Four portraits

1 Kwapong (1971, p.385).
2 Plato (1930, 1935). The quoted sentence is a translation by Popper (1966, p.52).
3 Darwin, C. (1901, pp.98, 99, 148).
4 Fisher, R.A. (1930, p.192). This book was reprinted in 1958 with minor changes. The passage quoted is unchanged from the first edition.
5 Fisher, R.A. (1930, p.205).
6 Wilson, E.O. (1975, p.575).
7 Skinner (1973, p.15).

Chapter 2 The pessimistic tradition

1 Popper (1966, p.35). A.E. Taylor (1948) gives an account of Plato's work, and G.C. Field (1948) of his life. For introductions to ancient philosophy, see Guthrie (1950), Russell, B. (1961), Wild (1960).
2 Russell, B. (1961, p.55).
3 Bury (1932, p.17).
4 Graves (1957, vol. I, p.313).
5 Bury (1932, pp.18–19). Not all the Greeks of that time were so strongly opposed to change; see Passmore (1970, p.196).
6 Russell, B. (1961, p.122).
7 Plato (1930, pp.306–7).
8 Plato (1930, p.221; on misology, p.227). This is the translation by Benjamin Jowett.
9 Plato (1903, pp.59–63).
10 Mazzeo (1965, part II) provides a good introduction.
11 Bock (1980, p.29). I use the translation of *The Prince* by George Bull (Machiavelli, 1961), whose introduction discusses Machiavelli's supposed depravity.

12 A translation by L.J. Walker has been valuably introduced and edited by Bernard Crick (Machiavelli, 1974).
13 Machiavelli (1974, p.112).
14 Machiavelli (1961, p.99).
15 Bury (1932, p.32).
16 *Leviathan*, first published in 1651, has been edited by John Plamenatz (Hobbes, 1962). For interpretations of Hobbes, Gauthier (1969), Hill (1958), Peters (1956).
17 See Chapter 6 of the present work.
18 Hobbes (1962, p.59). On Descartes, see Scott, J.F. (1952).
19 Hobbes (1962, p.123).
20 Peters (1956, p.44).
21 See Roy Pascal's introduction to *Thus Spake Zarathustra* (Nietzsche, 1960).
22 Stern (1980). See also Spengler (1926, 1928, 1934); Hughes (1952).

Chapter 3 Animals and analogy

1   Throughout this book, for lexical definitions I use the *Shorter Oxford English Dictionary* (OED) (2 vols, 1964).
2   Max-Müller (1864, p.358).
3   Bock (1980); Hiatt (1970); Lovejoy & Boas (1935); Ucko & Rosenfeld (1967).
4   Klingender (1971).
5   Bulmer (1978, p.3).
6   Evans-Pritchard (1956).
7   Douglas (1963).
8   Lovejoy & Boas (1935, especially pp. 207, 391, 393, 401).
9   Montaigne (1958).
10 Janson (1952).
11 Curley (1979); Janson (1952).
12 Singer (1917).
13 Hume (1975, p.104).
14 Maynard Smith (1961) and McFarland (1971) for general discussions. See also Edge (1973) on metaphor.
15 Ben Shaul (1962); Bernal & Richards (1973); Blurton Jones (1972).
16 Wolff (1965); Hayes & Watson (1981); Bruner (1969).
17 On common sense, Joynson (1974); and Chapter 12 of the present work.
18 Edge (1973, p.37).
19 Bowlby (1969, p.183).
20 Leyhausen (1965).
21 Kemp (1982); for a review, Boyden (1973).
22 Owen (1843); for a modern account, Alberts *et al.* (1983, ch.15).
23 Kemp (1982).
24 Beer (1974) gives a critical, ethological discussion.
25 Toynbee (1935, vol. 3, pp.88–9).
26 MacKenzie, W.J.M. (1967, p.46).
27 Cole (1923, pp.13–14). See also Macrae (1958).
28 Neither Marx nor Engels, however, used Darwinism consistently to support their sociology; and Engels explicitly criticized the crude argument by analogy. For sources and discussion, Venable (1946, especially pp. 62–66); Zirkle (1959, especially p. 456).
29 Ferri (1906, p.62; for Macdonald, pp.vii–viii).
30 Kropotkin (1910).
31 Berggren (1962, 1963); Haynes (1975); Hesse (1963); Nash (1963); Ortony (1975); Temkin (1949); Wagner (1968).
32 Young, J.Z. (1964, p.15).
33 Cheyne (1978).

34 Temkin (1949; quotations on pp.189–91).
35 Edge (1973, pp.41–2).
36 Hoffman (1980; quotation on p.396).

Chapter 4 Communication and instinct
1 Huxley, T.H. (1904, p.250).
2 Lorenz (1966, p.x; 1971, p.192).
3 Tinbergen (1968, pp.1414, 1415).
4 For fuller accounts, Barnett (1981a, especially ch.13); Halliday & Slater (1983a).
5 Thompson (1917, p.671).
6 McClintock (1971); Russell, M.J. *et al.* (1980); Veith, *et al.* (1983). The examples I take from animals are *signal* (or releaser) pheromones: they have an immediate effect on the receiver's behavior. The presumed pheromones of human beings are *primer* pheromones: they change the physiology of the receiver, but the effect on behavior occurs later.
7 For references, Barnett (1981a, ch. 12).
8 Beer (1972).
9 Tschanz (1968).
10 See, for example, Frisch (1967); Gould, J.L. (1976); Lindauer (1961); Menzel, R., *et al.* (1974).
11 Wolff (1969).
12 Jelliffe & Jelliffe (1978).
13 Note 10, above.
14 Williams, R. (1976, p.251).
15 Price (1975, p.108).
16 Lorenz (1952, pp.186–7).
17 Huxley, J.S. (1964).
18 Schenkel (1967).
19 Schjelderup-Ebbe (1922). For a more detailed account, Barnett (1981a, chs. 10, 11).
20 Dewsbury (1982).
21 Jay (1965).
22 Jolly, A. (1966).
23 Hall, A. (1977, p.177).
24 Wynne-Edwards (1962, p.187).
25 For sources, Barnett (1981a, chs. 10, 11).
26 Sollas (1924, p.126).
27 Keith (1948, p.29).
28 Godelier (1979).
29 Meggitt (1972).
30 Fortes (1970b, especially p.104). Consult Fortes for an extended review of kinship.
31 Long (1971, p.266). Consult Long for more information about Australiforms.
32 Godelier (1979) gives an instructive short survey.
33 Stahl (1980).
34 On feudalism, Gibbs (1949) for an introduction; Boutruche (1970).
35 For example, Sharma (1965).
36 Storr (1968, p.26).
37 Tawney (1961, p.50; see also p.27).
38 Tawney (1938, pp.45, 72).
39 Davis (1966, pp.3, 38). For a general review of slavery, Davis (1984).
40 Michener & Michener (1951, pp.184–7).
41 Wilson, E.O. (1971, pp.364–71).
42 Watson, J.L. (1980).
43 Carcopino (1956, pp.69–78); Finley (1973, ch. 2).
44 Furneaux (1974).
45 Watson, J.L. (1980, p.8).

46 Mackie (1978).
47 Wilson, E.O. (1975, p.246).
48 Barnett (1981a, chs. 10, 11).
49 Bertram (1975); Schaller (1972).
50 Angst & Thommen (1977) review infanticide among Primates. On langurs, see Dolinow (1977) and Hrdy (1977); on chimpanzees, Goodall (1986).
51 Dickeman (1975) reviews human infanticide. Swift J. (1934, pp.512–21).
52 Storr (1968, p.118). For unexpected deaths, Barnett (1979, 1988).
53 Freedman (1978, p.104). See also Factor & Waldron (1973) on crowding and health.
54 Barnett (1981b, pp.134–6, 241–2; 1988).
55 Mitchell, R.E. (1971). For other examples, Barnett (1979).
56 Coelho & Stein (1980, p.32).
57 Freedman (1975, 1978).
58 Draper (1973).
59 McCarthy *et al.* (1975). See also Factor & Waldron (1973); Webb & Collette (1975).
60 Thorpe (1963, p.29).
61 Wilm (1925, especially p.82).
62 Eibl-Eibesfeldt (1970, p.327). The second edition omits this statement, but retains references to aggressive and other drives (1975, p.339 and elsewhere).
63 Santayana (1954, pp.128–9).
64 Menninger (1948, p.343).
65 Lorenz (1966, pp.241–3).
66 Goldstein (1975, especially pp. 46–51); see also Clinard (1964).
67 Le Quesne (1983). And even cricket, now that it is big business, is going in for aggro.
68 Sipes (1973). See also Abel (1941).
69 Bandura (1969, p.159).
70 Hokanson (1970); Hokanson *et al.* (1968).

Chapter 5 The aggression labyrinth

1 Zimbardo (1978, p.167).
2 See, for example, Eibl-Eibesfeldt (1979); Moyer (1976); Wilson, E.O. (1975, especially pp.242–3).
3 For a valuable, balanced survey, Goldstein (1975).
4 Lion & Penna (1974); Madden & Lion (1981).
5 Rudé (1964, pp.255–257) gives figures on the high rate of serious violence by authority in Europe in the eighteenth and nineteenth centuries. See also Storch (1975) for police action viewed by rebellious wage-earners. On recent violent crime by police, S. Box (1983, ch. 3).
6 Bremer *et al.* (1973).
7 Durant (1981).
8 Abel (1941).
9 Tedeschi, *et al.* (1981, p.29). See also Berkowitz (1981).
10 Huxley, T.H. (1911, p.204).
11 Gould, S.J. (1977). See also Jones, G. (1980) for examples of racist and recapitulationary excesses.
12 For interpretations of Freud, with full references to sources, Fromm (1959, 1974); Gould, S.J. (1977, pp.155–61); Sulloway (1979).
13 Freud (1949a, pp.65–6).
14 Ardrey (1967, 1969, 1976).
15 Dart (1953, p.209). See also Dart (1959, pp.191–202).
16 For recent descriptions, Brain (1981).
17 Brain (1981). See also Isaac (1981) on the facts of our fossil history.
18 Ardrey (1969, p.30).
19 Lorenz (1966, p.205). For the influence of Lorenz, and a critical analysis, Kim (1976).
20 For Tomalin, *New Statesman*, Sept. 15, 1967. For Whitehorn, *Observer*, Oct. 29, 1967.

21 On Peckinpah, *New York Times*, 31 Aug. 1969, p.D9; on Kubrick, *NYT*, 30 Jan. 1972, p.D1. See also Montagu, A. (1976, pp.27–31).
22 Roper (1969); A. Walker (1981).
23 Jolly, C.J. (1970).
24 Leroi-Gourhan (1968); Marshak (1972).
25 See, for example, Hofman (1983); Pilbeam (1972); Stephan (1972).
26 On *Homo erectus*, J.D. Clark (1976); Pilbeam & Gould (1974).
27 Lee & DeVore (1968); Zihlman & Tanner (1978).
28 Schaller & Lowther (1969).
29 Isaac (1981).
30 Washburn & Avis (1958, p.433); Washburn & Lancaster (1968, p.299).
31 Fromm (1974, p.132).
32 Eibl-Eibesfeldt (1974, pp.41–3).
33 Clark, C.M.H. (1978, pp.210–18); Roth (1899); Thirkell (1874).
34 Clark, C.M.H. (1978, p.212). On whites seen from the other side, Blackburn (1979); Reynolds (1981).
35 Turnbull (1961).
36 Lee (1979, especially pp. 396–400; quotation on p.461).
37 Gardner, R. & Heider, K.G. (1969).
38 Chagnon (1968; quotation on p.118).
39 Levy (1969).
40 Dentan (1968).
41 Nance (1975).
42 Turnbull (1973).
43 Bohannan (1967).
44 Wolfgang (1967, especially pp. 15–28).
45 Proverbs, 22:6. On teaching, Chapter 14 of the present work.
46 Bandura (1973); Goldstein (1975); Stohl (1976).
47 Goldstein (1975; pp.29–30).
48 Goldstein (1975, pp.46–51).
49 Dicks (1972, p.55; p.234).
50 Adorno *et al.* (1950).
51 Arendt (1963).
52 Sutherland & Tanenbaum (1980); on Freud, Birnbach (1962).
53 Buchan (1921, p.467); Cruttwell (1982, pp.108–9).
54 Von der Mehden (1973, p.7).
55 Delgado (1969); Mark & Ervin (1970).
56 On Cameron, Marks (1979). For surveys of psychosurgery and allied topics, Breggin (1975); Scheflin & Opton (1978, especially ch.7). Two decades after Cameron's death, more revelations were still appearing, as victims claimed compensation (*New Scientist*, 6 November 1986, p.28).
57 Gross (1980, p.259).
58 Gross (1980, p.280).
59 Gross (1980, p.246). On the computer state, Burnham (1980).
60 Donner (1980, pp.xii, 452). See also Lasch (1969, ch. 3); Marks (1979) on the CIA; Bamford (1982) on extra-legal activities of the US National Security Agency.
61 Postman (1985).

Chapter 6 Evolution and natural selection

1 Ridley (1982).
2 Erwin, T.L. (1982).
3 Cuffey (1984); Godfrey (1984); K.R. Miller (1984). On creationism in Britain, also Howgate & Lewis (1984); Jukes (1981). Louisiana still has an 'equal time' creationist law (*Science* 225, p.36, 6 July 1984).

4 Fisher R.A. (1936, p.58). For an example of a good textbook which discusses progress in these terms, Dobzhansky *et al.* (1977, especially p.507).

5 Lovejoy (1960, p.59) slightly shortened.

6 Bateson (1973, p.424).

7 Bateson (1973, pp.313–14).

8 Froude (1883, pp.20–1).

9 Williams, G.C. (1966; pp.6–7).

10 Dobzhansky *et al.* (1977, p.96).

11 Karn & Penrose (1951).

12 Mitton (1975); Métral (1981).

13 Bishop (1981) for a review.

14 Lees (1981).

15 Allison (1969); Brewer (1979).

16 For introductions, Maynard Smith (1975); Sheppard (1975).

17 White (1978).

18 Mayr (1982).

19 Gould, S.J. (1980, 1982); Stanley (1979).

20 Falconer (1973).

21 For comprehensive introductions to human genetics, Bodmer & Cavalli-Sforza (1976); Levitan & Montagu (1977).

22 Bohm (1969) discusses this difficult concept.

23 WHO (1976).

24 For a review of the modern position, Fitch (1982).

25 Grene (1974) discusses the logic of teleology.

26 For arguments, Harris, H. (1976); Jones J.S. (1980); Kimura (1983); Lewontin (1980); Neal (1976).

27 Bodmer & Cavalli-Sforza (1976, ch. 12); Levitan & Montagu (1977, ch. 19).

28 Discussed by Maynard Smith (1982).

29 Woolf & Dukepoo (1969).

30 Genovés (1976, p.25).

31 Waddington (1960, p.385).

32 Brady, R.H. (1979, especially p.604).

33 Quoted by Heckhausen (1973, p.219).

34 For instance, Dunbar (1982, p.612). The biological concept of fitness contains a logical difficulty, as we can see from the variety of measures said to estimate fitness. Many well defined phenomena may be measured in more than one way - for example, the rate at which an animal uses oxygen. In such a case, whatever method is used, there is no doubt about the reality of the process recorded – the combination of oxygen and carbon atoms inside the animal's cells. When experimenters talk of measures of fitness, they may seem similarly to imply the existence of a single underlying phenomenon, of which the quantity is estimated by various techniques. The same implication is contained in the paradoxical assertion that fitness cannot be measured; for it signifies that fitness exists, but is inaccessible to quantitative enquiry. In fact, however, no single process or relationship corresponds to the word fitness. A defender of the concept may reply: but the *essence* of fitness (or *true* fitness) is survival – or reproduction (of genes, or genotypes, or phenotypes . . .) in the next (or later) generations; hence the measures we record are only shadowy representations of some remote but authentic fitness. But in natural science such 'essentialism' is in practice rejected as fruitless (*cf.* Popper, 1966, especially pp. 31–3). Instead, we try to name and to measure what can be observed, in principle, by anybody. We are therefore obliged, as stated in the text, to declare, for each investigation, just what is being recorded: this may be the number of females inseminated by a male, the number of eggs produced by a female, the longevity of either . . . Every such feature can, no doubt, have some effect on the character of later generations. 'Fitness' then becomes the name for a set of general, even trite – though important – statements: measurable features of organisms vary; some of the variation is genetically determined;

correlations can sometimes be found, or reasonably surmized, between the features of a given generation and the genotypic or phenotypic characteristics of later generations.

35   See Brady, R.H. (1979); Orians & Pearson (1977) for the original research; Pyke *et al.* (1977) for a review.
36   Gould S.J. & Lewontin (1979).
37   Weiner (1954).
38   Fisher, R.A. (1930, p.ix).
39   See, for example, Maynard Smith (1981); for critical comment, Etkin (1981).
40   Maynard Smith (1981).
41   Maynard Smith (1981, p.5).
42   John Maynard Smith, in a letter, writes: Haldane made the remark in a conversation. To be precise, he said it in 'The Orange Tree' [a London pub].
43   Maynard Smith (1964). For a review, Hamilton (1972).
44   Bertram (1978); Packer & Pusey (1982).
45   Riedman (1982).
46   Trivers (1971).
47   Hamilton (1964, p.42). Two questions unanswered by current theory are those of (i) group selection and (ii) the evolution of sexual reproduction. (i) On group selection, see Wilson, D.S. (1980), Wade (1982). Modern theory states that natural selection acts on individuals, not on groups. Neo-Darwinian calculations allow the possibility of behavior that injures the actor, provided that it sufficiently benefits kin; but they do not allow self-sacrificial actions of a kind familiar in human affairs, as when a person risks loss, injury or death on behalf of a group, such as a village or a nation. Formerly, such conduct was often supposed to have arisen, by natural selection, also in other species; but, as far as is known, this can happen, if at all, only if there are many small populations (demes) at least partly isolated from each other, and if demes reproduce themselves by division as they enlarge. Suppose that in some demes, by chance variation, there came to be a majority of self-sacrificing individuals; these demes would then have an advantage over the others, that is, greater chances of surviving and reproducing. Hence, in such a calculation, each deme (with tens or hundreds of members) is treated as an individual.

Attempts, both experimental and mathematical, to demonstrate group selection have, however, so far led to no consensus. (ii) This enigma has a bearing on the origin of sex. (For the biology of sex, Alberts *et al.*, 1983, ch. 14; on the origin of sex, Rose, M.R. 1983.) In biology, sexual reproduction refers to the fusion of two nuclei, usually from two cells; the cell, usually an egg, in which fusion takes place, develops into an adult of which most cells receive chromosomes from each parent. The principal effect of sexual reproduction is to enhance variation within the species: if a mutant gene survives, it can spread and become associated with a variety of other genes: the number of genotypes is thereby increased. Alternatives to sex include vegetative propagation, exemplified by growing a plant from a cutting, and parthenogenesis in which an egg develops unaided by a sperm. The rarity of such asexual reproduction among complex animals is conventionally explained by the advantages of the genetical diversity conferred by sex: the survival of a population is aided by the existence of varieties that can cope with changing environmental demands.

The origin of sex may have been a single event, a 'quantum leap' thousands of millions of years ago. How could sex then defeat the asexual in the 'struggle for existence'? Sex entails fusion, not division and so, in itself, is the opposite of reproduction. Of two types, one sexual and one asexual, the latter should multiply more rapidly and oust its rival. Moreover, each sexual organism transmits only half its genes to its offspring, while an asexual transmits all. The contribution of a sexual organism to the next generation is therefore diluted; and it is not clear how this can, in its immediate effects, be anything but disadvantageous. Granted, sex confers adaptability on populations that possess it; but for sex to survive in the beginning requires prevision on the part of the organisms concerned: the benefits of sex (in the Darwinian sense) become evident only when the environment changes and new demands are put on the population. But, as

Etkin (1981, p.55) remarks, 'genes are not . . . guided by future consequences'; they have no foresight. Hence sex seems to be a theoretical impossibility. Williams, G.C. (1980, p.382) therefore calls it 'a major unresolved mystery'; and Hamilton *et al.* (1981, p.376) ambiguously state that 'we still do not know what sex is for'. If, however, sex originated in a population of organisms divided into demes, it could have conferred an advantage in the same way as self-sacrificial behavior. As it is, with our present knowledge, we have to resign ourselves to sex without being able to explain it. The same, of course, applies to the altruism (in its primary sense) of human beings.

48 Brady, R.H. (1979, pp. 615–16; see also p. 610).

49 Birch & Ehrlich (1967) recommend their fellow ecologists not to write of evolutionary change when in fact they are describing observations on present, not past, events. The same could be urged in other divisions of biology, especially ethology.

50 Grene (1969, p.65; see also 1981).

51 Darwin, C. (1872, ch. 2). See also Manier (1978); R.M. Young (1971).

52 Carneiro (1967, p.x). See also Greene (1959).

53 Darlington (1958, p.239). For the extreme expression of 'geneticism', Darlington (1969), an interpretation of human society entirely in terms of imagined genetical differences between people and groups.

54 Huxley, J.S. (1953, pp.41, 38).

55 Huxley, T.H. & Huxley, J.S. (1947, p.217).

56 Huxley, J.S. (1953, pp.150, 81, 37).

57 Waddington (1960). Compare Grene (1974, ch. 12); Raphael (1958).

58 Lorenz (1966, p.257).

59 Lorenz (1966, p.17).

60 Williams, G.C. (1966, pp.254–5).

61 Durham (1976).

62 Gibbon (1976, p.181). This is an abbreviated version, edited by D.M. Low, of the work originally published in six volumes in 1776–1788.

63 Flew (1974, p.231).

64 Greene (1981, p.163).

65 Paley (1830, p.221; see also p.52). On the decline of religion, Hobsbawm (1977a, ch. 12).

66 Jacob (1977).

67 Medawar (1960, p.100); Medawar gives other examples. For an outline of modern immunology, Alberts *et al.* (1983, ch.17).

68 Dunbar (1982, pp.11–12).

69 Monod (1974, pp.137, 110, 36).

70 Bohm (1969, p.36).

71 Bohm (1969, pp.92, 38).

72 Quoted by Gillespie (1979, p.103).

73 See Popper (1966, ch. 5); Quinton (1966).

74 Popper (1966, p.71).

Chapter 7 Environment and heredity

1 Hsia (1967).

2 Simoons (1979).

3 Dubos & Dubos (1953).

4 Bodmer & Cavalli-Sforza (1976, p.493).

5 For a review, Wing (1973).

6 Shields & Gottesman (1973); for an extended, critical review, Rose, S.P.R. *et al.* (1984, ch. 8).

7 Crow (1983); Torrey & Peterson (1973).

8 Feldman & Lewontin (1975).

9 Block & Dworkin (1976); Eaves & Jinks (1972); Lewontin (1975); McAskie & Clarke (1976).

10 Suzuki (1969). See also Winner (1982, ch. 8).

11 Taken, with changes, from Barnett (1981a, pp.531–2). For a review of the meanings of 'race', Banton (1987).
12 Bodmer & Cavalli-Sforza (1976); Levitan & Montagu (1977).
13 Falconer (1960, p.36).
14 These figures are from Penrose (1963). See also Levitan & Montagu (1977, pp. 838–40).
15 Luria (1976).
16 Scribner & Cole (1973).
17 Hallowell (1955).
18 McKay *et al.* (1978).
19 Skodak & Skeels (1949).
20 Schiff *et al.* (1982). For the debate on IQ, Block & Dworkin (1976).
21 Tizard (1974, p.316).
22 Heyneman & Loxley (1983). On the USA, Postman (1985); also US Department of Education (1983) The Nation at Risk: the Imperative Need for Educational Reform.
23 Lazar *et al.* (1982, p.65).
24 Etkin (1981, p.88). For a summary, Lewontin (1983).
25 Barnett & Dickson (1984) and references therein.

Chapter 8 Stories of human evolution

 1 van den Berghe (1978). For versions of this self-portrait, Alexander (1979), Barash (1980), Symons (1979), Wilson, E.O. (1979).
 2 Fisher, E. (1979); Shapiro (1979).
 3 Goodale (1971) on Tiwi; Estioko-Griffin & Griffin (1981) on Agta. See also Lee & DeVore (1968).
 4 Discussed by Festinger (1983); Isaac (1983).
 5 Isaac (1983, p.562).
 6 Short (1976, 1979, 1981).
 7 Festinger (1983, pp.67–8).
 8 Short (1983, p.35). On breast-feeding, Jelliffe & Jelliffe (1978).
 9 Short (1981, p.336).
10 Short (1976, p.3).
11 Short (1981, p.338). Evidence for the universality of these features is not given. Hallowell (1955) describes a group of Amerindians, the Ojibwa, among whom female breasts arouse no erotic interest. The Mangaians (Pacific islanders) express surprise at Western interest in breasts: for them, it is an infantile feature (Marshall & Suggs, 1972, p.110).
12 See Barnett (1981a, pp.615–17).
13 Cant (1981).
14 See Symons (1980), and the discussion that follows.
15 Symons (1979, pp.12–14).
16 Baldwin & Baldwin (1980).
17 Alexander (1979, pp.56, 112).
18 Wilson, E.O. (1975, p.120).
19 Dawkins (1976, pp.2–3).
20 Romans, 3:9–17.
21 Augustinus (1945). This translation is edited by Ernest Barker. *De Civitatis Dei* was first published in AD 426.
22 Essock-Vitale & McGuire (1980; quotations from pp.233, 234, 242).
23 van den Berghe (1979, pp.46, 47).
24 van den Berghe (1979, pp.62, 63).
25 In the discussion of the paper by Symons (1980); see p.188.
26 Alexander (1975, pp.96–7).
27 Trivers (1981).
28 Young, J.Z. (1978, p.71).
29 van den Berghe (1978, p.39).
30 Bok (1984, p.61).

31 Bok (1984, pp.64–5).
32 Trivers (1981, p.32).
33 van den Berghe (1979, p.217).
34 Alexander (1979); Wilson, E.O. (1979).
35 Dawkins (1982, pp.37–8).
36 Alexander (1979, pp.82–3).
37 *Time*, Aug 1 1977, pp.18–23.
38 Midgley (1980, p.26).
39 *Daily Mail*, April 7 1978.
40 Bacon (1906, p.20). (First published in 1597.)
41 van den Berghe (1978, p.52).
42 Lotka (1945, p.167).
43 Medawar (1960, p.99).
44 Goldberg (1977).
45 Boyd & Richerson (1980); Richerson & Boyd (1978).
46 For the distinctive role of history, Bock (1980).
47 Caplan (1979, p.28).
48 My list of loaded terms in sociobiology is seriously incomplete. Gowaty (1982) documents a number of sexist terms, including rape, coyness, adultery and homosexuality, used for conduct only remotely analogous to that of human beings.
49 For example, Nagel, T. (1970).
50 Fortes (1970b); Keesing (1972); Schneider (1972).
51 Keesing (1972, pp.23, 24).
52 Leach (1982, p.107; see also pp.134–40); Leach (1961).
53 Brady, I. (1976, p.99).
54 van den Berghe (1981, pp.36, 27).
55 Huxley, T.H. (1911, pp.36–7).
56 Wilson, E.O. (1980, p.61).
57 Wilson, E.O. (1975, p.564).
58 Dawkins (1976, p.215). For critical comment, Daley (1980), Wade (1978).
59 van den Berghe (1978, p.52; 1979, p.69).
60 Alexander (1979, pp.59, 72, 153–4, 93).
61 Ghiselin (1974, p.247).
62 See Carr (1981, especially p.91); Mackenzie, W.J.M. (1975).
63 Grene (1978, p.216).
64 Meltzer (1981).
65 Harris, M. (1983, pp.18–19).

Chapter 9 Darwinism, genetics and politics

1 Huxley, T.H. (1911, p.83).
2 Mackenzie, D.A. (1981).
3 Box, J.F. (1978).
4 Fisher, R.A. (1918, p.433).
5 Norton (1980; quotation on p.489).
6 Nigel Lawson, Chancellor of the Exchequer, in an interview (*Guardian Weekly*, Sept. 18, 1983, p.19).
7 Darwin (1901, p.206).
8 Blacker (1952; quotations on p.82).
9 Searle (1976, especially p.53).
10 Galton (1908) gives his own account in an autobiography. See also Mackenzie, D.A. (1981); Searle (1976, 1981).
11 Darwin, L. (1926, p.138).
12 Haldane (1938, p.112).
13 Mackenzie D.A. (1981, especially pp.73–9); Searle (1981).
14 Adami (1922, p.185).

15 Mackenzie, D.A. (1981, p.197).
16 For a recent account, Geary (1981).
17 Stone, N. (1983, p.91).
18 Hobsbawm (1977b, ch. 6).
19 N. Stone (1983, especially ch. I, 4).
20 N. Stone (1983, especially p.142).
21 Norton (1981). On the history of the intelligence quotient, Gould, S.J. (1981).
22 Brim & Kagan (1980); Wohlwill (1980).
23 Tanner (1966).
24 Flynn (1984; quotation on p.47); Postman (1985).
25 Kagan (1979, p.181). See also Loehlin *et al.* (1975).
26 *Royal Commission on the Distribution of Income and Wealth.* Cmnd 6171, 7175. HMSO: London 1978.
27 Jencks *et al.* (1972). See also Bowles & Gintis (1973); Bowles & Nelson (1974).
28 Jencks *et al.* (1972, p.254).
29 Chomsky (1976, p.290).
30 Gardner (1983) outlines hypotheses on 'intelligences' of these kinds.
31 Haller (1963); Hofstadter (1955); Kamin (1974); Ludmerer (1972).
32 Bannister (1979, p.170); Scheinfeld (1944).
33 Dugdale (1910); Haller (1963).
34 Kamin (1974, p.11). For a review, Allen (1975).
35 Ludmerer (1972, pp.20, 60).
36 Allen (1975).
37 Haller (1963, p.124).
38 Haldane (1938).
39 Haldane (1938, p.97).
40 Haller (1963).
41 Jensen (1969).
42 Jensen (1980, especially pp. 175, 248).
43 See Halliday & Slater (1983b, especially chs. 5, 6).
44 Hearnshaw (1979), in a thoroughly researched biography, gives the full, and now accepted, evidence that Burt invented the findings he published on twins.
45 Halsey (1978, especially ch. 6).
46 Cox & Dyson (1971).
47 McDaniel (1985; quotation, p.38). For the debate on these questions, Block & Dworkin (1976).
48 Pearson (1900, p.310).
49 Galton (1865, 1869, 1908).
50 Galton (1865, p.325).
51 In an unsigned, dogmatically written article (*Anthropol. Rev.* 4, p.120, 1866).
52 Jones, G. (1980, pp.150–1). For the scale and pervasiveness of imperial and racist propaganda in Britain, see MacKenzie, J.M. (1984).
53 For detailed documentation, Billig (1981, especially chs 2, 4).
54 Pearson & Moul (1925).
55 Searle (1976, especially pp.40–2).
56 Compare Thoday (1969).
57 Haller (1963, especially ch. 10); Hofstadter (1955, ch. 9).
58 Ludmerer (1972, ch. 5).
59 Kamin (1974, p.21).
60 Kamin (1974, pp.23–4).
61 Ludmerer (1972, p.92).
62 Kamin (1974, p.25).
63 Boas (1911, 1940).
64 Klineberg (1935; quotation on p.189).
65 For a masterly, documented summary of race and crime, see Jencks (1987). On Jews, see Bonger (1943).

66 Klineberg (1935).

67 Gasman (1971). For a summary of Haeckel's life, Nordenskiöld (1928, ch. 14). See also Haeckel (1913).

68 Gasman (1971, p.15).

69 Gasman (1971, p.22).

70 Haeckel (1879, p.93).

71 Gasman (1971, p.157). See also Billig (1982, especially p.69).

72 Gasman (1971, p.99).

73 Spengler (1926, vol. 1, p.153).

74 Spengler (1934, p.21).

75 Bullock (1962, p.36; see also p.398).

76 Speer, A. (1970, p.588).

77 Gasman (1971, p.173).

78 Ludmerer (1972, pp.116–17).

79 Reitlinger (1961); for the sickening details, Gilbert (1986).

80 Morse (1968). See also Wyman (1984).

81 Eysenck (1971, p.130).

82 Scarr-Salapatek (1976, pp.114, 116). WASP: white anglo-saxon protestant.

83 Eysenck (1973, pp.111–12).

84 *Fortune*, October 1972, pp.132–48.

85 Edson (1969); Neary (1970); *Newsweek*, 19 March 1973; *Time*, 16 April 1973. For essays by 'hereditarians' (dedicated to Galton), Osborne *et al.* (1978); contributors include C.D. Darlington and A.R. Jensen.

86 *Fortune*, October 1972, p.142.

87 See the Eugenics Society's *Symposia*; also *Eug.Soc.Bull.* **15** (iii), p.85, 1983. The Society is now concerned principally with 'environmentalist' studies.

88 Boca & Giovana (1970, p.428). See also Billig (1978); Montagu, I. (1967); Wilkinson (1981). For American fascism, Epstein & Forster (1966) and the well documented Schoenberger (1969).

89 *Spearhead*, March 1979, pp.10–11; *New Nation*, Autumn 1980, p.18.

90 Cohn (1967, p.18).

91 For an example of the thinking of modern conservatism, Scruton (1980); for an encyclopedic description of neo-conservatism *etc.* in the USA, Peele (1984); for well documented critical analyses, Bosanquet (1983), Levitas (1986). Neo-fascism and the radical right are also represented in journals evidently addressed to academic readers (Billig, 1981). The most widely circulated, the British *Mankind Quarterly*, has had H.J. Eysenck as an adviser. It has propagated the race 'science' of the Nazis and has supported *apartheid* in South Africa. Anthropologists have condemned it as discrediting anthropology (*Current Anthropology* 3, 154–5, 1962). In Germany, *Neue Anthropologie* publishes similar matter and has A.R. Jensen among its contributors. In France, *Nouvelle École* shares members of its editorial board with *Mankind Quarterly* and has, as a leading theme, genetically determined differences between races in intelligence. This journal also provides a link with *Homo pugnax*, for K.Z. Lorenz has been a member of its editorial board and a contributor (Lorenz, 1976). The life and work of Lorenz illustrate the influence of Haeckel's social Darwinism. As a young man he was influenced by Haeckel, and he has persistently expressed anxiety about genetical degeneration supposed to result from civilization. This led him both to eugenics and to supporting Nazism (Kalikow, 1983). Correspondingly, his writings, like those of the more recent 'New Right', show no understanding of modern genetics.

92 Cowling (1978; quotations on pp. 49, 156, 183). For further examples of separateness, with criticisms, Barker (1981).

93 Scruton (1980, p.68).

94 Epistle to the Colossians, 3:11. But, until late in the eighteenth century, the dominant attitude of the Christian churches was to support slavery (Kahl, 1971).

95 Knox (1972; first published in 1558).

96 Tannahill (1980).

97 Fraser (1984).
98 Tuchman (1978, p.214).
99 Beard (1962, ch. 4) provides an extensive discussion.
100 In his *Politics* 1252 *b*.
101 Harrison (1981); Shields, S.A. (1980).
102 Harrison (1981); on brain size, see Gould, S.J. (1981), Sayers (1982, ch. 6).
103 Hobsbawm (1977b, p.295).
104 Spencer (1910). See also Carneiro (1967, especially p.x); Quinton (1966).
105 Spencer (1910).
106 In his *Principles of Biology* (1866). See also Bannister (1979, especially p.55); and Dupree (1977), who shows that 'social Darwinism' *preceded* Darwin!
107 Spencer (1866, pp.240–1).
108 Shields, S.A. (1980). See also Sayers (1982, especially ch. 1).
109 Spencer (1910).
110 See, for example, Rowbotham (1973, especially p.85); Longford (1981).
111 Tannahill (1980, p.371). See also Klein (1946) for some glimpses of history.
112 October 12 1975.
113 *She*, January 1979.
114 *Vogue*, April 1977.
115 Alper *et al.* (1978).
116 Patrick Jenkin, in a radio interview, quoted by Coote & Campbell (1982, p.87). For Patrick Buchanan, *Guardian Weekly*, May 19, 1985, p.17.
117 *Business Week*, April 10 1978.
118 University of Toronto Calendar.
119 Gastonguay (1975).
120 Judd (1978).
121 Some academics, said to be responsible for this material, have repudiated it (DeVore, 1977). Did the news of the repudiation reach the schools? In high schools in the USA, sociobiology is a regular component in the teaching of biology (Lowe, 1978).
122 Gould, S.J. (1980, p.263). See *New York Times*, November 30 1977.
123 Dinnerstein (1976, especially pp.24–5).
124 Galbraith (1975, ch. 23).
125 Mill (1970, p.22).
126 Cowling (1978, p.9).

Chapter 10 Conditioning and improvisation
1 Wittgenstein (1968, p.232).
2 Boring (1950, p.642).
3 Watson, J.B. (1913).
4 For example, Skinner (1974, p.220).
5 Blackman (1981, p.25).
6 Sechenov (1965). For historical reviews, Corson (1976); Kazdin (1978); Kimble (1967).
7 Kimble (1967, p.6).
8 Boring (1950, p.635).
9 Loeb (1964). This edition contains a valuable historical introduction by Donald Fleming.
10 Loeb (1901).
11 The standard work on animal orientations remains that of Fraenkel & Gunn (1961). See also Barnett (1981a, ch. 2).
12 Kazdin (1978).
13 Babkin (1951, p.75).
14 Pavlov (1928).
15 The traditional term 'conditioned reflex' is a mistranslation.
16 Feather (1965); Razran (1961). There is, however, some doubt on whether even such findings represent true CRs (Brewer, W.F. 1974).

17 Russell, B. (1958, p.149; 1927, pp.51–5). See also Russell's *History of Western Philosophy* (1961, p.741).

18 Needham (1946, p.116).

19 In the fourth edition of *The Physiological Basis of Medical Practice*, by C.H. Best & N.B. Taylor (London: Baillière, 1945), we read: 'In the training and education of the child conditioned reflexes also play a prominent rôle' (p.909); and on p.912 the student is told that Pavlov's findings on experimental neurosis 'have important psychiatric implications', but not what these implications are. In the third edition of *Human Physiology*, by F.R. Winton, & L.E. Bayliss (London: Churchill, 1948) seven pages are given to conditional reflexes; but at the end a single paragraph warns the reader 'not to generalise too hastily from Pavlov's results' (p.447); and exploratory behavior, and 'attention, insight and initiative' are mentioned.

20 Shaw (1944, p.202).

21 Pavlov (1927, p.395).

22 Anokhin (1968).

23 Corson (1971).

24 Dickinson (1980); Rescorla (1978).

25 Miller, N.E. (1959).

26 For example, Bridger (1967); Grings (1973).

27 Kaufmann *et al.* (1966).

28 Franks (1969, p.2)

29 Pavlov (1927, p.395).

30 Broadhurst (1973); Hebb (1947).

31 Thomas & O'Callaghan (1981, p.138).

32 See, for example, Deci (1980); Erwin E. (1978).

33 For documentation, Thomas & O'Callaghan (1981).

34 Davey (1981).

35 Elkins (1980, p.68).

36 Thomas & O'Callaghan (1981); Lick & Bootzin (1975). Contrast the account given by Blackman (1981).

37 See Dews (1981) on the absence of differences in *practise* among therapists who espouse different theories.

38 Elkins (1980).

39 Davey (1981); Deci (1980).

40 Erwin, E. (1978). See also a critical review by Grenander (1981).

41 Watson, J.B. (1913, p.158; 1930, p.2).

42 Herrnstein (1969); Woodworth (1931, pp.96–7).

43 Peterson *et al.* (1972).

44 Skinner (1938); see also Ferster & Skinner (1957); Skinner (1956).

45 Skinner (1966, p.14).

46 Barnett (1981b, p.43).

47 Darwin (1901, p.108).

48 Barnett (1981a, pp.201–22); Barnett & Cowan (1976).

49 Bowra (1959, p.173).

50 Genesis, 3:6–7.

51 Kagan (1979).

52 Watson & Ramey (1972).

53 Ross (1974).

54 Berlyne (1960); Schultz (1965); Vernon (1963).

55 Heyduk (1975).

56 Blackman (1981, p.13).

57 Skinner (1974, especially pp.189, 225).

58 Barnett (1981a, pp.216–22).

59 Barnett (1981a, pp.571–3).

60 Ganz (1968); Valvo (1971).

61 Gardner, H. (1983, ch. 8).
62 Gardner, H. (1983, p.171).
63 Gardner, H. (1983, pp.176–7).
64 Tuan (1974, p.74).
65 Gladwin (1970).
66 Blackman (1981, p.11). See also three volumes edited by Ulrich *et al.* (1966, 1970, 1974).
67 Skinner (1974, p.189).
68 Skinner (1966, p.16).
69 Skinner (1974, p.48).
70 Chomsky (1971, p.152).
71 For an explicit statement on these lines, Skinner (1953, pp.228–229).
72 Malcolm (1964, p.151).
73 Shotter (1981, p.159).
74 Malcolm (1964, p.153).
75 See, for example, Antaki (1981, chs. 3, 4).
76 Skinner (1973, p.113).
77 Skinner (1970, pp.20, 8).
78 Black (1973, p.125).
79 Ulrich *et al.* (1966, unpaginated preface).
80 Skinner (1973, p.200). For criticism of Skinner on control, Fromm (1974, ch. 2).
81 Skinner (1961, p.541).
82 Stolz (1978, p.96).
83 Fromm (1974, pp.39–40).
84 Skinner (1948).
85 Skinner (1970).
86 Skinner (1948, p.296).
87 Fromm (1974, p.41).
88 Popper (1966, vol. 1 , p.134–5). For criticism of Skinner's Utopia, Passmore (1970, chs. 8, 9).
89 Ulrich (1973, p.5).
90 Ulrich (1975, p.98).
91 Ulrich (1975, p.99).
92 Ulrich (1975, p.100). For another authentic account, see Kinkade (1973); and for other kinds of community in America, see Hall, J.R. (1978).
93 Iyer (1987, pt 7); Kanter (1973).
94 For village economics, see Swaminathan (1973).
95 Leon (1969); Rosner (1973).

Chapter 11 Work and play

1 Huizinga (1950).
2 For example, Bensusan-Butt (1978); Sen (1979).
3 Herbert (1935, p.1).
4 Sen (1979, p.12).
5 Bensusan-Butt (1978, p.5).
6 Galbraith (1975, pp.21, 28, 242). See also Routh (1980) on the failure of simple-minded 'market' theories to match (for example) the phenomenon of different wage rates for the same job.
7 Hayek (1983, pp.39, 43).
8 Simon (1983, pp.13–14).
9 McLean (1987; quotations on pp.129–30, 182).
10 Bosanquet (1983, p.84). On the limitations of computer programs, Rosznak (1986).
11 Titmuss (1970, p.198).
12 Stone, N. (1983, p.82).
13 Taylor, F.W. (1907, pp.24–5); see also Taylor (1911).

14 Stone, N. (1983, p.82).
15 Gilbreth (1973, especially pp.274–5, 283).
16 Bowlby (1969, p.178).
17 Gilbreth (1973, especially pp.318–19, 329–30).
18 For review, Ullmann (1969).
19 *Guardian Weekly*, July 24, 1983.
20 Scitovsky (1976, p.91).
21 Hobsbawm (1964, p.349).
22 Geary (1981, especially pp. 34, 72–3).
23 Sen (1979, pp.13, 14).
24 Kaupinnen-Toropainen *et al.* (1983; quotations on pp.206, 293).
25 Yankelovich *et al.* (1983, p.24).
26 Cooley (1980; quotations on pp.15–17); for comments by foreman and by computer operator, *New Society*, 7 February 1985, pp.207–8 and 7 March 1985, p.364, respectively.
27 Pugh (1977); Packard (1960, p. 196). See also Rosznak (1986).
28 Yankelovich *et al.* (1983).
29 Ingelhart (1977, especially ch. 10).
30 Kazdin (1978, especially p.242); Skinner (1968) on the technology of teaching, especially pp.32–3 on control.
31 Holland (1960).
32 Skinner (1968, p.20).
33 Bandura (1980); Deci (1975); Deci & Ryan (1980); B. Harris & Harvey (1981); Phares (1976).
34 Lepper *et al.* (1973).
35 Deci & Ryan (1980).
36 Deci (1975, p.222).
37 Deci *et al.* (1981; quotation on p.5).
38 Hebb (1955, p.246).
39 Einstein (1971, p.245).
40 Csikszentmihalyi (1975).
41 Csikszentmihalyi (1975, p.143).
42 Davidson (1976); Rosa (1976).

Part 5 *Homo sapiens*

1 Lodge (1981, p.169).

Chapter 12 The reductionist imperative

1 Krebs & Shelley (1975, p.24).
2 Clark, A. (1980, p.3).
3 For the work of the principal founder of this concept, René Descartes, J.F. Scott (1952). See also La Mettrie (1927).
4 Loeb (1964, p.33).
5 Walter (1953).
6 Young, J.Z. (1951, p.8).
7 Young, J.Z. (1978, pp.9, 142).
8 Blakemore (1977, p.184).
9 Young, J.Z. (1978, p.4).
10 Milner *et al.* (1968).
11 Sperry *et al.* (1969); for a review, Zangwill (1976).
12 Penfield (1958).
13 For an interesting discussion, Weiskrantz (1973).
14 Breggin (1975); Scheflin & Opton (1978, ch. 6).
15 Shutts (1982).

16 Freeman & Watts (1950); Shutts (1982).
17 Breggin (1975, especially p.351); Scheflin & Opton (1978, especially p.272).
18 Brady, J.V. (1958); Kling (1975).
19 Scheflin & Opton (1978, p.283).
20 Scheflin & Opton (1978, p.263).
21 It would be instructive to know how widespread is this presumption. See Barnett *et al.* (1983).
22 For example, by B. Russell (1948, pp. 38–9).
23 Lewis, R.A. (1970).
24 Rapoport (1966).
25 Grene (1969, 1974); Medawar (1974); Popper (1974).
26 Smart (1963; 1981).
27 Huby (1967).
28 Calvin (1961).
29 Malcolm (1964).
30 Hampshire (1975, p.113); see also Hollis (1977).
31 Popper (1982).
32 Popper (1982, p.127).
33 Popper (1972, ch. 6; quotation on p.207).
34 Popper (1972, p.213).
35 Popper (1972, p.222).
36 Popper (1972, pp.223–4).
37 Popper (1982, p.44).
38 Abelson (1977, p.7).
39 MacKenzie, B.D. (1977).
40 Gardner, H. (1983, p.287).
41 Ryle (1949, p.325).
42 Joynson (1974, pp.1–2).
43 Hargreaves (1980, pp.220–1).
44 Yourcenar (1959, p.41).
45 Schimmel (1979; the quotation below is from p.330). The importance of Schimmel's argument is brought out by J. Scott (1986) in a study of English medical students, many of whom evidently think poorly of psychiatry because it seems to lack a scientific foundation.
46 For the original works, begin with Freud (1933, 1949b). For perceptive analysis, Fromm (1959, 1970). For the idea of the unconscious, a poorly documented account by MacIntyre (1958). For a hatchet job on psychoanalysis as a science, Cioffi (1970). See also Chapter 5, note 12.
47 R.D. Laing (personal communication).
48 Freud (1933, pp.37–8).
49 Gardner, H. (1983, p.239).
50 In philosophy, for instance, a long-standing debate is between dualism, which states that there are two modes of being, the material and the mental, and monism, which states that there is only one. (This one may be material, *or* mental. When it is held to be mental, we have idealism.) K.R. Popper, in a well known proposal, describes *three* modes of being ('worlds'). The first, the material world, consists of observable objects – the world of the natural sciences. The second consists of 'mental states' – thoughts, memories, feelings, intentions, beliefs; we know these directly only in ourselves, but infer them in others. The third contains 'ideas in the objective sense', or objective knowledge; this includes the principles of logic and mathematics and the theories of science, also arguments. The third 'world' allows for truths which exist whether they have been thought of or not (Popper, 1972). This system has been much debated (Lakatos & Musgrave, 1970). I suggest that it is useful to regard it as a way of *classifying* kinds of knowledge. Compare Russell on dualism: 'I think that both mind and matter are merely convenient ways of grouping events' (1961, p.787).
51 Thoresen *et al.* (1981).
52 Lewis, C.S. (1943, p.209).

53 Shotter (1980, p.20). See also Gauld & Shotter (1977).
54 Arieti (1972).
55 Shotter (1980, p.21).
56 Popper (1974, p.284). See also Nagel, E. (1961).

Chapter 13 Human communication

1 Russell, B. (1962, p.21).
2 Ekman & Oster (1979).
3 Field T.M. *et al.* (1982).
4 Ekman & Oster (1979); Shimoda *et al.* (1978); Speer, D.C. (1972).
5 Ekman (1977) in fact suggests three. The third (called 'illustrators') includes scratching and licking lips.
6 Farb (1973, p.190).
7 For example, by Hewes (1973, 1977).
8 Farb (1973, pp.18–19).
9 Menzel (1971); Menzel & Halperin (1975).
10 Savage-Rumbaugh *et al.* (1978).
11 Lieberman *et al.* (1969).
12 Gardner, B.T. & Gardner, R.A. (1971).
13 Ristau & Robbins (1982); Umiker-Sebeok & Sebeok (1981).
14 Terrace (1979; quotation from Nim, p.210).
15 Ristau & Robbins (1982, p.177).
16 For a review, Passingham (1981).
17 Chomsky (1967, p.81).
18 Chomsky (1959, 1967, 1972); Hockett & Altmann (1968).
19 Chomsky (1967, p.81).
20 Johnson-Laird (1983; quotation on p.205).
21 Deighton (1966, p.76).
22 Passmore (1970, pp.272–3).
23 *Out of sight, out of mind.*
24 O'Donnell (1982, p.13).
25 Dreyfus (1972) – a philosopher's critique. For a recent summing up, Waldrop (1984).
26 Johnson (1968, p.309).
27 Kipling (1898, pp.173–4).
28 Popper (1972, pp.120–1).
29 Ryan (1974).
30 Bruner (1975); Church (1971).
31 Miller, G.A. (1981, p.119).
32 Miller, G.A. (1981, especially pp.118–19, 110).
33 Bruner (1975, pp.258–9).
34 Chomsky (1972, p.12).
35 Thorpe *et al.* (1972). See also Barnett (1981a, pp.384–95).
36 For a review, Winner (1982, chs 6–8).
37 Blacking (1973; quotation, p.54). See also Merriam (1964).
38 McPhee (1955).
39 Koranda (1980).
40 Chang & Trehub (1977).
41 Kessen *et al.* (1979).
42 For a documented discussion, Winner (1982, especially pp.210–15).
43 *Poetics*, 6.1449*b*.

Chapter 14 Teaching and tradition

1 For original sources not given in this chapter, Barnett (1973, 1977, 1981a).
2 Marrou (1964, pp.142–3).
3 Smalley (1974, p.16). And, on fifteenth-century France, Ariès (1962).

4   Passmore (1980, ch. 3. For definitions, ch. 2). My definition, *for the purposes of the present book*, is narrower than one previously used.
5   Lawick-Goodall (1973).
6   Stephens (1963).
7   Weisner & Gallimore (1977).
8   Rogoff *et al.* (1975).
9   Fortes (1970a, p.25).
10 Piaget (1932).
11 Opie & Opie (1959).
12 Blurton Jones (1972); Hartup & Keller (1960); Maccoby & Jacklin (1974); Mussen & Eisenberg-Berg (1977).
13 See also Chittenden (1942); Meister (1956); Pitkänen (1974).
14 For a review, Clark, J.D. (1976).
15 Gombrich (1966, pp.41–2, 33).
16 Gombrich (1966, p.53).
17 Gombrich (1966, p.30).
18 Clark, K. (1973, ch. 11; 1976, p.147).
19 Clark, K. (1969, p.289); for discussion, Clark, K. (1976, ch. 5), Gombrich (1966, ch. 25).
20 Chen *et al.* (1982).
21 Richerson & Boyd (1978).

Chapter 15 The question

1   Stone, N. (1983, p.15).
2   Popper (1970, p.57).
3   C. Darwin (1872, penultimate paragraph, unchanged from the first edition of 1859).
4   Passmore (1970, especially ch. 8); Sampson, R.V. (1956, especially ch. 3).
5   Passmore (1970, p.263). For a practical example of moderate optimism, W.H.O. (1982) – a program of health for all by the year 2000.
6   Jeremiah: 2:7. See also Gibbon (1976, p.456).
7   Davis (1969, ch. 6; quotation, p.175). See also Williams, E. (1944).
8   Anstey (1975). See also Davis (1984).
9   Furneaux (1974).
10 Hobsbawm (1977b, p.313).
11 Ali (1949, p.111). Note for social historians: this author, a Muslim, was a Commander of the Indian Empire and a member of His [British] Majesty's Judicial Committee of the Privy Council.
12 Westermarck (1917, pp.743–4).
13 UNESCO (1952).
14 Quoted by Hutt (1941, p.19). For the condition of the poor in Western Europe, S. Woolf (1986).
15 For introductions, Caute (1966, ch. 9); Hobsbawm (1977b, ch. 6); Stone, N. (1983).
16 Engels (1892); Hammond & Hammond (1925, 1936); Kuttner (1984, p.140); Piven & Cloward (1979); Torr (1948); Ware (1935).
17 Amato (1987) gives the Australian findings and reviews the evidence.
18 Stone, L. (1975, p.36). The history of the family, and attitudes to children, in Europe, are also reviewed.
19 Stone, L. (1975, p.40).
20 Ariès (1962). See also DeMause (1976).
21 For an introduction, Moore (1967); see also Macpherson (1962).
22 See, for example, Galbraith (1974, 1975), especially for the situation in the USA.
23 Elster (1985, p.521); see also Easton & Guddat (1967).
24 Hobbes (1962, p.59).
25 Keynes (1933, p.369).
26 Galbraith (1975, pp.319–20).
27 Galbraith (1975, p.297).

28 Galbraith (1974, p.194).
29 Eisenhower (1961, p.616).
30 For example, Noel-Baker (1936); Thayer (1970).
31 Quoted by A. Sampson (1978, p.327).
32 Cochran *et al.* (1983); Independent Commission (1982).
33 Sampson, A. (1974, pp.188–9); see also Sampson, A. (1980), Box, S. (1983).
34 On the food industry, Turner (1970); on the pharmaceutical industry, Braithwaite (1984). For the morals of violence by neglect, for instance of safety precautions. Harris, J. (1980).
35 Mintz (1985).
36 Taylor, P. (1984); British Medical Association (1986). On the third world, Nath (1987).
37 Box, S. (1983, p.57).
38 Galbraith (1975, p.259–60).
39 Schuettinger (1976, p.6).
40 Bury (1913, p.14).
41 Kozol (1985).
42 Postman (1985). Postman's account deals only with the USA, but countries in which television and its advertisements are less pervasive can also benefit from his warnings. The criticism is of television not as a source of entertainment but as a medium that smothers democracy and distorts education.
43 Smith (1987). For unemployment and death rates, Occupational Mortality: the Registrar General's Decennial Supplement for Great Britain, 1979–80, 1982–83 London: HMSO (1986). For unemployment and mental ill health, Layton (1986); Warr (1987).
44 See Sampson, R.V. (1956, ch. 6).
45 The title of a book by Schumacher (1973).
46 Skocpol (1979, especially ch. 7).
47 For the moral foundations of scientific practice, Edge (1974).
48 Russell, B. (1930, p.149).

References

Some of the titles of papers have been shortened.

Abel, T. 1941. The element of decision in the pattern of war. *Am. Soc. Rev.* 6, 853–9.

Abelson, R. 1977. *Persons*. London: Macmillan.

Adami, G. 1922. The true aristocracy. *Eug.Rev.* **14**, 174–86.

Adorno, T.W. *et al.* 1950. *The Authoritarian Personality*. New York: Harper.

Alberts, B. *et al.* 1983. *Molecular Biology of the Cell*. New York: Garland.

Alexander, R.D. 1975. The search for a general theory of behavior. *Behav. Sci.* **20**, 77–100.

Alexander, R.D. 1979. *Darwinism and Human Affairs*. Seattle: University of Washington Press.

Ali, A. 1949. *A Short History of the Saracens*. London: Macmillan.

Allen, G.E. 1975. Genetics, eugenics and class struggle. *Genetics* 79, 29–45.

Allison, A.C. 1969. Natural selection and population diversity. *J. Biosoc. Sci.*, Suppl. **1**, 15–30.

Alper, J., Beckwith, J. & Miller L.G. 1978. Sociobiology is a political issue. In: *The Sociobiology Debate*, ed. A.L. Caplan. New York: Harper & Row.

Amato, P.R. 1987. *Children in Australian Families*. Brookvale, NSW: Prentice-Hall.

Angst, W. & Thommen, D. 1977. Infant killing in monkeys and apes. *Folia Primatol.* **27**, 198–229.

Anokhin, P.K. 1968. Pavlov and psychology. In: *Historical Roots of Contemporary Psychology*, ed. B.B. Wolman. New York: Harper & Row.

Anstey, R. 1975. *The Atlantic Slave Trade*. London: Macmillan.

Antaki, C. (ed.) 1981. *The Psychology of Ordinary Explanations of Social Behaviour*. London: Academic.

Arblaster, A. 1984. *The Rise and Decline of Western Liberalism*. Oxford: Blackwell.

Ardrey, R. 1967. *The Territorial Imperative*. London: Collins.

Ardrey, R. 1969. *African Genesis*. London: Fontana.

Ardrey, R. 1976. *The Hunting Hypothesis*. New York: Atheneum.

Arendt, H. 1963. *Eichman in Jerusalem*. London: Faber.

Ariès, P. 1962. *Centuries of Childhood*. London: Cape.

Arieti, S. 1972. *The Will to be Human*. New York: Quadrangle.

Augustinus 1945. *The City of God* 2 vols. London: Dent. (First published in AD 426.)

Babkin, B.P. 1951. *Pavlov*. London: Gollancz.

Bacon, F. 1906. *Essays*. London: Dent. (First published in 1597.)

Baldwin, J.D. & Baldwin, J.I. 1980. Sociobiology or balanced biosocial theory? *Pacific Sociol. Rev.* **23**, 3–27.

Bamford, V.J. 1982. *Puzzle Palace.* Boston: Houghton Mifflin.

Bandura, A. 1969. *Principles of Behavior Modification.* New York: Holt, Rinehart & Winston.

Bandura, A. 1973. *Aggression: A Social Learning Analysis.* Englewood Cliffs, NJ: Prentice-Hall.

Bandura, A. 1980. The self and mechanisms of agency. In: *Social Psychological Perspectives on the Self*, ed. J. Suls. Hillsdale, NJ: Erlbaum.

Bannister, R.C. 1979. *Social Darwinism.* Philadelphia: Temple University Press.

Banton, M. 1987. *Racial Theories.* Cambridge: Cambridge University Press.

Barash, D.P. 1980. *The Whisperings Within.* London: Souvenir Press.

Barker, M. 1981. *The New Racism.* London: Junction Books.

Barnett, S.A. 1973. *Homo docens. J. Biosoc. Sci.* **5**, 393–403.

Barnett, S.A. 1977. The instinct to teach. Aggr. Behav. **3**, 209–29.

Barnett, S.A. 1979. Cooperation, conflict, crowding and stress. *Interdisc. Sci. Rev.* **4**, 106–31.

Barnett, S.A. 1981a. *Modern Ethology.* New York: Oxford University Press.

Barnett, S.A. 1981b. *The Rat: A Study in Behavior.* Chicago: Chicago University Press.

Barnett, S.A. 1988. Enigmatic death due to 'social stress'. *Interdisc. Sci. Rev.* **13**, 14–51.

Barnett, S.A. & Brown, V.A. 1981. Pull and push in educational innovation. *Stud. Higher Educ.* **6**, 13–22.

Barnett, S.A. & Brown, V.A. 1983. The education of biologists and the hazards of specialism. *Interdisc. Sci. Rev.* **8**, 294–6.

Barnett, S.A., Brown, V.A. & Caton, H. 1983. The theory of biology and the education of biologists. *Stud. Higher Educ.* **8**, 23–32.

Barnett, S.A. & Cowan, P.E. 1976. Activity, exploration, curiosity and fear. *Interdisc. Sci. Rev.* **1**, 43–62.

Barnett S.A. & Dickson, R.G. 1984. Milk production and consumption and growth of young of wild house mice in a cold environment. *J. Physiol.* **346**: 409–417.

Bateson, G. 1973. *Steps to an Ecology of Mind.* London: Paladin.

Beard, M.R. 1962. *Woman as a Force in History.* New York: Collier Books.

Beer, C.G. 1972. Individual recognition of voice in birds. *Proc. XV Intern. Ornithol. Congr.*, 339–56.

Beer, C.G. 1974. Comparative ethology and the evolution of behaviour. In: *Ethology and Psychiatry*, ed. N.F. White. University of Toronto.

Ben Shaul, D.M. 1962. The composition of the milk of wild animals. In: *Inter Zoo Year Book*, vol. 4, ed. C. Jarvis & D. Morris. London: Hutchinson.

Bensusan-Butt, D.M. 1978. *On Economic Man.* Canberra: Australian National University Press.

Berggren, D. 1962. The use and abuse of metaphor I. *Rev. Metaphysics* **16**, 237–58.

Berggren, D. 1963. The use and abuse of metaphor II. *Rev. Metaphysics* **16**, 450–72.

Berkowitz, L. 1981. The concept of aggression. In: *Multidisciplinary Approaches to Aggression Research*, ed. P.F. Brain & D. Benton, Amsterdam: Elsevier.

Berlyne, D.E. 1960. *Conflict, Arousal, and Curiosity.* New York: McGraw-Hill.

Bernal, J.F. & Richards, M.P.M. 1973. What can the zoologists tell us about human development? In: *Ethology and Development*, ed. S.A. Barnett. London: Heinemann Medical.

Bertram, B.C.R. 1975. Reproduction in wild lions. *J. Zool., Lond.* **177**, 463–82.

Bertram, B.C.R. 1978, Living in groups: predators and prey. *In: Behavioural Ecology: An Evolutionary Approach*, ed. J.R. Krebs & N.B. Davies. Oxford: Blackwell Scientific.

Billig, M. 1978. *Fascists*. London: Academic.

Billig, M. 1981. *L'Internationale Raciste*. Paris: Maspero.

Billig, M. 1982. *Ideology and Social Psychology*. Oxford: Blackwell.

Birch, L.C. & Ehrlich, P.R. 1967. Evolutionary history and population biology. *Nature, Lond.* **214**, 349–52.

Birnbach, M. 1962. *Neo-Freudian Social Philosophy*. London: Oxford University Press.

Bishop, J.A. 1981. A neodarwinian approach to resistance. In: *Genetic Consequences of Man-Made Change*, ed. J.A. Bishop & L.M. Cook. London: Academic.

Black, M. 1973. Some aversive responses to a would-be reinforcer. In: *Beyond the Punitive Society*, ed. H. Wheeler. London: Wildwood House.

Blackburn, J. 1979. *The White Men*. London: Orbis.

Blacker, C.P. 1952. *Eugenics: Galton and After*. London: Duckworth.

Blacking, J. 1973. *How Musical is Man?* Seattle: University of Washington Press.

Blackman, D. 1981. The experimental analysis of behaviour. In: *Applications of Conditioning Theory*, ed. G. Davey. London: Methuen.

Blakemore, C. 1977. *Mechanics of the Mind*. Cambridge: Cambridge University Press.

Block. N.J. & Dworkin, G. (eds.) 1976. *The IQ Controversy*. New York: Pantheon.

Blurton Jones, N. 1972. Comparative aspects of mother-child contact. In: *Ethological Studies of Child Behaviour*, ed. N. Blurton Jones. Cambridge: Cambridge University Press.

Boas, F. 1911. *The Mind of Primitive Man*. New York: Macmillan.

Boas, F. 1940. *Race, Language and Culture*. New York: Macmillan.

Boca, A. del & Giovana, M. 1970. *Fascism Today*. London: Heinemann.

Bock, K.E. 1980. *Human Nature and History*. New York: Columbia University Press.

Bodmer, W.F. & Cavalli-Sforza, L.L. 1976. *Genetics, Evolution, and Man*. San Francisco: Freeman.

Bohannan, P. 1967. Patterns of homicide among tribal societies in Africa. In: *Studies in Homicide*, ed. M. Wolfgang. New York: Harper.

Bohm, D. 1969. Some remarks on the notion of order. In: *Towards a Theoretical Biology 2. Sketches*, ed. C.H. Waddington. Edinburgh: Edinburgh University Press.

Bok, S. 1984. *Secrets*. Oxford: Oxford University Press.

Bonger, W.A. 1943. *Race and crime*. London: Oxford University Press.

Boring, E.G. 1950. *A History of Experimental Psychology*. New York: Appleton-Century-Crofts.

Bosanquet, N. 1983. *After the New Right*. London: Heinemann.

Boutruche, R. 1970. *Seigneurie et Féodalité*. Paris: Aubier.

Bowlby, J. 1969. *Attachment and Loss*, vol. 1. London: Hogarth.

Bowles, S. & Gintis, H. 1973. *Social Policy*, November 1972–February 1973, pp. 65–96.

Bowles, S. & Nelson, V.I. 1974. The 'inheritance of IQ'. *Rev. Econ. Stats.* **56**, 39–51.

Bowra, C.M. 1959. *The Greek Experience*. London: Weidenfeld & Nicolson.

Box, J.F. 1978. *R.A. Fisher: The Life of a Scientist*. New York: Wiley.

Box, S. 1983. *Power, Crime, and Mystification*. London: Tavistock.

Boyd R. & Richerson P.J. 1980. Sociobiology, culture and economic theory. *J. Econ. Behav. Organization* **1**, 97–121.

Boyden, A. 1973. *Perspectives in Zoology*. Oxford: Pergamon.

Brady, I. (ed.) 1976. *Transactions in Kinship*. Honolulu: University Press of Hawaii.

Brady, J.V. 1958. The paleocortex and behavioral motivation. In: *Biological and Biochemical Bases of Behavior*, ed. H.F. Harlow & C.N. Woolsey. Madison: University of Wisconsin Press.

Brady, R.H. 1979. Natural selection and the criteria by which a theory is judged. *Syst. Zool.* **28**, 600–21.

Brain, C.K. 1981. *The Hunters or the Hunted*. Chicago: Chicago University Press.

Braithwaite, J. 1984. *Corporate Crime in the Pharmaceutical Industry*. London: Routledge.

Breggin, P.R. 1975. Psychosurgery for the control of violence. In: *Neural Bases of Violence and Aggression*, ed. W.S. Fields & W.H. Sweet, St Louis, Miss.: Green.

Bremer, S., Singer, J.D. & Luterbacher, U. 1973. The population density and war proneness of European nations, 1816–1965. *Comp. Pol. Stud.* **6**, 329–48.

Brewer, G.J. 1979. Treatment of sickle cell anemia. *Persp. Biol. Med.* **22**, 250–72.

Brewer, W.F. 1974. No convincing evidence for operant or classical conditioning in adult humans. In: *Cognition and the Symbolic Processes*, ed. W.B. Weimer & D.S. Palermo. Hillsdale, NJ: Erlbaum.

Bridger, W.H. 1967. Contributions of conditioning principles to psychiatry. In: *Foundations of Conditioning and Learning*, ed. G.A. Kimble. New York: Appleton-Century-Crofts.

Brim, O.G. & Kagan, J. 1980. *Constancy and Change in Human Development*. Cambridge, MA: Harvard University Press.

British Medical Association. 1986. *Smoking Out the Barons*. Chichester: Wiley.

Broadhurst, P.L. 1973. Animal studies bearing on abnormal behavior. In: *Handbook of Abnormal Psychology*, ed. H.J. Eysenck. London: Pitman.

Bruner, J.S. 1969. On voluntary action and its hierarchical structure. In: *Beyond Reductionism*, ed. A. Koestler & J.R. Smythies. London: Hutchinson.

Bruner, J.S. 1975. From communication to language. *Cognition* **3**, 255–87.

Buchan, J. 1921. *A History of the Great War*, vol. 1. London: Nelson.

Bullock, A. 1962. *Hitler: A Study in Tyranny*. Harmondsworth: Penguin.

Bulmer, R.N.H. 1978. Totems and taxonomy. In: *Australian Aboriginal Concepts*, ed. L.R. Hiatt. New Jersey: Humanities Press.

Burnham, D. 1980. *The Rise of the Computer State*. New York: Random House.

Bury, J.B. 1913. *A History of Freedom of Thought*. New York: Holt.

Bury, J.B. 1932. *The Idea of progress*. New York: Macmillan.

Calvin, J. 1961. *The Institutes of the Christian Religion*. London: SCM Press. (First published in 1536.)

Cant, J.G.H. 1981. The evolution of human breasts and buttocks. *Am. Nat.* **117**, 199–204.

Caplan, A.L. 1979. Sociobiology, human nature, and psychological egoism. *J. Soc. Biol. Struct.* **2**, 27–38.

Carcopino, J. 1956. *Daily Life in Ancient Rome*, Harmondsworth: Penguin.

Carneiro, R.L. 1967. In: *The Evolution of Society, H. Spencer*. Chicago: Chicago University Press.

Carr, E.H. 1981. *The Twenty Years' Crisis 1919–1939*. London: Macmillan.

Caute, D. 1966. *The Left in Europe*. London: Weidenfeld & Nicolson.

Chagnon, N.A. 1968. *Yanomamö: the Fierce People*. New York: Holt, Rinehart & Winston.

Chang, H.W. & Trehub, S.E. 1977. Auditory processing by young infants. *J. Exp. Child Psychol.* **24**, 324–31.

Chen, K.H., Cavalli-Sforza, L.L. & Feldman, M.W. 1982. Cultural transmission in Taiwan. *Hum. Ecol.* **10**, 365–82.

Cheyne, J.A. 1978. Communication, affect, and social behavior. In: *Aggression, Dominance and Individual Spacing*, ed. L. Krames, P. Pliner & T. Alloway. New York: Plenum Press.

Chittenden, G.E. 1942. Measuring and modifying assertive behavior in young children. *Monog. Soc. Res. Child Dev.* **7** (1, Serial No. 31).

Chomsky, N. 1959. Verbal behavior. *Language,* 35, 26–58.

Chomsky, N. 1967. The general properties of language. In: *Brain Mechanisms Underlying Speech and Language,* ed. F.L. Darley. New York: Grune and Stratton.

Chomsky, N. 1971. The case against B.F. Skinner. *New York Review of Books,* 18–24.

Chomsky, N. 1972. *Language and Mind.* New York: Harcourt Brace Jovanovich.

Chomsky, N. 1976. The fallacy of Richard Herrnstein's IQ. In: *The IQ Controversy,* ed. N.J. Block & G. Dworkin. New York: Pantheon.

Church, J. 1971. The ontogeny of language. In: *The Ontogeny of Vertebrate Behavior,* ed. H. Moltz. New York: Academic.

Cioffi, F. 1970. Freud and the idea of a pseudo-science. In: Explanation in the Behavioural Sciences, ed. R. Borger & F. Cioffi. Cambridge: Cambridge University Press.

Clark, A. 1980. *Psychological Models and Neural Mechanisms.* Oxford: Clarendon.

Clark, C.M.H. 1978. *A History of Australia,* vol. 4. Melbourne: Melbourne University Press.

Clark, J.D. 1976. African origins of man. In: *Human Origins,* ed. G.L. Isaac & E.R. McCown. New York: Benjamin.

Clark, K. 1969. *Civilization.* London: Murray.

Clark, K. 1973. *The Romantic Rebellion.* London: Murray.

Clark, K. 1976. *Landscape into Art.* London: Murray.

Clinard, M.B. 1964. *Anomie and Deviant Behavior.* London: Collier-Macmillan.

Cochran, T.B., Arkin W.M. & Hoenig, M.M. 1983. *The Nuclear Weapons Data Book.* Cambridge, MA: Ballinger.

Coelho, G.V. & Stein, J.J. 1980. Change, vulnerability, and coping. In: *Uprooting and Development,* ed. G.V. Coelho & P.I. Ahmed. New York: Plenum.

Cohn, N. 1967. *Warrant for Genocide.* London: Eyre & Spottiswoode.

Cole, G.D.H. 1923. *Social Theory.* London: Methuen.

Cooley, M.J.E. 1980. *Architect or Bee?* London: Langley Technical Services.

Coote, A. & Campbell, B. 1982. *Sweet Freedom.* London: Pan Books.

Corson, S.A. 1971. Conditioning techniques in the study of psychosocial relationships. In: *Society, Stress and Disease,* vol. 1, ed. L. Levi. London: Oxford.

Corson, S.A. (ed.) 1976. *Psychiatry and Psychology in the USSR.* New York: Plenum.

Cowling, M. 1978. *Conservative Essays.* London: Cassell.

Cox, C.B. & Dyson, A.E. 1971. *The Black Papers on Education.* London: Davis-Poynter.

Crow, T.J. 1983. Is schizophrenia an infectious disease? *Lancet,* **1983** (i), 173–5.

Cruttwell, C.R.M.F. 1982. *A History of the Great War 1914–1918.* London: Granada. (First published in 1934.)

Csikszentmihalyi, M. 1975. *Beyond Boredom and Anxiety.* San Francisco: Jossey-Bass.

Cuffey, R.J. 1984. Paleontologic evidence and organic evolution. In: *Science and Creationism,* ed. A. Montagu. Oxford: Oxford University Press.

Curley, M.J. 1979. *Physiologus.* Austin: University of Texas Press.

Daly, M. 1980. Contentious genes. *Soc. Biol. Struct.* 3, 77–81.

Darlington, C.D. 1958. *The Evolution of Genetic Systems.* Edinburgh: Oliver & Boyd.

Darlington, C.D. 1969. *The Evolution of Man and Society.* London: Allen & Unwin.

Dart, R.A. 1953. The predatory transition from ape to man. *Int. Anthrop. Ling. Rev.* **1**, 201–18.

Dart, R.A. 1959. *Adventures with the Missing Link.* New York: Harper.

Darwin, C. 1872. *On the Origin of Species.* London: Murray.

Darwin, C. 1901. *The Descent of Man and Selection in Relation to Sex.* London: Murray.

Darwin, L. 1926. *The Need for Eugenic Reform.* London: Murray.

Davey, G. 1981. Conditioning and behaviour therapy. In: *Applications of Conditioning Theory*, ed. G. Davey. London: Methuen.

Davidson, J.M. 1976. The physiology of meditation. *Persp. Biol. Med.* **19**, 345–79.

Davis, D.B. 1969. *The Problem of Slavery in Western Culture*. Ithaca, NY: Cornell University Press.

Davis, D.B. 1984. *Slavery and Human Progress*. New York: Oxford University Press.

Dawkins, R. 1976. *The Selfish Gene*. London: Oxford University Press.

Dawkins, R. 1982. *The Extended Phenotype*. Oxford: Freeman.

Deci, E.L. 1975. *Intrinsic Motivation*. New York: Plenum.

Deci, E.L. 1980. *The Psychology of Self-Determination*. Lexington, MA: Lexington.

Deci, E.L., Nezlek, J. & Sheinman, L. 1981. The rewarder and intrinsic motivation. *J. Personal. Soc. Psychol.* **40**, 1–10.

Deci, E.L. & Ryan, R.M. 1980. Intrinsic motivational processes. *Adv. Exp. Soc. Psychol.* **13**, 39–80.

Deem, R. 1978. *Women and Schooling*. London: Routledge.

Deighton, L. 1966. *Funeral in Berlin*. Harmondsworth: Penguin.

Delgado, J.M.R. 1969. *Physical Control of the Mind: Toward a Psychocivilized Society*. New York: Harper & Row.

DeMause, W.L. 1976. *The History of Childhood*. London: Souvenir Press.

Dentan, R.K. 1968. *The Semai: A Nonviolent People of Malaya*. New York: Holt, Rinehart and Winston.

DeVore, I. 1977. DeVore explains sociobiology film interviews. *Anthropol. Newsletter,* **18**, 2, 14.

Dews, P.B. 1981. Pavlov and psychiatry. *J. Hist. Behav. Sci.* **17**, 246–50.

Dewsbury, D.A. 1982. Dominance rank, copulatory behavior, and differential reproduction. *Q. Rev. Biol.* **57**, 135–59.

Dickeman, M. 1975. Demographic consequences of infanticide. *Ann. Rev. Ecol. Syst.* **6**, 15–37.

Dickinson, A. 1980. *Contemporary Animal Learning Theory*. Cambridge: Cambridge University Press.

Dicks, H.V. 1972. *Licensed Mass Murder*. London: Heinemann.

Dinnerstein, D. 1976. *The Rocking of the Cradle and the Ruling of the World*. New York: Harper & Row.

Dobzhansky, T. *et al.* 1977. *Evolution*. San Francisco: Freeman.

Dolhinow, P. 1977. Normal monkeys? *Am. Sci.* **65**, 266–7.

Donner, F.J. 1980. *The Age of Surveillance*. New York: Knopf.

Douglas, M. 1963. *The Lele of the Kasai*. London: Oxford University Press.

Draper, P. 1973. Crowding among hunter-gatherers. *Science, NY* **182**, 301–3.

Dreyfus, H.L. 1972. *What Computers Can't Do*. New York: Harper & Row.

Dubos, R. & Dubos, J. 1953. *The White Plague*. London: Gollancz.

Dugdale, R.L. 1910. *The Jukes*. New York: Putnam.

Dunbar, R.I.M. 1982. Adaptation, fitness and the evolutionary tautology. In: *Current Problems in Sociobiology*, ed. King's College Sociobiology Group. Cambridge: Cambridge University Press.

Duncan, R.L. 1982. *Brimstone*. London: Sphere.

Dupree, A.H. 1977. Sociobiology and the natural selection of scientific disciplines. *Minerva,* **15**: 94–101.

Durant, J.R. 1981. The beast in man. In: *The Biology of Aggression*, ed. P.F. Brain & D. Benton. Rockville, MD: Sijthoff & Noordhoff.

Durham, W.H. 1976. A review of primitive war. *Q. Rev. Biol.* **51**, 385–415.

Easton, L.D. & Guddat, K.H. (ed.) 1967. *Writings of the Young Marx*. New York: Doubleday.

Eaves, L.J. & Jinks, J.L. 1972. Insignificance of evidence for differences in heritability of IQ. *Nature, Lond.* **240**, 84–8.

Edge, D.O. 1973. Technological metaphor. In: *Meaning and Control*, ed. D.O. Edge & J.N. Wolfe. London: Tavistock.

Edge, D.O. 1974. Technological metaphor and social control. *New Literary History* **6**, 135–47.

Edson, L. 1969. Jensenism. *New York Times Magazine*, Aug. 31, pp. 10–46.

Eibl-Eibesfeldt, I. 1970. *Ethology*. New York: Holt, Rinehart & Winston.

Eibl-Eibesfeldt, I. 1974. Determinants of aggressive behavior in man. In: *Determinants and Origins of Aggressive Behavior*, ed. J. de Wit & W.W. Hartup. The Hague: Mouton.

Eibl-Eibelsfeldt, I. 1979. *The Biology of Peace and War*. London: Thames & Hudson.

Einstein, A. 1971. *Mozart*. London: Panther.

Eisenberg, L. 1972. The *human* nature of human nature. *Science, NY* **176**, 123–8.

Eisenhower, D.D. 1965. *Waging Peace*. New York: Doubleday.

Ekman, P. 1977. Biological and cultural contributions to body and facial movement. In: *The Anthropology of the Body*, ed. J. Blacking. London: Academic.

Ekman, P. & Oster, H. 1979. Facial expressions of emotion. *Ann. Rev. Psychol.* **30**, 527–54.

Elkins, R.L. 1980. Covert sensitization. *Addictive Behaviors* **5**, 67–89.

Elster, J. 1985. *Making Sense of Marx*. Cambridge: Cambridge University Press.

Engels, F. 1892. *The Condition of the Working Class in England in 1844*. London: Allen & Unwin.

Epstein, B.R. & Forster, A. 1966. *The Radical Right*. New York: Random House.

Erwin, E. 1978. *Behavior Therapy*. Cambridge: Cambridge University Press.

Erwin, T.L. 1982. Tropical forests: their richness in Coleoptera. *Coleopterists Bull.* **36**, 74–5.

Essock-Vitale, S.M. & McGuire, M.T. 1980. Theories of kin selection and reciprocation. *Ethol. Sociobiol.* **1**, 233–43.

Estiokio-Griffin, A. & Griffin, P.B. 1981. Woman the hunter: the Agta. In: *Woman the Gatherer*, ed. F. Dahlberg. New Haven: Yale University Press.

Etkin, W. 1981. A biological critique of sociobiological theory. In: *Sociobiology and Human Politics*, ed. E. White. Lexington, MA: Heath.

Evans-Pritchard, E.E. 1956. *Nuer Religion*. Oxford: Clarendon Press.

Eysenck, H.J. 1971. *Race, Intelligence and Education*. London: Temple Smith.

Eysenck, H.J. 1973. *The Inequality of Man*. London: Temple Smith.

Factor, R.M. & Waldron, I. 1973. Population densities and human health. *Nature, Lond.* **243**, 381–4.

Falconer, D.S. 1960. *Introduction to Quantitative Genetics*. Edinburgh: Oliver & Boyd.

Falconer, D.S. 1973. Selection for body weight in mice. *Genet. Res., Camb.* **22**, 291–321.

Farb, P. 1973. *Word Play*. London: Hodder & Stoughton.

Feather, B.W. 1965. Semantic generalization of classically conditioned responses. *Psychol. Bull.* **63**, 425–41.

Feldman, M.W. & Lewontin, R.C. 1975. The heritability hang-up. *Science, NY* **190**, 1163–8.

Ferri, E. 1906. *Socialism and Positive Science*. London: Independent Labour Party.

Ferster, C.B. & Skinner, B.F. 1957. *Schedules of Reinforcement*. New York: Appleton-Century-Crofts.

Festinger, L. 1983. *The Human Legacy*. New York: Columbia University Press.

Field, G.C. 1949. *The Philosophy of Plato*. London: Oxford University Press.

Field, T.M. *et al.* 1982. Imitation of facial expressions by neonates. *Science, NY* 218, 179–81.

Finley, M.I. 1973. *The Ancient Economy*. Berkeley: University of California Press.

Fisher, E. 1979. *Woman's Creation*. New York: Doubleday.

Fisher, R.A. 1918. The correlation between relatives on the supposition of Mendelian inheritance. *Trans. R. Soc. Edin.* **52**, 399–433.

Fisher, R.A. 1930. *The Genetical Theory of Natural Selection*. Oxford: Clarendon Press.

Fisher, R.A. 1936. The measurement of selective intensity. *Proc. R. Soc. B* **121**, 58–62.

Fitch, W.M. 1982. The challenges to Darwinism. *Evolution* 36, 1133–43.

Flew, A. 1974. Evolutionary ethics. In: *New Studies in Ethics*, vol. 2, ed. W.D. Hudson. London: Macmillan.

Flynn, J.R. 1984. The mean IQ of Americans: massive gains 1932 to 1978. *Psychol. Bull.* **95**, 29–51.

Ford, D. 1985. *The Button: the Pentagon's Strategic Command and Control System*. New York: Simon & Schuster.

Fortes, M. 1970a. In: *From Child to Adult*, ed. J. Middleton. New York: Natural History Press. *Education in Taleland*.

Fortes, M. 1970b. *Kinship and the Social Order*. London: Routledge.

Fox, R.L. 1973. *Alexander the Great*. Harmondsworth: Allen Lane.

Fraenkel, G.S. & Gunn, D.L. 1961. *The Orientation of Animals*. New York: Dover.

Franks, C.M. (ed.) 1969. *Behavior Therapy: Appraisal and Status*. New York: McGraw-Hill.

Fraser, A. 1984. *The Weaker Vessel*. New York: Knopf.

Freedman, J.L. 1975. *Crowding and Behavior*. New York: Viking.

Freedman, J.L. 1978. The effects of high density on human behavior and emotions. In: *Aggression, Dominance, and Individual Spacing*, ed. L. Krames, P. Pliner & T. Alloway. New York: Plenum.

Freeman, W.J. & Watts, J. 1950. *Psychosurgery*. Oxford: Blackwell.

Freud, S. 1933. *New Introductory Lectures on Psycho-analysis*. London: Hogarth.

Freud, S. 1949a. *Instincts and their Vicissitudes*. London: Hogarth.

Freud, S. 1949b. *An Outline of Psychoanalysis*. London: Hogarth.

Frisch, K. von 1967. *The Dance Language and Orientation of Bees*. Cambridge, MA: Harvard University Press.

Fromm, E. 1959. *Sigmund Freud's Mission*. New York: Harper.

Fromm, E. 1966. *Marx's Concept of Man*. New York; Ungar.

Fromm, E. 1970. *The Crisis of Psychoanalysis*. London: Cape.

Fromm, E. 1974. *The Anatomy of Human Destructiveness*. London: Cape.

Froude, J.A. 1883. *Letters and Memorials of Jane Welsh Carlyle*. London: Longman.

Furneaux, R. 1974. *William Wilberforce*. London: Hamish Hamilton.

Galbraith, J.K. 1974. *The New Industrial State*. Harmondsworth: Penguin.

Galbraith, J.K. 1975. *Economics and the Public Purpose*. Harmondsworth: Penguin.

Galton, F. 1865. Hereditary talent and character. *Macmillan's Magazine*, **12**, 157–66, 318–27.

Galton, F. 1869. *Hereditary Genius*. London: Macmillan.

Galton, F. 1908. *Memories of My Life*. London: Methuen.

Ganz, L. 1968. Generalization in the stimulus deprived organism. In: *Early Experience and Behavior*, ed. G. Newton & S. Levine. Springfield: Thomas.

Gardner, B.T. & Gardner, R.A. 1971. Two-way communication with an infant chimpanzee. In: *Behavior of Nonhuman Primates*, ed. A.M. Schrier & F. Stollnitz. New York: Academic.

Gardner, H. 1983. *Frames of Mind*. New York: Basic Books.

Gardner, R. & Heider, K.G. 1969. *Gardens of War*. London: Deutsch.

Gasman, D. 1971. *The Scientific Origins of National Socialism*. London: MacDonald.

Gastonguay, P.R. 1975. A sociobiology of man. *Am. Biol. Teacher*, **37**, 481–6.

Gauld, A. & Shotter, J. 1977. *Human Action and its Psychological Investigation*. London: Routledge & Kegan Paul.

Gauthier, D.P. 1969. *The Logic of Leviathan*. Oxford: Clarendon Press.

Geary, D. 1981. *European Labour Protest 1848–1939*, London: Croom Helm.

Genovés, S. 1976. Behavior and violence. *Persp. Biol. Med.* **20**, 20–9.

Ghiselin, M.T. 1974. *The Economy of Nature and the Evolution of Sex*, Berkeley: University of California Press.

Gibbon, E. 1976. *The Decline and Fall of the Roman Empire* (abridged by D.M. Low). London: Chatto & Windus. (First published in 1776–1788.)

Gibbs, M. 1949. *Feudal Order*. London: Cobbett Press.

Gilbert, M. 1986. *The Holocaust*. London: Collins.

Gilbreth, M.L. 1973. *The Psychology of Management*. Easton: Hive. (First published in 1914.)

Gillespie, N.C. 1979. *Charles Darwin and the Problem of Creation*. Chicago: University of Chicago Press.

Gladwin, T. 1970. *East is a Big Bird*. Cambridge, MA: Harvard University Press.

Godelier, M. 1979. Territory and property in primitive society. In: *Human Ethology*, ed. M. von Cranach *et al.* Cambridge: Cambridge University Press.

Godfrey, L.R. 1984. Scientific creationism: the art of distortion. In: *Science and Creationism*, ed. A. Montagu. Oxford: Oxford University Press.

Goldberg, S. 1977. *The Inevitability of Patriarchy*. London: Temple Smith.

Goldstein, J.H. 1975. *Aggression and Crimes of Violence*. New York: Oxford University Press.

Gombrich, E.H. 1962. *Art and Illusion*, London: Phaidon.

Gombrich, E.H. 1966. *The Story of Art*. London: Phaidon.

Goodale, J.C. 1971. *Tiwi Wives*. Seattle: University of Washington Press.

Goodall, J. 1986. *The Chimpanzees of Gombe*. Cambridge, MA: Harvard University Press.

Gould, J.L. 1976. The dance-language controversy. *Q. Rev. Biol.* **51**, 211–44.

Gould, S.J. 1977. *Ontogeny and phylogeny*. Cambridge, MA: Harvard University Press.

Gould, S.J. 1980. Sociobiology and the theory of natural selection. In: *Sociobiology: Beyond Nature/Nurture?* ed. G.W. Barlow & J. Silverberg. Boulder, CO: Westview Press.

Gould, S.J. 1981. *Mismeasure of Man*. New York: Norton.

Gould, S.J. 1982. Darwinism and the expansion of evolutionary theory. *Science, NY* **216**, 380–7.

Gould, S.J. & Lewontin, R.C. 1979. The spandrels of San Marco and the Panglossian paradigm: a critique of the adaptationist programme. *Proc. R. Soc. Lond.* B **205**, 581–98.

Gowaty, P.A. 1982. Sexual terms in sociobiology. *Anim. Behav.* **30**, 630–1.

Graves, R. 1957. *The Greek Myths*. Harmondsworth: Penguin.

Greene, J.C. 1959 Biology and social theory in the nineteenth century. In: *Critical Problems in the History of Science*, ed. M. Clagett. Madison: University of Wisconsin Press.

Greene, J.C. 1981. *Science, Ideology, and World View*. Berkeley: University of California Press.

Grenander, M.E. 1981. *Br. J. Philos. Sci.* **32**, 85–94. A cautious overview of behaviour therapy.

Grene, M. 1969. Bohm's metaphysics and biology. In: *Towards a Theoretical Biology 2. Sketches*, ed. C.H. Waddington. Edinburgh: Edinburgh University Press.

Grene, M. 1974. *The Understanding of Nature*. Dordrecht, Holland: Reidel.

Grene, M. 1978. Sociobiology and the human mind. In: *Sociobiology and Human Nature*, ed. M.S. Gregory, A. Silvers & D. Sutch. San Franciso: Jossey-Bass.

Grene, M. 1981. Changing concepts of Darwinian evolution. *Monist*, **64**, 195–213.

Grings, W.W. 1973. Consciousness and cognition in autonomic behavior change. In: *The Psychophysiology of Thinking*, ed. F.J. McGuigan & R.A.Schoonover. New York: Academic.

Gross, B. 1980. *Friendly Fascism*. New York: Evans.

Guthrie, W.K.C. 1950. *The Greek Philosophers*. London: Methuen.

Haeckel, E. 1879. *Freedom in Science and Teaching*. New York: Appleton.

Haeckel, E. 1913. *The Riddle of the Universe*. London: Watts.

Haldane, J.B.S. 1938. *Heredity and Politics*. London: Allen & Unwin.

Hall, A. 1977. *The Mandarin Cypher*. London: Fontana.

Hall, J.R. 1978. *The Ways Out*. London: Routledge.

Haller, M.H. 1963. *Eugenics*. New Brunswick: Rutgers University Press.

Halliday, T.R. & Slater, P.J.B. (eds.) 1983a. *Communication*. Oxford: Blackwell.

Halliday, T.R. & Slater, P.J.B. (eds.) 1983b. *Genes, Development and Learning*. Oxford: Blackwell.

Hallowell, A.I. 1955. *Culture and Experience*. Philadelphia: University of Pennsylvania Press.

Halsey, A.H. 1978. *Change in British Society*. Oxford: Oxford University Press.

Hamilton, W.D. 1964. The genetical evolution of social behavior. *J. Theoret. Biol.* **27**, 1–52.

Hamilton, W.D. 1972. Altruism and related phenomena. *Ann. Rev. Ecol. Syst.* **3**, 193–232.

Hamilton, W.D., Henderson, P.A. & Moran, N.A. 1981. Factors in the maintenance of sex. In: *Natural Selection and Social Behavior*, ed. R. Alexander & D.W. Tinkle. New York:Chiron.

Hammond, J.L. & Hammond, B. 1925. *The Town Labourer*. London: Longman, Green.

Hammond, J.L. & Hammond, B. 1936. *The Village Labourer 1760–1832*. London: Longman, Green.

Hampshire, S. 1975. *Freedom of the Individual*. London: Chatto & Windus.

Hargreaves, D.H. 1980. Common-sense models of action. In: *Models of Man*, ed. A.J. Chapman & D.M. Lones. Leicester: British Psychological Society.

Harris, B. & Harvey, J.H. 1981. Attribution theory. In: *The Psychology of Ordinary Explanations of Social Behaviour*, ed. C. Antaki. London: Academic.

Harris, H. 1976. Enzyme variants in human populations. *Johns Hopkins Med. J.* **138**, 245–52.

Harris, J. 1980. *Violence and Responsibility*. London: Routledge.

Harris, M. 1983. *The Doctor Who Technical Manual*. London: Severn House.

Harrison, B. 1981. Women's health and the women's movement. In: *Biology, Medicine and Society 1840–1940*, ed. C. Webster. Cambridge: Cambridge University Press.

Hartup, W.W. & Keller, E.D. 1960. Nurturance in preschool children. *Child Devel.* **31**, 691–689.

Hayek, F.A. 1983. *Knowledge, Evolution, and Society*. London: Adam Smith Institute.

Hayes, L.A. & Watson, J.S. 1981. Facial orientation by parents and smiling by infants. *Infant Behav. Devel.* **4**, 333–40.

Haynes, F. 1975. Metaphor as interactive. *Educ. Theory,* **25**, 272–7.

Hearnshaw, L.S. 1979. *Cyril Burt, Psychologist.* London: Hodder & Stoughton.

Hebb, D.O. 1947. Spontaneous neurosis in chimpanzees. *Psychosom. Med.* **9**, 3–19.

Hebb, D.O. 1955. The mammal and his environment. *Am. J. Psychiat.* **111**, 826–31.

Heckhausen, H. 1973. Intervening cognitions in motivation. In: *Beyond Pleasure, Reward, Preference,* ed. D.E. Berlyne & K.B. Madsen. New York: Academic.

Herbert, A.P. 1935. *Uncommon Law.* London: Methuen.

Herrnstein, R.J. 1969. Behaviorism. In: *Schools of Psychology,* ed. D.L. Krantz. New York: Appleton-Century-Crofts.

Hesse, M.B. 1963. *Models and Analogies in Science.* London: Sheed & Ward.

Hewes, G.W. 1973. Primate communication and the gestural origin of language. *Curr. Anthropol.* **14**, 5–24.

Hewes, G.W. 1977. Language origin theories. In: *Language Learning by a Chimpanzee,* ed. D.M. Rumbaugh. New York: Academic.

Heyduk, R.G. 1975. Rated preference for musical compositions. *Percep. Psychophys.* **17**, 84–90.

Heyneman, S.P. & Loxley, W.A. 1983. The effect of primary school quality on academic achievement. *Am. J. Sociol.* **88**, 1162–94.

Hiatt, L.R. 1970. Men and other animals. *Quadrant,* **14**, 43–50.

Hill, J.E.C. 1958. *Puritanism and Revolution.* London: Secker & Warburg.

Hobbes, T. 1962. *Leviathan.* Glasgow: Fontana. (First published in 1651.)

Hobsbawm, E.J. 1964. *Labouring Men.* London: Weidenfeld & Nicolson.

Hobsbawm, E.J. 1977a. *The Age of Capital 1848–1875.* London: Sphere.

Hobsbawm, E.J. 1977b. *The Age of Revolution.* London: Sphere.

Hockett, C.F. & Altmann, S.A. 1968. A note on design features. In: *Animal Communication,* ed. T.A. Sebeok. Bloomington: Indiana University Press.

Hoffman, R.R. 1980. Metaphor in science. In: *Cognition and Figurative Language,* ed. R.P. Honeck & R.R. Hoffman. Hillsdale NJ: Erlbaum.

Hofman M.A. 1983. Encephalization in hominids. *Brain Behav. Evol.* **22**, 102–17.

Hofstadter, R. 1955. *Social Darwinism in American Thought.* New York: Braziller.

Hokanson, J.E. 1970. Psychophysiological evaluation of the catharsis hypothesis. In: *The Dynamics of Aggression,* ed. E.I. Megargee & J.E. Hokanson. New York: Harper & Row.

Hokanson, J.E., Willers, K.R. & Koropsak, E. 1968. Autonomic responses during aggressive interchange. *J. Personal.* **36**, 386–404.

Holland, J.G. 1960. Teaching machines. *J. Exp. Anal. Behav.* **3**, 275–87.

Hollis, M. 1977. *Models of Man.* Cambridge: Cambridge University Press.

Holton, G. 1973. *Thematic Origins of Scientific Thought.* Cambridge, MA: Harvard University Press.

Howgate, M. & Lewis, A. 1984. The case of Miocene Man. *New Scientist,* **29** March, 44–45.

Hrdy, S.B. 1977. Infanticide as a primate reproductive strategy. *Am. Sci.* **65**, 40–9.

Hsia, D.Y.-Y. 1967. The hereditary metabolic diseases. In: *Behavior-Genetic Analysis,* ed. J. Hirsch. New York: McGraw-Hill.

Huby, P. 1967. The first discovery of the freewill problem. *Philosophy,* **42**, 353–62.

Hughes, H.S. 1952. *Oswald Spengler.* New York: Scribner.

Huizinga, J. 1950. *Homo Ludens.* Boston: Beacon.

Hume, D. 1975. *Enquiries Concerning Human Understanding.* Oxford: Clarendon Press. (First published in 1777.)

Hutt, A. 1941. *British Trade Unionism*. London: Lawrence & Wishart.

Huxley, J.S. 1953. *Evolution in Action*. London: Chatto & Windus.

Huxley, J.S. 1964. Psychometabolism: general and Lorenzian. *Persp. Biol. Med.* 7, 399–432.

Huxley, T.H. 1904. *Methods and Results*. New York: Macmillan.

Huxley, T.H. 1911. *Evolution and Ethics*. London: Macmillan.

Huxley, T.H. & Huxley, J.S. 1947. *Evolution and Ethics*. London: Pilot Press.

Independent Commission on Disarmament. 1982. *Common Security*. New York: Simon & Schuster.

Ingelhart, R. 1977. *The Silent Revolution*. Princeton, NJ: Princeton University Press.

Isaac, G.L. 1981. Emergence of human behaviour. *Phil. Trans. R. Soc. Lond.* B 292, 177–88.

Isaac, G.L. 1983. Aspects of human evolution. In: *Evolution from Molecules to Men*, ed. D.S. Bendall. Cambridge: Cambridge University Press.

Israel, J. 1971. *Alienation*. Boston: Allyn & Bacon.

Iyer, R. (ed.) 1986. *The Moral and Political Writings of Mahatma Gandhi*, Vol. II. Oxford: Clarendon.

Jacob, F. 1977. Evolution and Tinkering. *Science*, NY **196**, 1161–6.

Janson, H.W. 1952. *Apes and Ape Lore*. London: Warburg Institute.

Jay, P. 1965. The common langur of north India. In: *Primate Behavior*, ed. I. DeVore. New York: Holt, Rinehart & Winston.

Jelliffe, D.B. & Jelliffe, E.F.P. 1978. *Human Milk in the Modern World*. Oxford: Oxford University Press.

Jencks, C. 1987. Genes and crime. *New York Review of Books*, **34** (ii), Feb. 12; 33–41.

Jencks, C. *et al.* 1972. *Inequality*. New York: Basic Books.

Jensen, A.R. 1969. How much can we boost IQ and scholastic achievement? *Harvard Educ. Rev.* **39**, 1–123.

Jensen, A.R. 1980. *Bias in Mental Testing*. London: Methuen.

Johnson, S. 1968. *Selected Writings*. Harmondsworth: Penguin.

Johnson-Laird, P.N. 1983. *Mental Models*. Cambridge, MA: Harvard University Press.

Jolly, A. 1966. Lemur social behavior and primate intelligence. *Science*, NY **153**, 501–6.

Jolly, C.J. 1970. The seed-eaters. *Man*, **5**, 5–26.

Jones, G. 1980. *Social Darwinism and English Thought*. Brighton: Harvester.

Jones, J.S. 1980. How much genetic variation? *Nature, Lond.* **288**, 10–11.

Joynson, R.B. 1974. *Psychology and Common Sense*. London: Routledge.

Judd, T. 1978. Naturizing what we do. *Science for the People*, Jan/Feb, pp.16–19.

Jukes, T.H. 1981. Creating problems. *Nature, Lond.* **289**, 218, 335.

Kalikow, T.J. 1983. Konrad Lorenz's ethological theory. *J. Hist. Biol.* **16**, 39–73.

Kamin, L.J. 1974. *The Science and Politics of I.Q.* Hillsdale, NJ: Erlbaum.

Kanter, R.M. 1973. *Communes*. New York: Harper.

Karn, M.N. & Penrose, L.S. 1951. Birth weight and gestation time. *Ann. Eugen., Lond.* **16**, 147–64.

Kaufman, A., Baron, A. & Kopp, R.E. 1966. Some effects of instructions on human operant behavior. *Psychol. Monogr. Suppl.* **1**, 243–250.

Kauppinen-Toropainen, K., Kandolin, T. & Mutanen, P. 1983. Job dissatisfaction and work-related exhaustion. *J. Occup. Behav.* **4**, 193–207.

Kazdin, A.E. 1978. *History of Behavior Modification*. Baltimore: University Park Press.

Keesing, R.M. 1972. The lure of kinship. In: *Kinship Studies in the Morgan Centennial Year*, ed. P. Reining. Washington DC: Anthropological Society of Washington.

Keith, A. 1948. *A New Theory of Human Evolution*. London: Watts.

Kemp, T.S. 1982. *Mammal-like Reptiles and the Origin of Mammals*. London: Academic.

Kessen, W., Levine, J. & Wendrich, K. 1979. Imitation of pitch in infants. *Infant Behav. Devel.* **2**, 93–9.

Keynes, J.M. 1933. *Essays in Persuasion*. London: Macmillan.

Kim, S.S. 1976. The Lorenzian theory of aggression. *J. Peace Res.* **13**, 253–76.

Kimble, G.A. 1967. Sechenov and the anticipation of conditioning theory. In: *Foundations of Conditioning and Learning*, ed. G.A. Kimble. New York: Appleton-Century-Crofts.

Kimura, M. 1983. *The Neutral Theory of Molecular Evolution*. Cambridge: Cambridge University Press.

Kinkade, K. 1973. *A Walden Two Experiment*. New York: Morrow.

Kipling, R. 1898. *The Day's Work*. London: Macmillan.

Klein, V. 1946. *The Feminine Character*. London: Kegan Paul.

Klineberg, O. 1935. *Race Differences*. New York: Harper.

Kling, A. 1975. Brain lesions and aggressive behavior. In: *Neural Bases of Violence and Aggression*, ed. W.S. Fields & W.H. Sweet. St Louis: Green.

Klingender, F. 1971. *Animals in Art and Thought to the End of the Middle Ages*, ed. E. Antal & J. Harthan. London: Routledge.

Knox, J. 1972. *The First Blast of the Trumpet against the Monstrous Regiment of Women*. Amsterdam: Theatrvm Orbis Terrarvm. (First published in 1558.)

Koranda, L.D. 1980. Music of the Alaskan Eskimos. In: *Musics of Many Cultures*, ed. E. May. Berkeley: University of California Press.

Kozol, J. 1985. *Illiterate America*. Garden City, New York: Doubleday.

Krebs, H.A. & Shelley, J.H. 1975. *The Creative Process in Science and Medicine*. Amsterdam: Excerpta Medica.

Kropotkin, P.A. 1910. *Mutual Aid*. London: Heinemann.

Kuttner, R. 1984. *The Economic Illusion*. New York: Houghton Mifflin.

Kwapong, A.A. 1971. Epilog. In: *Man and Beast*, ed. J.F. Eisenberg & W.S. Dillon. Washington: Smithsonian.

Lakatos, I. & Musgrave, A. (eds.) 1970. *Criticism and the Growth of Knowledge*. Cambridge: Cambridge University Press.

La Mettrie, J.O. de 1927. *Man a Machine*. Chicago: Open Court. (First published 1747.)

Lasch, C. 1969. *Agony of the American Left*. New York: Knopf.

Lawick-Goodall, J. van 1973. Cultural elements in a chimpanzee community. *Symp. IVth Congr. Primat.* **1**, 144–84.

Layton, C. 1986. Unemployment and health. *Psychol. Rep.* **58**: 807–810.

Lazar, I. *et al.* 1982. Lasting effects of early education. *Monog. Soc. Res. Child Devel.* **47**, ser.195.

Leach, E.R. 1961. *Pul Eliya: a Village in Ceylon*. Cambridge: Cambridge University Press.

Leach, E.R. 1982. *Social Anthropology*. Glasgow: Fontana.

Lee, R.B. 1979. *The !Kung San*. Cambridge: Cambridge University Press.

Lee, R.B. & DeVore, I. 1968. *Man the Hunter*. Chicago: Aldine.

Lees, D.R. 1981. Industrial melanism. In: *Genetic Consequences of Man Made Change*, ed. J.A. Bishop & L.M. Cook. London: Academic.

Leon, D. 1969. *The Kibbutz*. Oxford: Pergamon.

Lepper, M.R., Greene, D.T. & Nisbett, R.E. 1973. Undermining children's intrinsic interest. *J. Personal. Soc. Psychol.* **28**, 129–37.

Le Quesne, L. 1983. *The Bodyline Controversy*. London: Secker & Warburg.

Leroi-Gourhan, A. 1968. *The Art of Prehistoric Man in Western Europe*. London: Thames and Hudson.

Levitan, M. & Montagu, A. 1977. *Textbook of Human Genetics*. New York: Oxford University Press.

Levitas, R. (ed.) 1986 *The Ideology of the New Right*. Cambridge: Polity Press.

Levy, R.I. 1969. On getting angry in the Society Islands. In: *Mental Health Research in Asia and the Pacific*. Honolulu: East-West Center Press.

Lewis, C.S. 1943. *Perelandra*. London: Bodley Head.

Lewis, R.A. 1970. *Sickle States*. Accra: Ghana Universities Press.

Lewontin, R.C. 1975. Genetic aspects of intelligence. *Ann. Rev. Genet.* **9**, 387–405.

Lewontin, R.C. 1980. Sociobiology: another biological determinism. *Int. J. Hlth. Serv.* **10**, 347–63.

Lewontin, R.C. 1983. Gene, organism and environment. In: *Evolution from Molecules to Men*, ed. D.S. Bendall. Cambridge: Cambridge University Press.

Leyhausen, P. 1965. The sane community – a density problem? *Discovery*, **26**, 27–33.

Lick, J. & Bootzin, R. 1975. Expectancy in the treatment of fear. *Psychol. Bull.* **82**, 917–31.

Lieberman, P.H., Klatt, D.H. & Wilson, W.H. 1969. Vocal tract limitations on the vowel repertoires of primates. *Science, NY* **164**, 1185–7.

Lindauer, M. 1961. *Communication among Social Bees*. Cambridge, MA: Harvard University Press.

Lion, J.R. & Penna, M. 1974. Human aggression. In: *The Neuropsychology of Aggression*, ed. R.E. Whalen. New York: Plenum.

Lodge, D. 1981. *How Far Can You Go?* Harmondsworth: Penguin.

Loeb, J. 1901. *Comparative Physiology of the Brain*. London: Murray.

Loeb, J. 1964. *The Mechanistic Conception of Life*. Cambridge, MA: Harvard University Press. (First published 1912.)

Loehlin, J.C., Lindzey, G. & Spuhler, J.N. 1975. *Race Differences in Intelligence*. San Francisco: Freeman.

Long, J.P.M. 1971. The Pintubi. In: *Aboriginal Man and Environment in Australia*, ed. D.J. Mulvaney & J. Golson. Canberra: Australian National University Press.

Longford, E. 1981. *Eminent Victorian Women*. London: Weidenfeld & Nicolson.

Lorenz, K.Z. 1952. *King Solomon's Ring*. London: Methuen.

Lorenz, K.Z. 1966. *On Aggression*. London: Methuen.

Lorenz, K.Z. 1971. *Studies in Animal and Human Behaviour*, vol. 2. London: Methuen.

Lorenz, K.Z. 1976. Pathologie de la civilisation. *Nouvelle École*, 9–30.

Lotka, A.J. 1945. The law of evolution as a maximal principle. *Hum. Biol.* **17**, 167–94.

Lovejoy, A.O. 1960. *The Great Chain of Being*. New York: Harper.

Lovejoy, A.O. & Boas, G. 1935. *Primitivism and Related Ideas in Antiquity*. Baltimore: Johns Hopkins Press.

Lowe, M. 1978. Sociobiology and sex differences. *Signs*, **4**, 118–25.

Ludmerer, K.M. 1972. *Genetics and American Society: A Historical Appraisal*. Baltimore: Johns Hopkins University Press.

Luria, A.R. 1976. *Cognitive Development*. Cambridge, MA: Harvard University Press.

Maccoby, E.E. & Jacklin, C.N. 1974. *The Psychology of Sex Differences*. Stanford: Stanford University Press.

Machiavelli, N. 1961. *The Prince*. Harmondsworth: Penguin. (First published about 1513.)

Machiavelli, N. 1974. *The Discourses*. Harmondsworth: Penguin. (First published in 1513–1519).

MacIntyre, A.C. 1958. *The Unconscious*. London: Routledge.

Mackenzie, B.D. 1977. *Behaviourism and the Limits of Scientific Method*. London: Routledge.

Mackenzie, D.A. 1981. Sociobiologies in competition. In: *Biology, Medicine and Society 1840–1940*, ed. C. Webster. Cambridge: Cambridge University Press.

Mackenzie, J.M. 1984. *Propaganda and Empire*. Manchester: Manchester University Press.

Mackenzie, W.J.M. 1967. *Politics and Social Science*. Harmondsworth: Penguin.

Mackenzie, W.J.M. 1975. *Power, Violence, Decision*. Harmondsworth: Penguin.

Mackie, J.L. 1978. The law of the jungle. *Philosophy* 53, 455–64.

Macpherson, C.B. 1962. *The Political Theory of Possessive Individualism*. Oxford: Clarendon Press.

Macrae, D.B. 1958. Darwinism and the social sciences. In: *A Century of Darwin*, ed. S.A. Barnett. London: Heinemann.

Madden, D.J. & Lion, J.R. 1981. Clinical management of aggression. In: *Multidisciplinary Approaches to Aggression Research*, ed. P.F. Brain & D. Benton. Amsterdam: Elsevier.

Malcolm, N. 1964. Behaviorism as a philosophy of psychology. In: *Behaviorism and Phenomenology*, ed. T.W. Wann. Chicago: University of Chicago Press.

Manier, E. 1978. *The Young Darwin and his Cultural Circle*. Dordrecht-Holland: Reidel.

Mark, V.H. & Ervin, F.R. 1970. *Violence and the Brain*. New York: Harper & Row.

Marks, J. 1979. *The Search for the 'Manchurian Candidate': the CIA and Mind Control*. New York: Times Books.

Marrou, H.I. 1956. *A History of Education in Antiquity*. New York: Sheed & Ward.

Marshack, A. 1972. *The Roots of Civilization*. New York: McGraw-Hill.

Marshall, D.S. & Suggs, R.C. 1972. *Human Sexual Behavior*. Englewood Cliffs, NJ: Prentice-Hall.

Max-Müller, M. 1864. *Lectures on the Science of Language*. Second Series. London: Longman.

Maynard Smith, J. 1961. Evolution and history. In: *Darwinism and the Study of Society*, ed. M. Banton. London: Tavistock.

Maynard Smith, J. 1964. Group selection and kin selection. *Nature, Lond.* 201, 1145–7.

Maynard Smith, J. 1975. *The Theory of Evolution*. Harmondsworth: Penguin.

Maynard Smith, J. 1981. Evolutionary games. In: *Evolution Today*, ed. G.G.E. Scudder & J.L. Reveal. Pittsburgh: Carnegie-Mellon.

Maynard Smith, J. 1982. Unsolved evolutionary problems. In: *Genome Evolution*, ed. G.A. Dover & R.B. Flavell. London: Academic.

Mayr, E. 1982. *The Growth of Biological Thought*. Cambridge, MA: Harvard University Press.

Mazzeo, J.A. 1965. *Renaissance and Revolution*. New York: Pantheon.

McAskie, M. & Clarke, A.M. 1976. Parent-offspring resemblances in intelligence. *Br. J. Psychol.* 67, 243–73.

McCarthy, J.D., Galle, O.R. & Zimmern, W. 1975. Population density, social structure, and interpersonal violence. *Am. Behav. Sci.* 18, 771–91.

McClintock, M.K. 1971. Menstrual synchrony and suppression. *Nature, Lond.* 229, 244–5.

McDaniel, C.G. 1985. The tyranny of testing. *Med. J. Aust.* 142, 38–40.

McFarland, D.J. 1971. *Feedback Mechanisms in Animal Behaviour*. London: Academic.

McKay, H. *et al.* 1978. Improving cognitive ability in chronically deprived children. *Science*, NY **200**, 270–8.

McLean, I.S. 1987. *Public Choice*. Oxford: Blackwell.

McPhee, C. 1955. Children and music in Bali. In: *Childhood in Contemporary Cultures*, ed. M. Mead & M. Wolfenstein. Chicago: University of Chicago Press.

Medawar, P.B. 1960. *The Future of Man*. London: Methuen.

Medawar, P.B. 1974. A geometric model of reduction and emergence. In: *Studies in the Philosophy of Biology*, ed. F.J. Ayala & T. Dobzhansky. London: Macmillan.

Meggitt, M.J. 1972. Australian aboriginal society. In: *Kinship Studies in the Morgan Centennial Year*, ed. P. Reining. Washington: Anthropological Society.

Meister, A. 1956. Perception and acceptance of power relations in children. *Group Psychotherapy & Psychodrama* **9**, 153–63.

Mellow, J.R. 1974. *Charmed Circle, Gertrude Stein and Company*. London: Phaidon.

Meltzer, A. 1981. Doomed subspecies. *Nature, Lond.* **291**, 608.

Menninger, W.C. 1948. Recreation and mental health. *Recreation* **42**, 340–6.

Menzel, E.W. 1971. Communication about the environment in young chimpanzees. *Folia Primat.* **15**, 220–32.

Menzel, E.W. & Halperin, S. 1975. Objective communication between chimpanzees. *Science*, NY **189**, 652–4.

Menzel, R., Erber, J. & Masuhr, T. 1974. Learning and memory in the honeybee. In: *Experimental Analysis of Insect Behaviour*, ed. L.B. Browne. Berlin: Springer-Verlag.

Merriam, A.P. 1964. *The Anthropology of Music*. Bloomington: Northwestern University Press.

Métral, G. 1981. The action of natural selection on the menstrual cycle. *J. Biosoc. Sci.* **13**, 337–43.

Michener, C.D. & Michener, M.H. 1951. *American Social Insects*. Toronto: Van Nostrand.

Midgley, M. 1980. Rival fatalisms. In: *Sociobiology Examined*, ed. A. Montagu. New York: Oxford University Press.

Mill, J.S. 1970. *The Subjection of Women*. Cambridge, MA: MIT Press. (First published in 1869.)

Miller, G.A. 1981. *Language and Speech*. San Francisco: Freeman.

Miller, K.R. 1984. Scientific creationism versus evolution. In: *Science and Creationism*, ed. A. Montagu. Oxford: Oxford University Press.

Miller, N.E. 1959. Liberalization of basic S-R concepts. In: *Psychology: A Study of a Science*, vol. 2, ed. S. Koch. New York: McGraw-Hill.

Milner, B., Corkin, S. & Teuber, H.L. 1968. Hippocampal amnesic syndrome. *Neuropsychologia*, **6**, 215–34.

Mintz, M. 1985. *At Any Cost: Corporate Greed, Women, and the Dalkon Shield*. New York: Pantheon.

Mitchell, R.E. 1971. Some implications of high density housing. *Am. Sociol. Rev.* **36**, 18–29.

Mitton, J.B. 1975. Normalizing selection for height. *Hum. Biol.* **47**, 189–200.

Monod, J. 1974. *Chance and Necessity*. Glasgow: Fontana.

Montagu, A. 1976. *The Nature of Human Aggression*. London: Oxford University Press.

Montagu, I. 1967. *Germany's New Nazis*. London: Panther.

Montaigne, M. de 1958. *Essays*. Harmondsworth: Penguin.

Moore, B. 1967. *Social Origins of Dictatorship and Democracy*. London: Allen Lane.

Morse, A.D. 1968. *While Six Million Died*. London: Secker & Warburg.

Moyer, K.E. 1976. *The Psychobiology of Aggression*. New York: Harper & Row.

Mussen, P. & Eisenberg-Berg, N. 1977. *Roots of Caring, Sharing and Helping*. San Francisco: Freeman.

Nagel, E. 1961. *The Structure of Science*. London: Routledge.

Nagel, T. 1970. *The Possibility of Altruism*. Oxford: Clarendon Press.

Nance, J. 1975. *The Gentle Tasaday*. New York: Harcourt Brace Jovanovich.

Nash, H. 1963. The role of metaphor in psychological theory. *Behavl Sci*. 8, 336–45.

Nath, V.R. 1987. *Smoking: Third World Alert*. Oxford: Oxford University Press.

Neal, A.G. (ed.) 1976. *Violence in Animal and Human Societies*. Chicago: Nelson Hall.

Neary, J. 1970. A scientist's variations on a disturbing racial theme. *Life Magazine*, 68, 5813–65.

Needham, J. 1946. *History is on our Side*. London: Allen & Unwin.

Nietzsche, F.W. 1960. *Thus Spake Zarathustra*. London: Dent. (Originally published in 1883–1891.)

Noel-Baker, P. 1936. *The Private Manufacture of Armaments*. London: Gollancz.

Nordenskiöld, E. 1928. *The History of Biology*. New York: Knopf.

Norton, B.J. 1980. Fisher and the neo-Darwinian synthesis. *Proc. Int. Congr. Hist. Sci.* 15, 481–94.

Norton, B.J. 1981. Psychologists and class. In: *Biology, Medicine and Society 1840–1940*, ed. C. Webster. Cambridge: Cambridge University Press.

O'Donnell, P. 1982. *The Xanadu Talisman*. London: Pan.

Opie, I. & Opie, P. 1959. *The Lore and Language of Schoolchildren*. London: Oxford University Press.

Orians, G.H. & Pearson, N.E. 1977. On the theory of central place foraging. In: Analysis of Ecological Systems, ed. D.J. Horn *et al.* Columbus: Ohio State University Press.

Ortony, A. 1975. Why metaphors are necessary and not just nice. *Educ. Theory*, 25, 45–53.

Osborne, R.T., Noble, C.E. & Weyl, N. (eds.) 1978. *Human Variation*. New York: Academic.

Owen, R. 1843. *Lectures on the Comparative Anatomy and Physiology of the Invertebrate Animals*. London: Longman.

Packard, V. 1960. *The Hidden Persuaders*. Harmondsworth: Penguin.

Packer, C. & Pusey, A.E. 1982. Cooperation and competition within coalitions of male lions. *Nature, Lond.* 296, 740–2.

Paley, W. 1830. *The Works*, vol. 4. London.

Passingham, R.E. 1981. Broca's area and the origins of human vocal skill. *Phil. Trans. R. Soc. Lond.* B 292, 167–75.

Passmore, J. 1970. *The Perfectibility of Man*. London: Duckworth.

Passmore, J. 1980. *The Philosophy of Teaching*. London: Duckworth.

Pavlov, I.P. 1927. *Conditioned Reflexes*. London: Oxford University Press.

Pavlov, I.P. 1928. *Lectures on Conditioned Reflexes*. London: Lawrence & Wishart.

Pearson, K. 1900. *The Grammar of Science*. London: Black.

Pearson, K. & Moul, M. 1925. The problem of alien immigration. *Ann. Eugen.* 1, 5–127.

Peele, G. 1984. *Revival and Reaction: the Right in Contemporary America*. Oxford: Clarendon.

Penfield, W. 1958. *The Excitable Cortex in Conscious Man*. Liverpool: Liverpool University Press.

Penrose, L.S. 1963. *The Biology of Mental Defect*. London: Sidgwick & Jackson.

Peters, R.S. 1956. *Hobbes*. Harmondsworth: Penguin.

Peterson, G.B. *et al.* 1972. Conditioned behavior toward signals for reinforcement. *Science*, NY **177**, 1009–1011.

Phares, E.J. 1976. *Locus of Control in Personality*. Morristown, NJ: General Learning Press.

Piaget, J. 1932. *The Moral Judgment of the Child*. London: Routledge.

Pilbeam, D. 1972. *The Ascent of Man*. New York: Macmillan.

Pilbeam, D. & Gould, S.J. 1974. Size and scaling in human evolution. *Science*, NY **186**, 892–901.

Pitkänen, L. 1974. The control of aggressive behaviour in children. *Scand. J. Psychol.* **15**, 169–77.

Piven, F.F. & Cloward, R.A. 1979 *Poor People's Movements*. New York: Random House.

Plato 1903. *The Four Socratic Dialogues*. Oxford: Clarendon Press. (First published 4th century BC.)

Plato 1930. *The Republic* vol. 1. London: Heinemann. (First published about BC 393.)

Plato 1935. *The Republic*, vol. 2. London: Heinemann. (First published about BC 393.)

Popper, K.R. 1966. *The Open Society and Its Enemies*, vol. 1. London: Routledge.

Popper, K.R. 1970. Normal science. In: *Criticism and the Growth of Knowledge*, ed. I. Lakatos & A. Musgrave. Cambridge: Cambridge University Press.

Popper, K.R. 1972. *Objective Knowledge An Evolutionary Approach*. Oxford: Clarendon.

Popper, K.R. 1974. Scientific reduction. In: *Studies in the Philosophy of Biology*, ed. F.J. Ayala & T. Dobhansky. London: Macmillan.

Popper, K.R. 1982. *The Open Universe*. London: Hutchinson.

Postman, N. 1985. *Amusing Ourselves to Death*. New York: Viking.

Price, A. 1975. *October Men*. London: Hodder Paperbacks.

Prins, G. 1984. *The Choice: Nuclear Weapons versus Security*. London: Chatto & Windus.

Pugh, G.E. 1977. *The Biological Origin of Human Values*. New York: Basic Books.

Pyke, G.H., Pulliam, H.R. & Charnov, E.L. 1977. Optimal foraging. *Q. Rev. Biol.* **52**, 137–54.

Quinton, A. 1966. Ethics and the theory of evolution. In: *Biology and Personality*, ed. I.T. Ramsey. Oxford: Blackwell.

Raphael, D.D. 1958. Darwinism and ethics. In: *A Century of Darwin*, ed. S.A. Barnett. London: Heinemann.

Rapoport, A. 1966. Models of conflict. In: *Conflict in Society*, ed. A. de Reuck & J. Knight. London: Churchill.

Razran, G. 1961. Interoceptive conditioning, semantic conditioning, and the orienting reflex. *Psychol. Rev.* **68**, 81–147.

Reitlinger, G.R. 1961. *The Final Solution*. New York: Barnes.

Rescorla, R.A. 1978. A cognitive perspective on Pavlovian conditioning. In: *Cognitive Processes in Animal Behavior*, ed. S.H. Hulse *et al.* Hillsdale, NJ: Erlbaum.

Reynolds, H. 1981. *The Other Side of the Frontier*. Townsville: James Cook University.

Richerson, P.J. & Boyd, R. 1978. A dual inheritance model of the human evolutionary process. *J. Soc. Biol. Struct.* **1**, 127–54.

Ridley, M. 1982. How to explain organic diversity. *New Scientist*, 6 May, 359–61.

Riedman, M.L. 1982. The evolution of alloparental care and adoption. *Q. Rev. Biol.* **57**, 405–35.

Ristau, C.A. & Robbins, D. Language in the great apes. *Adv. Stud. Behav.* **12**, 141–255.

Rogoff, B. *et al.* 1975. Assignment of roles and responsibilities to children. *Hum. Dev.* **18**, 353–60.

Roper, M.K. 1969. A survey of the evidence for intrahuman killing in the Pleistocene. *Curr. Anthropol.* **10**, 427–59.

Rosa, K.R. 1976. *Autogenic training.* London: Gollancz.

Rose, M.R. 1983. The contagion mechanism for the evolution of sex. *J. Theor. Biol.* **101**, 137–46.

Rose, S.P.R., Kamin, L.J. & Lewontin, R.C. 1984. *Not in Our Genes.* Harmondsworth: Penguin.

Rosner, M. 1973. Direct democracy in the kibbutz. In: *Communes,* ed. R.M. Canter. New York: Harper & Row.

Ross, H.S. 1974. Exploratory behavior in 12-month-old infants. *J. Exp. Child Psychol.* **17**, 436–51.

Roszak, T. 1986. *The Cult of Information.* Cambridge: Lutterworth.

Roth, H.L. 1899. *The Aborigines of Tasmania.* Halifax: King.

Routh, G. 1980. *Occupation and Pay in Great Britain 1906–1979.* London: Macmillan.

Rowbotham S. 1973. *Hidden from History.* London: Pluto Press.

Rudé, G.F.E. 1964. *The Crowd in History.* New York: Wiley.

Russell, B. 1927. *An Outline of Philosophy.* London: Allen & Unwin.

Russell, B. 1930. *The Conquest of Happiness.* London: Allen & Unwin.

Russell, B. 1948. *Human Knowledge: Its Scope and Limits.* London: Allen & Unwin.

Russell, B. 1958. *Portraits From Memory and other Essays.* London: Allen & Unwin.

Russell, B. 1961. *History of Western Philosophy.* London: Allen & Unwin.

Russell, B. 1962. *The Scientific Outlook.* London: Allen & Unwin.

Russell, M.J., Switz, J.M. & Thompson, K. 1980. Olfactory influences on the human menstrual cycle. *Pharmacol. Biochem. Behav.* **13**, 737–8.

Ryan, J. 1974. Early language development. In: The *Integration of a Child into a Social World,* ed. M.P.M. Richards. Cambridge: Cambridge University Press.

Ryle, G. 1949. *The Concept of Mind.* London: Hutchinson.

Sampson, A. 1974. *The Sovereign State.* London: Hodder Fawcett.

Sampson, A. 1978. *The Arms Bazaar.* London: Hodder & Stoughton.

Sampson, A. 1980. *The Seven Sisters.* London: Hodder & Stoughton.

Sampson, R.V. 1956. *Progress in the Age of Reason.* London: Heinemann.

Santayana, G. 1954. *The Life of Reason.* London: Constable.

Savage-Rumbaugh, E.S., Rumbaugh, D.M. & Boysen, S. 1978. Linguistically mediated tool use by chimpanzees. *Behav. Brain Sci.* **4**, 539–54.

Sayers, J. 1982. *Biological Politics.* London: Tavistock.

Scarr-Salapatek, S. 1976. Unknowns in the IQ equation. In: *The IQ Controversy,* ed. N.J. Block & G. Dworkin. New York: Pantheon.

Schaller, G.B. 1972. *The Serengeti Lion.* Chicago: University of Chicago Press.

Schaller, G.B. & Lowther, G.R. 1969. The relevance of carnivore behavior to the study of early hominids. *Southwestern J. Anthropol.* **25**, 307–41.

Scheflin, A.W. & Opton, E.M. 1978. *The Mind Manipulators.* New York: Paddington Press.

Scheinfeld, A. 1944. The Kallikaks after thirty years. *J. Hered.* **35**, 259–64.

Schenkel, R. 1967. Submission in the wolf and dog. *Am. Zool.* **7**, 319–29.

Schiff, M. *et al.* 1982. How much *could* we boost scholastic achievement and IQ scores? *Cognition,* **12**, 165–96.

Schimmel, S. 1979. Anger in Graeco-Roman and modern psychology. *Psychiatry,* **42**, 320–37.

Schjelderup-Ebbe, T. 1922. Beiträge zur Sozialpsychologie des Haushuhns. *Z. Psychol.* **88**, 226–52.

Schneider, D.M. What is kinship all about? In: *Kinship Studies in the Morgan Centennial Year*, ed. P. Reining. Washington DC: Anthropological Society of Washington.

Schoenberger, R.A. (ed.) 1969. *The American Right Wing.* New York: Holt, Rinehart & Winston.

Schuettinger, R.L. 1976. *Lord Acton: Historian of Liberty.* La Salle, IL: Open Court.

Schultz, D.P. 1965. *Sensory Restriction Effects on Behavior.* New York: Academic.

Schumacher, E.F. 1973. *Small is Beautiful.* New York: Harper & Row.

Scitovsky, T. 1976. *The Joyless Economy.* New York: Oxford University Press.

Scott, J. 1986. What puts medical students off psychiatry? *Bull. R. Coll. Psychiat.* **10**, 98–100.

Scott, J.F. 1952. *The Scientific Work of René Descartes.* London: Taylor & Francis.

Scribner, S. & Cole, M. 1973. Cognitive consequences of formal and informal education. *Science, NY* **182**, 553–9.

Scruton, R. 1980. *The Meaning of Conservatism.* London: Macmillan.

Searle, G.R. 1976. *Eugenics and Politics in Britain 1900–1914.* Noordhof: Leyden.

Searle, G.R. 1981. Eugenics and class. In: *Biology, Medicine and Society 1840–1940*, ed. C. Webster. Cambridge: Cambridge University Press.

Sechenov, I.M. 1965. *Reflexes of the Brain*, Cambridge, MA: MIT Press. (First published in 1863.)

Seeman, M. 1959. On the meaning of alienation. *Am. Sociol. Rev.* **24**, 783–91.

Sen, A. 1979. Rational fools: a critique of the behavioural foundations of economic theory. In: *Scientific Models and Man*, ed. H. Harris. Oxford: Clarendon.

Shapiro, J. 1979. Cross-cultural perspectives on sexual differences. In: *Human Sexuality*, ed. H.A. Katchadourian. Berkeley, CA: University of California Press.

Sharma, R.S. 1965. *Indian Feudalism.* Calcutta: University of Calcutta.

Shaw, G.B. 1944. *Everybody's Political What's What.* London: Constable.

Sheppard, P.M. 1975. *Natural Selection and Heredity.* London: Hutchinson.

Shields, J. & Gottesman, I.I. 1973. Genetic studies of schizophrenia. *Biochem. Soc. Spec. Publ.* **1**, 165–74.

Shields, S.A. 1980. Evolutionary theory and male scientific bias. In: *Sociobiology: beyond Nature/Nurture?* ed. G.W. Barlow & J. Silverberg. Boulder, Colorado: Westview Press.

Shimoda, K., Argyle, M. & Bitti, P.R. 1978. Intercultural recognition of emotional expressions. *Europ. Soc. Psychol.* **8**, 169–79.

Short, R.V. 1976. The evolution of human reproduction. *Proc. R. Soc. B* **195**, 3–24.

Short, R.V. 1979. Sexual selection illustrated by man and the great apes. *Adv. Stud. Behav.* **9**, 131–58.

Short, R.V. 1981. Sexual selection in man and the great apes. In: *Reproductive Biology of the Great Apes*, ed. C.E. Graham. New York: Academic.

Short, R.V. 1983. The biological basis for the contraceptive effects of breast-feeding. *Adv. Internat. Matern. Child Hlth* **3**, 27–39.

Shotter, J. 1980. Men the magicians. In: *Models of Man*, ed. A.J. Chapman & D.M. Jones. Leicester: British Psychological Society.

Shotter, J. 1981. Telling and reporting. In: *The Psychology of Ordinary Explanations of Social Behaviour*, ed. C. Antaki. London: Academic.

Shutts, D. 1982. *Lobotomy: Resort to the Knife.* New York: van Nostrand Reinhold.

Simon, H.A. 1983. *Reason in Human Affairs.* Stanford, CA: Stanford University Press.

Simoons, F.J. 1979. Lactose malabsorption in Eurasia. *Anthropos* **74**, 61–80.

Singer, C. 1917. The medical literature of the dark ages. *Proc. R. Soc. Med. Hist. Med.* **10**, 110–60.

Sipes, R.G. 1973. War, sports and aggression. *Am. Anthropol.* **75**, 64–86.

Skinner, B.F. 1938. *The Behavior of Organisms*. New York: Appleton-Century.

Skinner, B.F. 1948. *Walden Two*. New York: Macmillan.

Skinner, B.F. 1953. *Science and Human Behavior*. New York: Macmillan.

Skinner, B.F. 1956. Freedom and the control of men. *Am. Scholar,* **25**, 47–65.

Skinner, B.F. 1961. The design of cultures. *Daedalus,* **90**, 534–46.

Skinner, B.F. 1966. The phylogeny and ontogeny of behavior. *Science, NY* **153**, 1205–13.

Skinner, B.F. 1968. *The Technology of Teaching*. New York: Appleton-Century-Crofts.

Skinner, B.F. 1970. An autobiography. In: *Festschrift for B.F. Skinner*, ed. P.B. Dews. New York: Appleton-Century-Crofts.

Skinner, B.F. 1973. *Beyond Freedom and Dignity*. Harmondsworth: Penguin.

Skinner, B.F. 1974. *About Behaviorism*. New York: Knopf.

Skocpol, T. 1978. *States and Social Revolutions*. Cambridge: Cambridge University Press.

Skodak, M. & Skeels, H.M. 1949. A final follow-up study of one hundred adopted children. *J. genet. Psychol.* **75**, 85–125.

Smalley, B. 1974. *Historians in the Middle Ages*. London: Thames & Hudson.

Smart, J.J.C. 1963. Materialism. *J. Philos.* **60**, 651–62.

Smart, J.J.C. 1981. Physicalism and emergence. *Neuroscience* **6**, 109–13.

Smith, R. 1987. *Unemployment and Health*. London: Oxford University Press.

Sollas, W.J. 1924. *Ancient Hunters*. London: Macmillan.

Speer, A. 1975. *Inside the Third Reich*. London: Sphere.

Speer, D.C. 1972. Nonverbal communication of affective information. *Comp. Group Stud.* **3**, 409–23.

Spencer, H. 1866. *Principles of Biology*. New York: Appleton.

Spencer, H. 1892. *Social Statics*. London: Williams & Norgate.

Spencer, H. 1910. *The Principles of Sociology*. London: Appleton.

Spengler, O. 1926, 1928. *The Decline of the West*. London: Allen & Unwin.

Spengler, O. 1934. *The Hour of Decision*. London: Allen & Unwin.

Sperry, R.W., Gazzaniga, M.S. & Bogen, J.E. 1969. Interhemispheric relationships. In: *Handbook of Clinical Neurology*, vol. 4, ed. P.H. Vinken & G.W. Bruyn. Amsterdam: North-Holland.

Stahl, H.H. 1980. *Traditional Romanian Village Communities*. Cambridge: Cambridge University Press.

Stanley, S.M. 1979. *Macroevolution: Pattern and Process*. San Francisco: Freeman.

Stephan, H. 1972. Evolution of primate brains. In: *The Functional and Evolutionary Biology of Primates*, ed. R. Tuttle. Chicago: Aldine.

Stephens, W.N. 1963. *The Family in Cross-cultural Perspective*. New York: Holt, Rinehart & Winston.

Stern, J.P. 1980. The *Weltangst* of Oswald Spengler. *Times Literary Supplement*, Oct. 10, pp. 1149–52.

Stohl, M. 1976. *War and Domestic Political Violence*. Beverly Hills: Sage Publications.

Stolz, S.B. 1978. *Ethical Issues in Behavior Modification*. San Francisco: Jossey-Bass.

Stone, L. 1975. The rise of the nuclear family. In: *The Family in History*, ed. C.E. Rosenberg. Philadelphia: University of Pennsylvania Press.

Stone, N. 1983. *Europe Transformed 1878–1919*. Glasgow: Fontana.

Storch, R.D. 1975. The plague of the blue locusts. *Int. Rev. Soc. Hist.* **20**, 61–90.

Storr, A. 1968. *Human Aggression*. London: Allen Lane.

Sulloway, F.J. 1979. *Freud: Biologist of the Mind*. New York: Basic Books.

Sutherland, S.L. & Tanenbaum, E.J. 1980. Submissive authoritarians. *Canad. Rev. Sociol. Anthropol.* **17**, 1–23.

Suzuki, S. 1969. *Nurtured by Love*. New York: Exposition Press.

Swaminathan, M.S. 1973. *Our Agricultural Future*. New Delhi: Sardar Patel Memorial Lectures.

Swift, J. 1939. *Selected Writings*. London: Nonesuch Press.

Symons, D. 1979. *The Evolution of Human Sexuality*. New York: Oxford University Press.

Symons, D. *et al.* 1980. Précis of *The Evolution of Human Sexuality*. *Behavl Brain Sci.* **3**, 171–214.

Tannahill, R. 1980. *Sex in History*. London: Sphere.

Tanner, J.M. 1966. Growth and physique in different populations. In: *The Biology of Human Adaptability*, ed. P.T. Baker & J.S. Weiner. Oxford: Clarendon Press.

Tawney, R.H. 1938. *Religion and the Rise of Capitalism*. Harmondsworth: Penguin.

Tawney, R.H. 1961. *The Acquisitive Society*. Glasgow: Collins.

Taylor, A.E. 1948. *Plato: the Man and his Work*. London: Methuen.

Taylor, F.W. 1907. On the art of cutting metals. *Trans. Am. Soc. Mech. Eng.* **28**, 31–279.

Taylor, F.W. 1911. *The Principles of Scientific Management*. New York: Harper.

Taylor, P. 1984. *Smoke Ring: the Politics of Tobacco*. London: Bodley Head.

Tedeschi, J.T. *et al.* 1981. Is the concept of aggression useful? In: *Multidisciplinary Approaches to Aggression Research*, ed. P.F. Brain & D. Benton. Amsterdam: Elsevier.

Temkin, O. 1949. Metaphors of human biology. In: *Science and Civilization*, ed. R.C. Stauffer. Madison: University of Wisconsin Press.

Terrace. H.S. 1980. *Nim*. London: Eyre Methuen.

Thayer, G. 1970. *The War Business*. London: Paladin.

Thirkell, R. 1874. The aborigines of Tasmania. *Proc. R. Soc. Tasmania for 1873*, **28**.

Thoday, J.M. 1969. Limitations to genetic comparisons of populations. *J. Biosoc. Sci., Suppl.* **1**, 3–14.

Thomas, G., O'Callaghan, M. 1981. Pavlovian principles and behaviour therapy. In: *Applications of Conditioning Theory*, ed. G. Davey. London: Methuen.

Thompson, D.W. 1917. *Growth and Form*. Cambridge: Cambridge University Press.

Thoresen, C.E., Telch, M.J. & Eagleston, J.R. 1981. Altering the Type A behavior pattern. *Psychosomatics* **22**, 472–82.

Thorpe, W.H. 1963. *Learning and Instinct in Animals*. London: Methuen.

Thorpe, W.H. *et al.* 1972. Duetting and antiphonal song in birds. *Behav. Suppl.* **18**, 1–197.

Tinbergen, N. 1968. On war and peace in animals and man. *Science, NY* **160**, 1411–1418.

Titmuss, R.M. 1970. *The Gift Relationship*. London: Allen & Unwin.

Tizard, B. 1944. IQ and race. *Nature, Lond.* **247**, 316.

Torr, D. (ed.) 1948. *History in the Making*, 3 vols. London: Lawrence & Wishart.

Torrey, E.F. & Peterson, M.R. 1973. Slow and latent viruses in schizophrenia. *Lancet*, **2**, 22–24.

Toynbee, A.J. 1935. *A Study of History*, vol 3. London: Oxford University Press.

Trivers, R.L. 1971. The evolution of reciprocal altruism. *Q. Rev. Biol.* **46**, 35–57.

Trivers, R.L. 1981. Sociobiology and politics. In: *Sociobiology and Human Politics*, ed. E. White. Lexington, MA: Heath.

Tschanz, B. 1968. Die Entstehung der persönlichen Bezeihung zwischen Jungvogel und Eltern. *Z. Tierpsychol.* **25**, 244–45.

Tuan, Y-F. 1974. *Topophilia*. Englewood Cliffs, NJ: Prentice-Hall.

Tuchman, B.W. 1978. *A Distant Mirror: the Calamitous 14th Century*. New York: Knopf.

Turnbull, C.M. 1961. *The Forest People*. London: Chatto & Windus.

Turnbull, C.M. 1973. *The Mountain People*. London: Cape.

Turner, J.A. 1970. *The Chemical Feast*. New York: Grossman.

Ucko, P.J. & Rosenfeld, A. 1967. *Paleolithic Cave Art*. London: Weidenfeld & Nicholson.

Ullman, L.P. 1969. Behavior therapy as a social movement. In: *Behavior Therapy*, ed. C.M. Franks. New York: McGraw-Hill.

Ulrich, R.E. 1973. Toward experimental living. *Behav. Modification Monogr.* **11**, 1–74.

Ulrich, R.E. 1975. Ethical implications of behavior modification. *Enero*, **1**, 97–106.

Ulrich, R.E., Stachnik, T. & Mabry, J. (eds.) 1966. *Control of Human Behavior*, vol. 1. Glenview, IL: Scott, Foresman.

Ulrich, R.E., Stachnik, T. & Mabry, J. (eds) 1970. *Control of Human Behavior*, vol. 2. Glenview, IL: Scott, Foresman.

Ulrich, R.W. Stachnik, T. & Mabry, J. (eds) 1974. *Control of Human Behavior*, vol. 3. Glenview, IL: Scott, Foresman.

Umiker-Sebeok, J. & Sebeok, T.A. 1981. Clever Hans and smart simians. *Anthropos* **76**, 89–165.

UNESCO 1952. *The Race Concept*. Paris: UNESCO.

Valvo, A. 1971. *Sight Restoration after Long-term Blindness*. New York: American Foundation for the Blind.

van den Berghe, P.L. 1978. Bridging the paradigms: biology and the social sciences. In: *Sociobiology and Human Nature*, ed. M.S. Gregory, A. Silvers & D. Sutch. San Francisco: Jossey-Bass.

van den Berghe, P.L. 1979. *Human Family Systems*. New York: Elsevier.

van den Berghe, P.L. 1981. *The Ethnic Phenomenon*. New York: Elsevier.

Veith, J.L. *et al.* 1983. Exposure to men influences ovulation. *Physiol. Behav.* **31**, 313–315.

Vernon, J.A. 1963. *Inside the Black Room*. London: Souvenir Press.

von der Mehden, F.R. 1973. *Comparative Political Violence*. Englewood-Cliffs, NJ: Prentice-Hall.

Waddington, C.H. 1960. *The Ethical Animal*. London: Allen & Unwin.

Wade, M.J. 1978. The selfish gene. *Evolution*, **32**, *220–1*.

Wade, M.J. 1982. Group selection. *Evolution*, **36**, 949–61.

Wagner, G. 1968. *On the Wisdom of Words*. London: Allen & Unwin.

Waldrop, M.M. 1984. Natural language understanding. *Science, NY* **224**, 372–4.

Walker, A. 1981. Diet and teeth. *Phil. Trans. R. Soc.* B **292**, 57–64.

Walter, W.G. 1953. *The Living Brain*. London: Duckworth.

Ware, N.J. 1935. *Labor in Modern Industrial Society*. New York: Russell & Russell.

Warr, P. 1987. *Work, Unemployment and Mental Health*. Oxford: Oxford University Press.

Washburn, S.L. & Avis V. 1958. Evolution of human behavior. In: *Behavior and Evolution*, ed. A. Roe & G.G. Simpson. New Haven: Yale University Press.

Washburn, S.L. & Lancaster, C.S. 1968. The evolution of hunting. In: *Man the Hunter*, ed. R.B. Lee & I. DeVore. Chicago: Aldine.

Watson, J.B. 1913. Psychology as the behaviorist views it. *Psychol. Rev.* **20**, 158–77.

Watson, J.B. 1930. *Behaviorism*. Chicago: University of Chicago Press.

Watson, J.L. (ed.) 1980. *Asian and African Systems of Slavery*. Oxford: Blackwell.

Watson, J.S. & Ramey, C.T. 1972. Reactions to response-contingent stimulation in early infancy. *Merrill-Palmer Q.* **18**, 219–27.

Webb, S.D. & Collette, J. 1975. Urban correlates of stress-alleviative drug use. *Am. Behav. Sci.* **18**, 750–70.

Weiner, J.S. 1954. Nose shape and climate. *Am. J. Phys. Anthropol.* **12**, 1–4.

Weiskrantz, L. 1973. Problems and progress in physiological psychology. *Br. J. Psychol.* **64**, 511–20.

Weisner, T.S. & Gallimore, R. 1977. Child and sibling caretaking. *Curr. Anthropol.* **18**, 169–90.

Westermarck, E. 1917. *The Origin and Development of the Moral Ideas*, vol. 1. London: Macmillan.

White, M.J.D. 1978. *Modes of Speciation*. San Francisco: Freeman.

WHO 1976. Resistance to pesticides. *WHO Tech. Rep. Ser.* **585**, 7–88.

WHO 1982. *Seventh General Programme of Work*. Geneva: World Health Organization.

Wild, J. 1960. The concept of man in Greek thought. In: *The Concept of Man*, ed. S. Radhakrishnan & P.T. Raju. London: Allen & Unwin.

Wilkinson, P. 1981. *The New Fascists*. London: Grant McIntyre.

Williams. E. 1966. *Capitalism and Slavery*. New York: Putnam.

Williams, G.C. 1966. *Adaptation and Natural Selection*. Princeton, NJ: Princeton University Press.

Williams, G.C. 1980. The paradox of sexuality. In: *Sociobiology: beyond Nature/Nurture?*, ed. G.W. Barlow & J. Silverberg. Boulder, CO: Westview Press.

Williams R. 1976. *Keywords*. Glasgow: Fontana.

Wilm, E.C. 1925. *The Theories of Instinct*. New Haven: Yale University Press.

Wilson, D.S. 1980. *The Natural Selection of Populations and Communities*. Menlo Park, CA: Benjamin/Cummings.

Wilson, E.O. 1971. *The Insect Societies*. Cambridge, MA: Harvard University Press.

Wilson, E.O. 1975. *Sociobiology*. Cambridge, MA: Harvard University Press.

Wilson, E.O. 1979. *On Human Nature*. Cambridge, MA: Harvard University Press.

Wilson, E.O. 1980. *Comparative Social Theory*. Michigan: University of Michigan.

Wing, J.K. 1973. Causation and treatment of schizophrenia. *Biochem. Soc. Spec. Publ.* **1**, 153–63.

Winner, E. 1982. *Invented Worlds*. Cambridge, MA: Harvard University Press.

Wittgenstein, L. 1968. *Philosophical Investigations*. Oxford: Blackwell.

Wohlwill, J.F. 1980. Cognitive development in childhood. In: *Constancy and Change in Human Development*, ed. O.G. Brim & J. Kagan. Cambridge, MA: Harvard University Press.

Wolff, P.H. 1965. The natural history of crying. In: *Determinants of Infant Behaviour*, ed. B.M. Foss. London: Methuen.

Wolff, P.H. 1969. What we must and must not teach our young children. In: *Planning for better learning*, ed. P.H. Wolff & R. MacKeith. London: Heinemann.

Wolfgang, M.E. (ed.) 1967. *Studies in Homicide*. New York: Harper & Row.

Woodworth, R.S. 1931. *Contemporary Schools of Psychology*. London: Methuen.

Woolf, C.M. & Dukepo, F.C. 1969. Inbreeding and albinism. *Science, NY* **164**, 30–7.

Woolf, S. 1986. *The Poor in Western Europe in the Eighteenth and Nineteenth Centuries*. London: Methuen.

Wyman, D.S. 1984. *The Abandonment of the Jews*. New York: Pantheon.

Wynne-Edwards, V.C. 1962. *Animal Dispersion*. Edinburgh: Oliver & Boyd.

Yankelovich, D. *et al.* 1983. *Work and Human Values*. New York: Aspen Institute.

Young, J.Z. 1951. *Doubt and Certainty in Science*. Oxford: Clarendon Press.

Young, J.Z. 1964. *A Model of the Brain*. Oxford: Clarendon Press.

Young, J.Z. 1978. *Programs of the Brain*, Oxford: Oxford University Press.

Young, R.M. 1971. Darwin's metaphor: does nature select? *Monist*, 55, 442–503.

Yourcenar, M. 1959. *Memoirs of Hadrian*, trans. G. Frick. Harmondsworth: Penguin.

Zangwill, O.L. Thought and the brain. *Br. J. Psychol.* 67, 301–14.

Zihlman, A.L. & Tanner, N.M. 1978. *Female Hierarchies*. Chicago: Beresford Book Service.

Zimbardo, P.G. 1978. The psychology of evil. In: *Aggression, Dominance, and Individual Spacing*, ed. L. Krames *et al*. New York: Plenum.

Zirkle, C. 1959. Commentary. In: *Critical Problems in the History of Science*, ed. M. Clagett. Madison: University of Wisconsin Press.

Name index

Subject index

Animals are listed under their vernacular names (used in the text); Latin names, except those of domestic species, are in parentheses.

Academy, The 7
achondroplasia 109
acne 123
'acquired characters', *see* Lamarck
adaptation 88–92
  defined 303
adopted children
  and intelligence 111–12
  and schizophrenia 105
  see also adoption
adoption 43
  and human genetics 111
  see also kin
Adventures of the Black Girl in her Search for God, The 182
advertising 109, 298
'aggression' 4, 56–76
  and Freud 59–60
  and politics 73–6
  as habit 70–2
  as instinct 33, 53–4
  evolution of 92–3
  meanings of 57–9, 303
  psychosurgery and 74
Agta 116
alarm calls 34, 93–4
albinism 89, 109
alcoholism 128, 186
alienation
  and neo-behaviorism 201
  and Taylorism 214
  and work 217, 292
  defined 303–4
altruism
  and evolution 124
  and giving blood 212
  defined 304
  disregarded 138

misused 134–5
of children 279
practical 301
see also bioaltruism
amygdala 234
analogy 16–28
  and persuasion 24–6
  defined 20
  medieval 20
  Platonic 1–2
  political 24–6
  reasoning by 304
  scientific 20–24, 27, 33
animalitarianism 17–20
anomie, *see* alienation
anorexia 185
anthropology , teaching of 154
ants (Formicidae) 46
apes (Pongidae)
  and imitation 271–2
  and language 255–7
  and reproduction 119, 120–1
  and social dominance 39, 40
  and territory 42, 120
Apology, The 10
aposematism 90, 128
armaments manufacture 293–4
arts
  and Darwinian fitness 130–1
  and tradition 279–80
  related to exploration 193, 221
asexual reproduction 322–3
ashram 205
Australian aborigines 17
  and food 64, 116
  and property 43
  and violence 66
Australopithecus 61–3, 82
Authoritarian Personality, The 72
autonomous human being 73, 187, 191, 198–9, 221–4
  and Marx 292